Reviews of

104 Physiology, Biochemistry and Pharmacology

Editors
E. Habermann, Giessen · E. Helmreich, Würzburg
H. Holzer, Freiburg · R. Jung, Freiburg
R. J. Linden, Leeds · P. A. Miescher, Genève
J. Piiper, Göttingen · W. Singer, Frankfurt/M
U. Trendelenburg, Würzburg · K. Ullrich, Frankfurt/M
W. Vogt, Göttingen

With 16 Figures

Springer-Verlag Berlin Heidelberg GmbH

ISBN 978-3-662-31044-1 ISBN 978-3-540-39672-7 (eBook)
DOI 10.1007/978-3-540-39672-7

Library of Congress-Catalog-Card Number 74-3674

Originally published by Springer-Verlag Berlin Heidelberg New York Tokyo in 1986
Softcover reprint of the hardcover 1st edition 1986

2127/3130-543210

Contents

Indexed in Current Contents

Rev. Physiol Biochem. Pharmacol., Vol. 104

Eicosanoids and the Coronary Circulation

THOMAS SIMMET and BERNHARD A. PESKAR

Contents

Institute of Pharmacology and Toxicology, Ruhr-Universität Bochum, Im Lottental, D–4630 Bochum 1, FRG

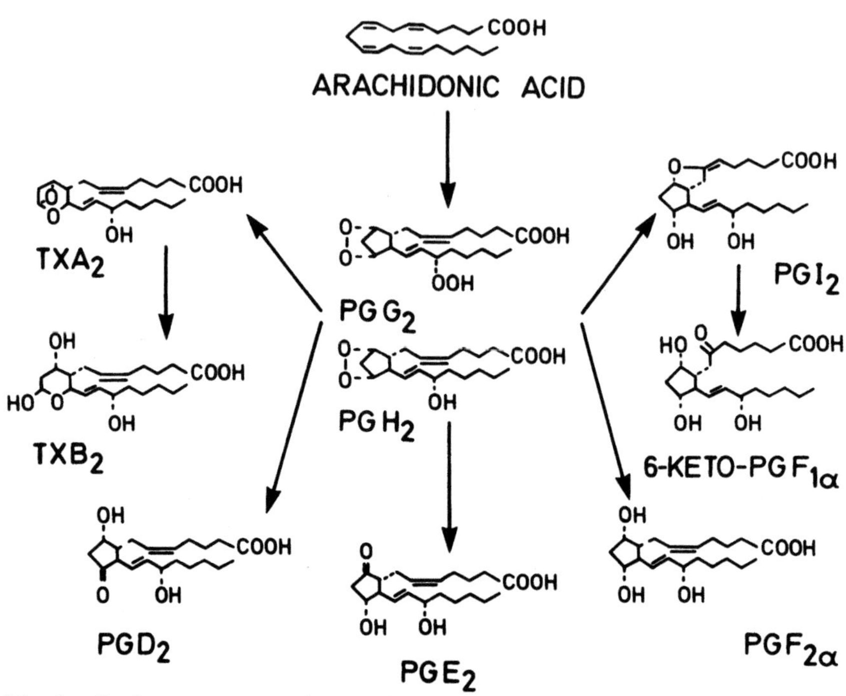

Fig. 1a. Cyclooxygenase pathway of arachidonic acid metabolism

1 Introduction

1.1 General Introduction

As early as 1930, *Kurzrock* and *Lieb,* gynecologists studying artificial insemination, reported that the human uterus reacted with either strong contractions or relaxation to the instillation of fresh human semen. These initial observations were extended by the studies of *von Euler* (1934) and *Goldblatt* (1935), who independently discovered and studied the strong activity in seminal plasma that stimulated smooth muscle. In 1935 *von Euler* termed the active principle of seminal plasma "prostaglandin." Lipid extracts prepared from the vesicular glands of different species revealed that the activity was associated with a fraction containing lipid soluble acids (*von Euler* 1934, 1935, 1937; *Goldblatt* 1935).

Bergström and coworkers isolated in pure cristalline form the first two prostaglandins (PG), now called PGE_1 and $PGF_{1\alpha}$ (*Bergström* and *Sjövall* 1960a,b) and soon thereafter elucidated the complete structure of these compounds (*Abrahamsson* et al. 1962; *Bergström* et al. 1963; *Nugteren* et al. 1966). Structurally, the first PG described had the same basic carbon skeleton as prostanoic acid (*Bergström* et al. 1968). However, the discovery of PGI_2 (prostacyclin) with a double-ring structure (*Johnson* et al.

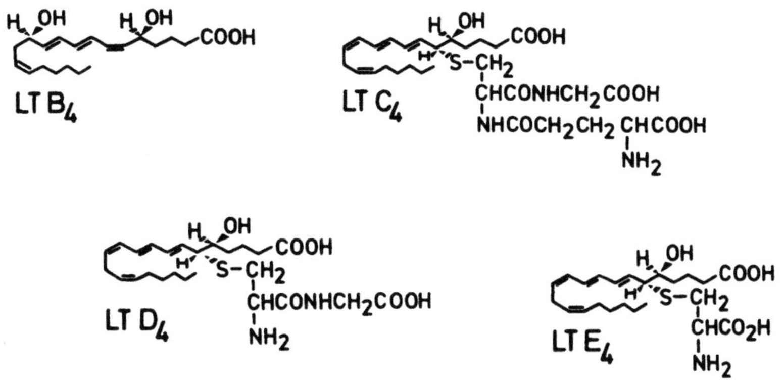

Fig. 1b. Chemical structure of leukotrienes

1976), thromboxane (TX) A_2 with an oxane ring (*Hamberg* et al. 1975), and the lipoxygenase products with the straight chain conformation (*Taylor* and *Morris* 1983) has led to the generic name "eicosanoids" proposed by *Corey* et al. (1980) to embrace cyclooxygenase and lipoxygenase products as well (Fig. 1a,b).

1.2 Eicosanoid Synthesis and Catabolism

1.2.1 Substrate Liberation

All eicosanoids, including PGs, TXs, and leukotrienes (LTs), are derived from certain polyunsaturated fatty acids. Dihomo-γ-linolenic acid, arachidonic acid, and 5,8,11,14,17-eicosapentaenoic acid were shown to be the precursor fatty acids of prostaglandins of the 1-, 2-, and 3-series, respectively, the series number indicating the number of double bonds in the side chains (*Bergström* et al. 1964; *van Dorp* et al. 1964; *Struijck* et al. 1966).

Arachidonic acid, which is the precursor of all bisenoic PGs, may be obtained directly from the diet or by anabolic desaturation and chain elongation from the dietary essential linoleic acid, the latter mechanism representing probably only a minor pathway (*Hassam* and *Crawford* 1976). In blood, arachidonic acid is largely bound to albumin and is incorporated mainly as part of phospholipids in all cell membranes of the body (*Ramwell* et al. 1977). The substrate, usually arachidonic acid, must be in the nonesterified form for the oxidative transformation to eicosanoids (*Lands* and *Samuelsson* 1968; *Vonkeman* and *van Dorp* 1968). However, intracellular levels of free fatty acids are extremely low (*Lands* and *Samuelsson* 1978; *Vogt* 1978). Therefore, the first step of eicosanoid synthesis requires hydrolytic cleavage of the esterified substrate at or near the sites of enzymatic peroxidation. Since unsaturated fatty acids, including all eicosanoid precursors, are generally esterified in the β position of the phosphatide, phospholipase A_2, which hydrolyzes this bond, becomes an important regulatory enzyme of the arachidonic acid cascade, even though phospholipase C and D in combination with lipases might also participate in the fatty acid liberation (*Kunze* and *Vogt* 1971; *Vogt* 1978; *Irvine* 1982; *Blackwell* and *Flower* 1983).

A number of different hormones seem to be involved in the activation of phospholipase A_2 (*Piper* and *Vane* 1971; *Vogt* 1978), but even simple mechanical stimulation can result in PG generation (*Gryglewski* and *Vane* 1972; *Flower* and *Blackwell* 1976). Furthermore, deprivation of oxygen availability to the heart is a stimulus for PG production. This has been reported after the induction of hypoxia in isolated heart preparations (*Block* et al. 1975; *Needleman* et al. 1975), or after ischemia created by coronary artery ligation in situ (*Berger* et al. 1976; *Kraemer* et al. 1976). Interestingly, glucocorticoids inhibit phospholipase A_2 activity by induction of phospholipase A_2 inhibitory proteins (*Blackwell* et al. 1980; *Hirata* et al. 1980).

Apart from binding to protein and other mechanisms which might influence the fate of the unsaturated fatty acids, liberated arachidonic acid can be reacylated by an apparently very efficient mechanism (*Isakson* et al. 1977; *Needleman* 1978), or can be further metabolized by either the cyclooxygenase or lipoxygenase pathway. The importance of the reacylation mechanism has recently been stressed (*Kröner* et al. 1981; *Körner* et al. 1984; *Schmidt* et al. 1984). Thus, in macrophages the reacylation capacity is about 15 times higher than the degradation rate of phospholipids, and the fast accumulation of arachidonic acid induced by the divalent cation ionophore A23187 was found to be due to the combined effects of activated phospholipase A_2 and inhibited reacylation (*Kröner* et al. 1981; *Schmidt* et al. 1984). Furthermore, in human platelets inhibition of the reincorporation of free fatty acids into phospholipids by the lysophosphatide acyltransferase inhibitor thimerosal (Merthiolate) was accompanied by a 39-fold increase in TX formation (*Körner* et al. 1984).

From experiments with ischemic myocardium it has recently been postulated that prolonged ischemia will cause a defect in the mechanism for the reacylation of arachidonate into phosphatidylcholine (*Chien* et al. 1984). Such a pathophysiologic mechanism not only increases the level of free arachidonic acid, but might also contribute to the net loss of membrane phospholipid during myocardial ischemia.

1.2.2 Metabolism of Arachidonic Acid via the Cyclooxygenase Pathway

Cyclooxygenase activity has been purified from bovine (*Miyamoto* et al. 1976) and sheep (*Hemler* et al. 1976; *van der Ouderaa* et al. 1977) vesicular glands. Cyclooxygenase is present in virtually all tissues and has a negative feedback mechanism based on self-catalyzed inactivation (*Smith* and *Lands* 1972; *Egan* et al. 1976; *Samuelsson* et al. 1978; *Lands* 1979). In the presence of heme and oxygen this microsomal glycoprotein catalyzes the bis-dioxygenation of arachidonic acid to form PGG_2, and further, the peroxidative reduction of PGG_2 to PGH_2 (*van der Ouderaa* et al. 1977; *Ohki* et al. 1979; *Kulmacz* and *Lands* 1983). These compounds are also called PG endoperoxides (*Hamberg* et al. 1974b). Nonsteroidal anti-inflammatory drugs such as aspirin and indomethacin inhibit cyclooxygenase activity and thereby PG production (*Vane* 1971; *Smith* and *Willis* 1971; *Ferreira* et al. 1971; *Flower* 1974).

PG endoperoxide E isomerase catalyzes the isomerization of PGH to PGE in the presence of the cofactor glutathione (*Ogino* et al. 1977). Another glutathione-dependent isomerase reaction catalyzes the formation of PGD_2 from PG endoperoxides (*Nugteren* and *Hazelhof* 1973; *Christ-Hazelhof* et al. 1976). PGD_2 and PGE_2 are also formed by nonenzymatic degradation of the endoperoxide PGH_2 (*Samuelsson* et al. 1978), a process which is additionally influenced by the presence of albumin (*Christ-Hazelhof* et al. 1976; *Hamberg* and *Fredholm* 1976). Reductase activity is involved in the synthesis of $PGF_{2\alpha}$ (*Lands* 1979), while a TX synthetase, which has been detected in platelets from several species (*Needleman* et al. 1976; *Hammarström* and *Falardeau* 1977; *Sun* 1977), converts PG endoperoxides to TX. The highly unstable TXA_2 is readily degraded to TXB_2 (*Hamberg* et al. 1975). The biosynthesis of TXA_2 is linked to that of 12L-hydroxy-5,8,10-heptadecatrienoic acid (HHT), a C17-hydroxy acid, since thromboxane synthetase catalyzes HHT production presumably by splitting off malondialdehyde (MDA) (*Hamberg* et al. 1974a; *Yoshimoto* et al. 1977; *Wlodawer* and *Hammarström* 1978). However, MDA and HHT can also be formed nonenzymatically by spontaneous degradation of PGH_2 (*Hammerström* and *Falardeau* 1977). PGI_2 synthetase converts PG endoperoxide to the unstable PGI_2, which then spontaneously hydrolyzes to 6-keto-$PGF_{1\alpha}$ (*Johnson* et al. 1976).

The rapid metabolism of prostaglandins is associated with both the enzymatic oxidation of the 15-hydroxyl group of the active PG (*Änggård* et al. 1971) and the spontaneous hydrolytic reactions of TXA_2 and PGI_2 mentioned above. The oxidation at C15 is followed by a reduction of the Δ^{13} double-bond, thereby forming biologically inactive 13,14-dihydro-15-keto derivatives, which are major metabolites in the peripheral circulation (*Änggård* et al. 1971; *Samuelsson* et al. 1975; *Bakhle* 1983). By β-oxidation of the carboxyl side chain and ω-oxidation of the alkyl side chain mono- and dicarboxylic acids with shorter chain length are finally produced; these are the main metabolites observed in urine (*Samuelsson* et al. 1975; *Bakhle* 1983).

1.2.3 Metabolism of Arachidonic Acid via the Lipoxygenase Pathway

Another major pathway of arachidonic acid metabolism is represented by the lipoxygenase system, which is responsible for the formation of several different hydroperoxy-eicosatetraenoic acids (HPETEs) (*Taylor* and *Morris* 1983). Blood platelets contain a 12-lipoxygenase (*Hamberg* and *Samuelsson* 1974; *Nugteren* 1975), and 12-HPETE is produced during platelet aggregation (*Hamberg* and *Samuelsson* 1974). By reduction the hydroperoxide is converted to a hydroxy acid (12-HETE). The presence of 12-lipoxygenase activity has also been described in several other tissues, such as the rabbit aorta and pericardium (*Greenwald* et al. 1979; *Herman* et al. 1979).

Leukocytes, on the other hand, are known to exhibit 5- and 15-lipoxygenase activity (*Borgeat* et al. 1976; *Narumiya* et al. 1981). The former pathway is of particular interest, since 5-lipoxygenation is the initial reaction for the synthesis of LTs, which are mediators of immediate hypersensitivity reactions and inflammation (*Hammarström* 1983; *Piper* 1983; *Samuelsson* 1983). Thus, 5-HPETE may be either reduced to 5-HETE or transformed to an unstable epoxide, LTA_4, by a dehydrase. By means of a hydrolase LTA_4 is converted to LTB_4 (5S12R-DHETE), which is a potent chemotactic factor (*Ford-Hutchinson* et al. 1980; *Hammarström* 1983; *Samuelsson* 1983). Glutathione-S-transferase catalyzes the transformation from LTA_4 to LTC_4 by the introduction of glutathione at C6 with a thioether linkage. LTC_4 can be metabolized to LTD_4 by the enzymatic elimination of glutamic acid by γ-glutamyl transpeptidase. A renal dipeptidase hydrolyzes the remaining peptide bond of LTD_4, and the product of this reaction is LTE_4. Finally, LTE_4 can be an acceptor for glutamic acid, giving rise to a γ-glutamyl cysteinylderivative, LTF_4 (*Hammarström* 1983; *Samuelsson* 1983; *Taylor* and *Morris* 1983).

It should be noted that tissues produce in general a certain pattern of eicosanoids which depends on the species, the tissue used, the substrate concentration, and the stimulus used for substrate liberation.

2 Thromboxane and Prostaglandin I_2 in the Coronary Circulation

2.1 Thromboxane and Prostaglandin I_2 Biosynthesis by Cardiac Tissue

PG synthesis in cardiac tissue was first reported by *van Dorp* (1966) and *Karim* et al. (1967), who demonstrated the presence of PG-synthesizing capacity in sheep and human cardiac muscle, as well as the occurrence of PGE_2 and $PGF_{2\alpha}$ in the hearts of several animal species (*Karim* et al. 1968). Although PGE_2 and $PGF_{2\alpha}$ are released from isolated rat, rabbit, and guinea pig hearts, PGI_2 is by far the major PG formed in these organs (*De Deckere* et al. 1977; *Schrör* et al. 1978). Similarly, 6-keto-$PGF_{1\alpha}$, the hydrolysis product of PGI_2, was found to be the predominant PG in human coronary sinus blood, alongside substantial amounts of PGD_2, PGE_2, and $PGF_{2\alpha}$ (*Nowak* et al. 1980). While some authors have concluded that the coronary vessels might be the major source of PG in perfused rabbit hearts (*Hsueh* and *Needleman* 1978; *Wennmalm* 1979), experiments with cultured rat and dog heart myocytes have shown that considerable amounts of PGI_2 were synthesized after incubation with radiolabeled arachidonic acid (*Bolton* et al. 1980). Others, however, using cells from neonatal rats, reported a lack of PGI_2 synthesizing activity, while PGE_2 and $PGF_{2\alpha}$ are generated in these cultured cells (*Ahumada* et al. 1980). It is not clear whether the discrepancy is due to contamination with vascular cells, which have been suggested to be the primary source of PGI_2 biosynthesis (*Bunting* et al. 1976; *Moncada* et al. 1976a), or to an age-dependent expression of enzyme activity.

Interestingly, in adult heart cell culture, TXB_2 release could be detected as well as PGI_2 (*Bolton* et al. 1980). These data indicate that heart myocytes, in addition to the vascular system, might also be able to contribute significantly to overall cardiac eicosanoid synthesis.

The discovery of PGI_2 (*Johnson* et al. 1976; *Moncada* et al. 1976a,b), which exerts biological effects just opposite to those of TXA_2 on platelets and vascular smooth muscle, led to the hypothesis of a TXA_2-PGI_2 balance (*Moncada* et al. 1976a,b). Although the interaction between these two eicosanoids might have been overemphasized occasionally, this concept has stimulated an abundance of basic and clinical research.

2.2 Prostaglandin I_2 and Its General Effects

PGI_2 is synthesized by blood vessel microsomes prepared from fresh tissue from the endoperoxide PGH_2 or arachidonic acid (*Bunting* et al. 1976; *Moncada* et al. 1976a; *Moncada* and *Vane* 1979). It is the main product of arachidonic acid metabolism in all arteries and veins so far tested (*Moncada*

1982). The capacity of the vessel wall to synthesize PGI_2 is greatest at the intimal surface and progressively decreases toward the adventitia (*Moncada* et al. 1977b). This observation is supported by data obtained from cultured vessel wall cells, showing that endothelial cells are the most active producers of PGI_2 (*Weksler* et al. 1977; *MacIntyre* et al. 1978), the production persisting even after numerous subcultures in vitro (*Christofinis* et al. 1979). The biosynthesis of PGI_2 is inhibited by 15-HPETE and other fatty acid hydroperoxides (*Gryglewski* et al. 1976).

PGI_2 is a general vasodilator, except in the case of pig coronary arteries and the venous part of some vascular beds, where it was found to be a weak vasoconstrictor or without effect on vascular tone (*Dusting* et al. 1977a; *Levy* 1978; *Moncada* and *Vane* 1979). In addition, PGI_2 is the most potent endogenous inhibitor of platelet aggregation yet discovered, being more than 1000 times more active than adenosine monophosphate (AMP; *Born* 1962). PGI_2 inhibits platelet aggregation by stimulating adenylate cyclase, thus leading to an increase in cyclic AMP (cAMP) levels in the platelets (*Best* et al. 1977; *Gorman* et al. 1977a; *Tateson* et al. 1977). It was found to be much more potent than either PGE_1, which supposedly shares the same receptor, or PGD_2, which acts on different receptors on the platelet membrane (*Whittle* et al. 1978). PGI_2 applied locally has been shown to inhibit thrombus formation due to adenosine diphosphate (ADP) in the microcirculation of the hamster cheek pouch (*Higgs* et al. 1977), and administered systemically to rabbits it increases bleeding time and prevents electrically induced thrombus formation in the carotid artery (*Ubatuba* et al. 1979). Furthermore, PGI_2 disaggregates platelets in vitro (*Moncada* and *Vane* 1979), as well as in the circulation of man (*Szczeklik* et al. 1978b). The substrance is chemically unstable with a half-life of 2–3 min, and is hydrolyzed to 6-keto-$PGF_{1\alpha}$. Albumins stabilize PGI_2 (*Wynalda* and *Fitzpatrick* 1980), and hydrolysis is delayed in blood (*Pifer* et al. 1981) and plasma (*Jørgensen* et al. 1979).

2.3 Thromboxane A_2 and Its General Effects

Hamberg et al. (1974b) have shown that the addition of PGG_2 or PGH_2 to a suspension of platelets causes rapid aggregation. On the other hand, aggregation induced by thrombin was accompanied by the release of PG endoperoxides, indicating the involvement of these compounds in platelet aggregation (*Hamberg* et al. 1974b). By trapping experiments, the formation of TXA_2 has been demonstrated, and it has been reported that the releasing and aggregating effects of arachidonic acid and PGG_2 on platelets declines with a half-life identical to that of TXA_2 (*Hamberg* et al. 1975). TXA_2 degrades to TXB_2 with a chemical and biological half-life of about

30 s, as compared to a half-life of 5 min for the PG endoperoxides (*Hamberg* et al. 1975; *Svensson* et al. 1975). Using potent inhibitors of TXA_2 synthesis, such as 9,11-azoprosta-5,13-dienoic acid or 1-*N*-butylimidazole, PGH_2-induced platelet aggregation and TXA_2 synthesis were inhibited in platelet-rich plasma, washed platelets, or whole blood, suggesting that conversion of the PG endoperoxide to TXA_2 might be necessary for platelet aggregation (*Blackwell* et al. 1978; *Fitzpatrick* and *Gorman* 1978).

Addition of PGG_2 to human platelet-rich plasma decreased cAMP (*Salzman* 1977). In platelet-rich plasma stimulated with PGI_2, the cAMP-lowering effect of TXA_2 generation could be inhibited by TX synthetase inhibitors. Therefore, it was concluded that TXA_2 is the compound responsible for the abolition of the increase in cAMP induced by PGI_2 (*Gorman* et al. 1977b). Furthermore, TXA_2 is a potent constrictor of vascular and airway smooth muscle, being more potent than the parent PG endoperoxides (*Moncada* and *Vane* 1979; *Bunting* et al. 1983). TX synthetase, which generates TXA_2 from PG endoperoxides was first identified in the high-speed particulate fraction of horse and human blood platelets (*Needleman* et al. 1976).

Polymorphonuclear leukocytes and macrophages also have TX-synthesizing capacity (*Higgs* et al. 1976; *Morley* et al. 1979), and might therefore contribute to the release of this compound during myocardial infarction. However, certain vascular tissues have also been reported to release TXB_2, including cultured bovine endothelial cells, the rabbit pulmonary artery, and some human arteries and veins (*Tuvemo* et al. 1976; *Salzman* et al. 1980; *Ingerman-Wojenski* et al. 1981; *Neri Serneri* et al. 1983).

The PG endoperoxides are the common precursors of compounds with opposing biological effects. TXA_2, produced mainly by platelets, strongly constricts blood vessels and induces platelet aggregation. On the other hand, PGI_2, synthesized from PGH_2 by the vessel wall, is a strong vasodilator and the most potent endogenous inhibitor of platelet aggregation. Both compounds have opposing effects on cAMP concentrations, resulting in a balanced control mechanism that might affect thrombus formation and eventually play an important role in hemostasis (*Moncada* and *Vane* 1979).

2.4 Effect of Thromboxane A_2 on Coronary Arteries

In isolated cat and porcine coronary artery strips, TXA_2 was found to be several times more potent than PG endoperoxides in causing contractions (*Ellis* et al. 1976; *Svensson* and *Hamberg* 1976; *Ogletree* et al. 1978). Also, bovine and canine coronary artery strips have been shown to be contracted by TXA_2 (*Dusting* et al. 1977b; *Needleman* et al. 1977; *Imai* et al. 1983).

Furthermore, a dose-dependent coronary vasospastic action of TXA_2 has been reported in isolated guinea pig hearts (*Terashita* et al. 1978). Other authors, using a platelet-perfused guinea pig heart preparation, have not been able to detect any influence on coronary vascular resistance at TX concentrations up to 25 ng/ml (*Schrör* et al. 1981c). This lack of action has been attributed to an increased counteracting PGI_2 synthesis.

In fact, similar results were reported for open-chest greyhounds (*Herbaczynska-Cedro* et al. 1982), while in mongrel dog heart-lung preparations TXA_2 was shown to be a coronary vasoconstrictor (*Imai* et al. 1983). *Wang* et al. (1980), using adult dogs and puppies, observed a vasoconstrictor activity of TXA_2 in the coronary arteries of puppies, which could not be detected in adult dogs. In rabbits, intracoronary infusion of TXA_2 resulted in ischemia accompanied by ECG changes and hypotension (*Marooka* et al. 1977), which could be prevented by a TXA_2 antagonist (*Shimamoto* 1978). Carbocyclic TXA_2, a stable analogue of TXA_2, contracted human, monkey, dog, cat, rabbit, and guinea pig coronary arteries, while sheep vessels were reported to be unresponsive (*Smith* et al. 1981a; *Burke* et al. 1983b; *Toda* 1984). Interestingly, this TX analogue exhibits a selective profile of activity in that it is a constrictor of coronary arteries yet fails to aggregate platelets (*Burke* et al. 1983b). The same compound has been shown to reduce coronary blood flow and cardiac output, as well as to aggravate myocardial ischemia induced by coronary artery ligation, in cats in terms of further reduction of myocardial cellular integrity (*Smith* et al. 1981b).

Another TXA_2 mimetic, U 44069, has also been reported to contract isolated canine coronary arteries, an effect potentiated by indomethacin or 5-hydroxytryptamine (*Mullane* et al. 1982). Infusion of the vasoconstrictor (*Dusting* et al. 1978; *Wang* et al. 1980) and proaggregatory endoperoxide analogue U 46619 in dogs was followed by PGI_2 release from the coronary arteries, accompanied by a decrease in vascular resistance, indicating that the release of PGI_2 might be a protective autoregulatory mechanism in canine vascular beds, thus maintaining blood flow in the presence of vasoconstrictor and platelet-aggregating stimuli (*Mehta* et al. 1984b).

2.5 Effect of Prostaglandin I_2 on Coronary Arteries

PGI_2 relaxed bovine coronary artery strips (*Dusting* et al. 1977b), but, as mentioned earlier, was found to be a weak contractor of coronary arteries of the pig (*Dusting* et al. 1977a). There is some controversy about the action of PGI_2 on cAMP levels in bovine coronary arteries. It has been reported that in isolated coronary artery strips the relaxation induced by

PGI_2 is accompanied by a reduced cAMP level (*Schrör* and *Rösen* 1979; *Schrör* et al. 1980b). Other authors found PGI_2 to increase cAMP in bovine coronary arteries in the presence or absence of a phosphodiesterase inhibitor (*Dembinska-Kiéc* et al. 1979; *Kukovetz* et al. 1979). Despite certain differences in the experimental procedure, the discrepancies cannot be readily explained. In the isolated perfused cat coronary artery, as well as in the isolated guinea pig and rabbit heart, PGI_2 produced a dose-dependent vasodilatation (*Lefer* et al. 1978; *Ogletree* et al. 1978; *Schrör* and *Moncada* 1979; *Allan* and *Levi* 1980).

Adenosine, formed by the breakdown of ATP in the myocardium, is considered to link the coronary flow rate to the metabolic state of the heart. In isolated rabbit hearts the dose-dependent increase in coronary flow produced by the infusion of adenosine was counteracted by inhibition of phospholipase, cyclooxygenase, or PGI_2 synthetase. Furthermore, in hearts radiolabeled with ^{14}C-arachidonic acid release of ^{14}C-PGI_2 has been observed upon exposure to adenosine, suggesting that the flow-promoting effect of adenosine might depend on the formation of PGI_2 (*Wennmalm* 1984). In the isolated rat heart, PGI_2 induced a dose-dependent coronary vasodilatation which did not alter myocardial energy-rich phosphate content, but diminished myocardial cAMP content and adenosine release (*De Deckere* 1979; *Schrör* et al. 1980a,b). *Karmazyn* et al. (1978), however, reported an increase in perfusion pressure when PGI_2 in concentrations smaller than 3×10^{-9} mol/l was added to the medium perfusing rat and rabbit hearts. Only higher concentrations were found to be vasodilatory. These authors also observed an increase in heart rate and a positive inotropic effect of PGI_2 in isolated rat hearts, while *Schrör* et al. (1980a,b) found no direct effect of PGI_2 on myocardial contractile force, oxygen consumption, or heart rate.

The discrepancies might be due to the different perfusion rates employed by both groups, thus resulting in differences in oxygen supply. Interestingly, PGI_2 produced a marked coronary dilator effect and an increase in myocardial contractility in normotensive NWK rats, whereas both effects were less pronounced in spontaneously hypertensive rats (*Smirnov* et al. 1980).

The vasodilator action of PGI_2 was further confirmed in canine coronary arteries in vitro (*Imai* et al. 1983; *Toda* 1984) and in vivo (*Armstrong* et al. 1977; *Dusting* et al. 1978; *Ito* et al. 1980; *Wang* et al. 1980; *Mehta* et al. 1981b; *Dusting* and *Angus* 1984), and in isolated heart preparations (*Scholtholt* et al. 1981; *Imai* et al. 1983). Given systemically, PGI_2 decreased peripheral and coronary resistance as well as left ventricular pressure, and increased coronary flow in anesthetized open-chest dogs (*Armstrong* et al. 1977; *Dusting* et al. 1978; *Ito* et al. 1980; *Mehta* et al. 1981b). Similar data were obtained in conscious dogs (*Hintze* et al. 1984).

After left intraventricular administration of PGI_2, coronary vasodilation was accompanied by a significant increase in coronary arterial cAMP tissue concentration, while the cyclic guanosine monophosphate (cGMP) level remained unchanged, resulting in a decrease in the cGMP/cAMP ratio (*Ito* et al. 1980).

Interestingly, PGI_2 and arachidonic acid cause bradycardia after systemic administration to dogs, despite the simultaneous induction of hypotension. The decrease in heart rate could be abolished or reversed to tachycardia by atropine, hexamethonium, or vagal section, suggesting that it is a reflex response which is mediated by intracardial receptors subserved by vagal fibers (*Chapple* et al. 1978; *Hintze* et al. 1978, 1979, 1981; *Kaley* et al. 1980). In addition to the direct vasodilatory activity of PGI_2, this reflex seems to contribute to peripheral vasodilation (*Chapple* et al. 1980) via activation of the left ventricular receptors, resulting in vasodilation in kidney and skeletal muscle (*Hintze* and *Kaley* 1984). Female beagles were less sensitive than males to the PGI_2-induced bradycardia, which seems to be dependent on both the dose and the basal heart rate (*Chiavarelli* et al. 1982).

Since coronary artery spasm has been associated with TX release (*Hirsh* et al. 1981; *Neri Serneri* et al. 1984a), the effect of PGI_2 on TX-induced vasoconstriction has recently been investigated. PGI_2 caused a concentration-related relaxation in human and dog coronary artery strips maximally precontracted with carbocyclic TXA_2. In monkey coronary artery strips precontracted with carbocyclic TXA_2, the contracted vessel did not relax upon PGI_2 administration (*Toda* 1984). In conscious dogs, PGI_2 was found to be able to reverse coronary constriction caused by a TXA_2 mimic, U 46619 (*Hintze* et al. 1984). In both studies, however, PGI_2 was shown to be less effective than nitroglycerin in relaxing precontracted coronary blood vessels (*Hintze* et al. 1984; *Toda* 1984).

A biphasic action of PGI_2 was observed in human coronary ring segments in vitro; at low concentrations it is a potent relaxant, while at concentrations higher than 10^{-6} mol/l it induced a contractile response of the vessel rings (*Davis* et al. 1980; *Ginsburg* et al. 1980; *Toda* 1984). In contrast to cats and dogs, where PGI_2 was found to cause vagal reflex bradycardia, most investigators observed a rise in heart rate during PGI_2 infusion in man, accompanying a decrease in peripheral vascular resistance (*Szczeklik* et al. 1978b; *FitzGerald* et al. 1979; *Warrington* and *O'Grady* 1980; *O'Grady* et al. 1980).

2.6 Thromboxane, Prostaglandin I_2, and Coronary Risk Factors

There is a well-known association between so-called risk factors and coronary heart disease. Though the pathophysiological mechanisms by which these risk factors promote ischemic heart disease are not fully understood, disturbance of TX and PGI_2 biosynthesis has repeatedly been reported.

A reduced PGI_2-synthesizing capacity has been shown in coronary arteries and cultured aortic smooth muscle cells from rabbits fed an atherogenic diet (*Dembinska-Kiéc* et al. 1977; *Larrue* et al. 1982). Similarly, surgical specimens of atherosclerotic arteries from patients released diminished amounts of PGI_2, as judged by bioassay, and no PGI_2-like activity could be detected in human atheromatous plaques (*D'Angelo* et al. 1978; *Sinzinger* et al. 1979). However, recent data demonstrating significantly higher excretion of a urinary PGI_2 metabolite in patients with severe atherosclerosis indicate that reduced PGI_2-synthesizing capacity of vascular tissue in vitro might coexist with an enhanced rate of PGI_2 biosynthesis in vivo (*G.A. FitzGerald* et al. 1984). In patients with atherosclerosis obliterans, moderately elevated 6-keto-$PGF_{1\alpha}$ plasma levels were accompanied by high concentrations of TXB_2 in the circulation (*Jouve* et al. 1984).

Hyperlipidemia has been associated with increased platelet aggregability (*Carvalho* et al. 1974; *Tremoli* et al. 1984). Platelets of type II hyperlipoproteinemic patients have further been demonstrated to release increased amounts of TXB_2, whereas simultaneously higher concentrations of PGI_2 are required to inhibit aggregation of such platelets (*Tremoli* et al. 1979, 1984; *Colli* et al. 1983). The inhibitory activity of low-density lipoproteins on PGI_2 biosynthesis in superfused bovine coronary arteries and in arterial microsomal preparations (*Beitz* and *Förster* 1980; *Szczeklik* and *Gryglewski* 1980) has in part been related to artificial generation of lipid peroxides during the experimental procedure (*Szczeklik* et al. 1981), lipid peroxides being known inhibitors of PGI_2 formation (*Gryglewski* et al. 1976).

Diabetic humans and rats with spontaneous or streptozotocin-induced diabetes mellitus have hyperreactive platelets with increased TX generation in response to arachidonic acid, collagen, ADP, and epinephrine (*Ziboh* et al. 1979; *Butkus* et al. 1980; *Gerrard* et al. 1980; *Subbiah* and *Deitemeyer* 1980; *Halushka* et al. 1981). In the other hand, vascular PGI_2 synthesis was found to be reduced (*Harrison* et al. 1978; *Silberbauer* et al. 1979; *Gerrard* et al. 1980; *Subbiah* and *Deitemeyer* 1980; *Rösen* et al. 1984).

Cigarette smoking has been reported to increase platelet aggregation in venous blood ex vivo (*Levine* 1973). A number of studies have indicated an inhibitory effect of cigarette smoke or nicotine on cardiovascular PGI_2

formation, whereas TX synthesis in platelets is not inhibited and might even be enhanced (*Neri Serneri* et al. 1981b; *Wennmalm* 1982; *Van der Giessen* et al. 1983; *Nadler* et al. 1984). Although there was an increase in arterial TXB_2 concentrations in patients with ischemic heart disease after smoking, no such phenomenon was observed in blood from the coronary sinus and no consistent smoking-induced changes were found in arterial 6-keto-$PGF_{1\alpha}$ concentrations (*Van der Giessen* et al. 1983). On the other hand, in isolated rabbit hearts nicotine inhibits synthesis of PGI_2-like material (*Wennmalm* 1980).

2.7 Involvement of Thromboxane and Prostaglandin I_2 in Ischemic Heart Disease

Even though coronary artery spasm without permanent occlusion might contribute to acute myocardial infarction (*Maseri* et al. 1979; *Luchi* and *Chahine* 1981; *Oliva* 1983), there remains a firm connection between thrombosis and the pathogenesis of myocardial infarction (*Schwartz* 1982). It is known that platelet activation occurs during myocardial ischemia, as evidenced by hyperaggregability, increases in circulating platelet aggregates and increased levels of β-thromboglobulin and platelet factor 4 (*Dreyfuss* and *Zahavi* 1973; *Guyton* and *Willerson* 1977; *Green* et al. 1980; *Neri Serneri* et al. 1981a; *Smitherman* et al. 1981).

Platelets from patients who have survived myocardial infarction generate more TXA_2 from given amounts of arachidonic acid and require significantly less TXA_2 to initiate their aggregation than platelets from healthy subjects (*Szczeklik* et al. 1978a). These results were confirmed by *Mehta* et al. (1980a) in patients with angina pectoris. Further observations demonstrated a decreased sensitivity of platelets from these patients to the antiaggregatory effect of PGI_2 and an increased sensitivity to the proaggregatory effect of U 46619, a PG endoperoxide analogue. The sensitivity of platelets to the antiaggregatory effects of PGI_2 and PGE_1 was described to be further decreased during an angina pectoris attack. In patients with maturity-onset diabetes plus coronary heart disease, platelet sensitivity to PGI_2 and PGE_1 was always lower than in patients with coronary heart disease without diabetes mellitus (*Sinzinger* et al. 1981). Platelets from patients with myocardial infarction, effort angina or spontaneous angina have been demonstrated to release significantly more TXB_2 after stimulation in vitro with thrombin or arachidonic acid than platelets from healthy controls (*Neri Serneri* et al. 1981a,b, 1984b).

Platelet aggregation ex vivo and platelet counts were significantly lower in the coronary venous blood than in the aortic blood of patients with coronary heart disease (*Mehta* et al. 1978). The latter effect has been attributed to platelet sequestration in the atherosclerotic myocardial vasculature. This phenomenon, as well as the increase in the number of platelet aggregates in coronary venous blood during tachycardia stress, could be inhibited by oral administration of aspirin (*Mehta* et al. 1978, 1980b). The platelet trapping in patients with coronary heart disease could not be confirmed by others (*Cortellaro* et al. 1983; *Thaulow* 1983). However, the restoration of the decreased platelet sensitivity to ADP or collagen in the coronary sinus blood of such patients by aspirin and dazoxiben, a selective TX synthetase inhibitor, has been confirmed (*Thaulow* 1983).

During atrial pacing in patients with stable angina, the coronary sinus TXB_2 concentration and the coronary sinus/aortic bulb TXB_2 ratio were significantly increased (*Tada* et al. 1981; *Cortellaro* et al. 1983). Sulfinpyrazone treatment reduced coronary sinus TXB_2 levels at rest and after atrial pacing, but not the ischemic threshold at atrial pacing (*Cortellaro* et al. 1983). Other authors described lower concentrations of TXB_2 in the coronary venous blood than in aortic blood, but also observed an increase in TXB_2 concentrations after atrial pacing (*Mehta* et al. 1984a).

The intracardiac formation of PGI_2 was found to be low or undetectable both in the basal state and during pacing-induced myocardial ischemia (*Edlund* et al. 1983). In one case, coronary artery occlusion after atrial pacing was preceded by the release of high amounts of TX into the coronary circulation, whereas no measurable 6-keto-$PGF_{1\alpha}$ levels could be detected in the coronary venous blood (*Mehta* et al. 1982). Exercise-induced myocardial ischemia was reported to cause a significantly higher rise of peripheral TXB_2 plasma levels, as compared to 6-keto-$PGF_{1\alpha}$ plasma levels. In contrast, in normal subjects a high increase in peripheral 6-keto-$PGF_{1\alpha}$ plasma levels was accompanied by only moderate increases in TXB_2 after exersice (*Mehta* et al. 1981a). Adrenergic stimulation, induced by cold stress, has been claimed to induce a sudden increase in plasma TXB_2 and in circulating PGI_2 levels as determined by bioassay. Under the same experimental conditions, PGI_2 production was reported to be significantly lower in patients with ischemic heart disease than in controls, whereas the increase in TXB_2 in plasma was higher (*Neri Serneri* et al. 1981b, 1984b). Also, in patients at rest with myocardial infarction, stable or unstable angina, levels of circulating PGI_2 have been found to be significantly decreased as compared to those in healthy subjects (*Neri Serneri* et al. 1981b, 1982).

During variant angina attacks, which are caused by coronary artery spasm (*Hillis* and *Braunwald* 1978), but not in the symptom-free interval, the number of platelet aggregates is increased in peripheral blood and in coronary sinus blood (*Robertson* et al. 1980). Elevated levels of TXB_2 in the peripheral blood of patients with Prinzmetal's angina were first described by *Lewy* et al. (1979a,b). An association between increased levels of TXB_2 in the peripheral or coronary sinus blood and variant or unstable angina has been confirmed by other groups (*Hirsh* et al. 1981; *Robertson* et al. 1981; *Tada* et al. 1981; *Mehta* et al. 1984a; *Neri Serneri* 1984). Unlike the observations in patients with variant angina, TXB_2 was undetectable in peripheral blood in patients with classical angina pectoris at rest. Furthermore, spontaneous angina attacks in variant angina patients or pacing-induced angina attacks in classical angina pectoris patients were both followed by increased TXB_2 plasma levels, although responses in the latter patient group were smaller (*Lewy* et al. 1979b). Other data also indicate that the release of TXB_2 into coronary sinus blood is much greater during spontaneous angina than during pacing stress in patients with coronary artery disease (*Mehta* et al. 1984a). Even at rest, peripheral and coronary sinus TXB_2 levels have been reported to be elevated in patients with effort angina, but these values were exceeded by those from patients with variant angina (*Tada* et al. 1981).

Tada et al. (1981) also demonstrated increased TXB_2 levels in two out of five patients with variant angina during a spontaneous attack or attacks induced by ergonovine or by atrial pacing. In patients with spontaneous angina, increased levels of TXB_2 in the coronary sinus were paralleled by decreased levels of 6-keto-$PGF_{1\alpha}$. In contrast to healthy controls, sympathetic stimulation did not induce a significant elevation of 6-keto-$PGF_{1\alpha}$ levels (*Neri Serneri* 1984). *Robertson* et al. (1981), although observing an approximately fivefold rise of coronary sinus TXB_2 levels during spontaneous ischemic attacks, did not observe this increase in samples taken early in ischemia (i.e., within 0.5–1.5 min after ST-segment elevation). Aspirin or indomethacin had no beneficial effect on frequency or duration of ischemia in their patients with vasotonic angina. Patients with unstable angina pectoris who reported chest pain within 24 h of study had higher coronary sinus/aortic blood TXB_2 ratios than patients whose most recent anginal pain attack was more than 96 h before study (*Hirsh* et al. 1981). During and especially after pacing, considerable increases in TXB_2 were observed in coronary sinus and arterial blood of patients suffering from spontaneous angina (*Lewy* et al. 1980).

In conclusion, these findings indicate that PG and TX biosynthesis is altered in ischemic heart disease. Basically, two different deviations in arachidonic acid metabolism can be observed – reduced synthesis of the vasodilator PGI_2 and/or an abnormal generation of TXA_2.

It should be noted that data on PGI_2 (or 6-keto-$PGF_{1\alpha}$) and TXB_2 levels in plasma should be interpreted with great caution, since physiologic concentrations of these compounds are extremely low, and unappropriate sample handling will rapidly stimulate synthesis of TX (*FitzGerald* et al. 1983b). In studies on PGI_2 and TX production in the human coronary circulation, samples are usually obtained after the insertion of catheters. Although some authors have claimed that blood sampling through long catheters is reliable with respect to thromboxane concentrations (*Hirsh* et al. 1982), others found endogenous PGI_2 biosynthesis to be stimulated by cardiac catheterization and angiography (*Roy* et al. 1983). It is also worth noting that heparin, an often used anticoagulant, was reported to potentiate synthesis of TXA_2 in human platelets in response to a variety of agents and to enhance cardiac PGI_2 release in patients suffering from coronary artery disease (*Anderson* et al. 1980; *Wallis* et al. 1982). On the other, increased urinary excretion of a TX metabolite has been shown by gas chromatography/mass spectroscopy after myocardial infarction (*Vesterqvist* and *Gréen* 1984), supporting the conclusions drawn from radioimmunoassay data on TXB_2 levels in plasma.

2.8 The Therapeutic Approach

2.8.1 Inhibition of Thromboxane Synthesis with Platelet-Active Drugs

On the basis of the fact that platelet aggregation is inhibited by aspirin, it was assumed that this drug might have a prophylactic effect in thromboembolic disorders, particularly in coronary artery thrombosis (*Elwood* et al. 1974). This view was supported by an epidemiologic survey of the *Boston Collaborative Drug Surveillance Group* (1974), demonstrating that the percentage of patients regularly using aspirin was significantly lower in a group hospitalized with acute myocardial infarction than in a group hospitalized for other reasons. Six randomized controlled multicenter trials involving the use of aspirin in the treatment of patients recovering from myocardial infarction have been reported (*Elwood* et al. 1974; *Coronary Drug Project Research Group* 1976; *Breddin* et al. 1979; *Elwood* and *Sweetnam* 1979; *Aspirin Myocardial Infarction Study Research Group* 1980; *Persantine-Aspirin Reinfarction Study Research Group* 1980). The dose of aspirin given varied between 300 mg and 1500 mg daily. The entry time after myocardial infarction and the observation time also differed considerably. In all but one study (*Aspirin Myocardial Infarction Study Research Group* 1980), the mortality rate was lower in the aspirin-treated group, but the difference never reached a conventional level of significance. In the *Persantine-Aspirin Reinfarction Study* (1980) patients were treated with 324 mg aspirin or 324 mg aspirin plus 75 mg dipyridamole three times daily. Dipyridamole is a vasodilator which is supposedly increasing vascular PGI_2 production, as well as inhibiting phosphodiesterase activity and adenosine uptake (*FitzGerald* and *Sherry* 1982). No significantly positive results were achieved in this study, either. In all six trials, the incidence of nonfatal myocardial reinfarction was reduced, although not significantly, in each trial. Combining all the evidence available so far, it seems that aspirin has a small beneficial effect in the secondary prevention of myocardial death, and a larger beneficial effect on nonfatal infarction. When the results from all six trials were statistically analyzed together, the effects of aspirin were found to be highly significant (*Editorial* 1980).

The results and the limitations of these trials have been discussed extensively (*Sherry* 1982, 1983; *Mustard* et al. 1983). A major point of criticism must be the dosage of aspirin, which is now known to inhibit not only TXA_2 synthesis in platelets, but also PGI_2 biosynthesis of the vessel wall. Although no dose of aspirin can be expected to be entirely selective on TX synthesis, recent data indicate that a dose of aspirin as low as 20 mg daily causes considerable inhibition of TX synthesis, but only slight inhibition of PGI_2 biosynthesis (*Patrignani* et al. 1982; *Davi* et al. 1983; *FitzGerald* et al. 1983a). An interesting new observation is that long-term daily administration of 1 mg aspirin renders human platelets more sensitive to the antiaggregatory effect of PGI_2 and PGE_1 (*Sinzinger* et al. 1984).

In the recently completed *Veterans Administration Cooperative Study* (*Lewis* et al. 1983), protective effects of aspirin against acute myocardial infarction and death in men with unstable angina have been demonstrated. In this multicenter, double-blind, placebo-controlled randomized trial, patients were treated with 324 mg aspirin daily for 12 weeks. The incidence of death or acute myocardial infarction was 51% lower in the aspirin group than in the placebo group, the difference being highly significant. The reduction in total mortality in the aspirin group was also 51%, though this figure was not statistically significant. This primary prevention trial is the first multicenter study demonstrating a significant protective effect of aspirin in patients with coronary heart disease.

The uricosuric drug sulfinpyrazone, a derivative of the nonsteroidal antiphlogistic drug phenylbutazone, has been employed in two randomized, double-blind multicenter trials (*Anturane Reinfarction Trial Research Group* 1978, 1980; *Anturane Reinfarction Italian Study Group* 1982). Sulfinpyrazone, a cyclooxygenase inhibitor like aspirin, inhibits the platelet release reaction and interferes with PG and TX synthesis in platelets (*FitzGerald* and *Sherry* 1982). Sulfinpyrazone (4 × 200 mg daily) reduced cardiac death by 48.5% and sudden cardiac death by 57.2% during the 1st year, and by 32% and 43% respectively at 24 months after first myocardial infarction. With the exception of the 32% reduction in cardiac death at the end of the 2nd year, differences were found to be significant (*Anturane Reinfarction Trial Research Group* 1978, 1980).

The benefit of sulfinpyrazone was attributable entirely to a reduction in sudden death during the 2nd through 7th month after first infarction.

The *Anturane Reinfarction Study* was given a great deal of publicity because it was the first study with a platelet-active drug claiming a major benefit in reducing reinfarction (*Mustard* et. al. 1983; *Sherry* 1983). On the other hand, in the *Anturane Reinfarction Italian Study* (1982) a significant 56% reduction of reinfarction was observed, whereas there was no apparent effect on sudden death. Nevertheless, a significant reduction of what were considered to be thromboembolic endpoints was reported. In contrast to the *Anturane Reinfarction Trial Research Group* (1978, 1980), no significant differences could be detected with respect to cardiac death within 24 months.

2.8.2 Dietary Manipulation of Thromboxane Synthesis

Greenland Eskimos, whose diet is rich in ω-3-polyunsaturated fatty acids, have a prolonged bleeding time related to the reduction of ex vivo platelet aggregation, and a very low incidence of cardiovascular death (*Dyerberg* et al. 1975, 1978; *Dyerberg* and *Bang* 1979). On the basis of these observations, it was suggested that 5,8,11,14,17-eicosapentaenoic acid (EPA), the main fatty acid available for PG biosynthesis in Greenland Eskimos, might be useful for preventing the thrombotic complications of cardiovascular disease (*Dyerberg* et al. 1978).

EPA metabolism via cyclooxygenase gives rise to trienoic PG and can be converted by vascular tissue in vitro into an antiaggregatory substance (*Dyerberg* et al. 1978). Bovine aortic microsomes were found to convert PGH_3 to PGI_3, which has an antiaggregatory effect similar to PGI_2. On the other hand, microsomes from aspirin-treated platelets synthesized TXA_3 from PGH_3. In contrast to PGH_2 and TXA_2, the trienoic compounds, by increasing platelet cAMP levels, inhibited aggregation of platelet-rich plasma by other agonists (*Needleman* et al. 1979). While EPA is readily acylated into platelet phospholipids, it has been shown to be a relatively poor substrate for platelet cyclooxygenase. It is, however, an inhibitor of arachidonic acid metabolism by platelet cyclooxygenase and lipoxygenase (*Needleman* et al. 1979). EPA inhibits thrombin-induced aggregation and aggregation of imidazole- and aspirin-treated platelets, as well as aggregation of platelet-rich plasma induced by collagen, ADP, arachidonic acid, and an endoperoxide analogue (*Gryglewski* et al. 1979; *Jakubowski* and *Ardlie* 1979; *Dyerberg* and *Jørgensen* 1980). Thus, it is obvious that PG-dependent and PG-independent pathways of platelet aggregation are inhibited by EPA.

Dyerberg et al. (1981) reported the release of Δ^{17}-6-keto-$PGF_{1\alpha}$, the degradation product of PGI_3, from human umbilical artery incubated with EPA. They also found trace amounts of this substance after incubation of rat aorta with EPA. Other investigators failed to detect vascular

PGI_3 formation after feeding rats with EPA or after incubation of rat aortic tissue with EPA (*Hornstra* et al. 1981; *Hamazaki* et al. 1982). *Hornstra* et al. (1981) reported inhibition of TXA_2 and PGI_2 production by rat platelets and aorta respectively after dietary administration of EPA. Others found an increased PGI_2 synthesis (*Hamazaki* et al. 1982) or decreased PGI_2 synthesis in the presence of unchanged TX production. This latter biochemical effect was associated with a rise in arterial blood pressure of the EPA-treated rats (*Scherhag* et al. 1982).

Studying healthy Caucasians who had been on a mackerel diet for 1 week, *Siess* et al. (1980) demonstrated decreased ex vivo platelet aggregation and TX production after low-dose collagen stimulation, as compared to control values obtained during normal diet. The change in platelet function was associated with an increase in the C20:5/C20:4 platelet phospholipid ratio. Platelet aggregation induced by exogenous arachidonic acid was unchanged, indicating normal cyclooxygenase activity.

Decreased platelet aggregation and TX formation induced by collagen were also observed during daily supplementation of 25 ml of cod liver oil to the diet of volunteers (*Brox* et al. 1981). Administration of cod liver oil equivalent to 1.8 g EPA daily for 6 weeks resulted in a significant increase in bleeding time, accompanied by an increase in EPA in the platelet phospholipids (*Sanders* et al. 1980). In contrast to rat platelets (*Hornstra* et al. 1981), TXA_3 biosynthesis could be demonstrated by gas chromatography/mass spectrometry in human platelet-rich plasma if the subjects had ingested cod liver oil equivalent to 4 g EPA daily for 25 days (*Fischer* and *Weber* 1983). Furthermore, formation of PGI_3 in man has been demonstrated after dietary EPA which increased with increasing EPA incorporation into plasma phospholipids (*Fischer* and *Weber* 1984). Release of immunoreactive TXB_2 and platelet aggregation ex vivo by collagen stimulation was simultaneously decreased. Interestingly, excretion of 2,3-dinor-6-keto-$PGF_{1\alpha}$, a urinary metabolite of PGI_2, was unchanged (cod liver oil diet) or even increased (mackerel diet) under the influence of dietary EPA (*Fischer* and *Weber* 1984), whereas in cultured human endothelial cells PGI_2 production was reported to be inhibited by EPA (*Brox* and *Nordøy* 1963; *Spector* et al. 1983).

A beneficial effect of EPA has been shown in experimental myocardial infarction in dogs receiving a dietary supplementation of fish oil rich in this fatty acid (*Culp* et al. 1980). In fish oil-fed dogs the size of myocardial infarction induced by electrical stimulation of the left circumflex coronary artery was reduced by 88% compared to control dogs, and treated animals had significantly fewer ectopic beats.

As discussed in Sect. 3, the synthesis of proinflammatory and chemotactic eicosanoids may be of considerable importance during myocardial infarction. Therefore, it is interesting to note that EPA is a good substrate for lipoxygenase enzymes (*Terano* et al. 1984a). In rats fed an EPA-rich diet, the EPA content of leukocyte phospholipids was significantly increased and the ratio of EPA/arachidonic acid correlated with the LTB_5/LTB_4 ratio produced after stimulation of the leukocytes (*Terano* et al. 1984b). These results are of great interest since LTB_5 is at least 30 times less active than LTB_4 with respect to the aggregation of rat neutrophils, the chemokinesis of human polymorphonuclear leukocytes, lysosomal enzyme release from human leukocytes, and potentiation of bradykinin-induced plasma exudation (*Terano* et al. 1984a).

2.8.3 Selective Thromboxane Synthetase Inhibition

Imidazole was the first compound reported to inhibit TX formation without affecting cyclooxygenase activity (*Moncada* et al. 1977a). A selective inhibitor of TX synthetase should be expected to be an antithrombotic agent superior to cyclooxygenase inhibitors such as aspirin, as it would not affect PGI_2 formation by vessel walls or other cells (*Bunting* et al. 1983). Recently, numerous TX synthetase inhibitors with in vivo activity have been described. These include imidazole derivatives such as UK-37248-01 (*Tyler* et al. 1981), UK-38385 (*Fischer* et al. 1983), OKY-046 (*Ogawa* et al. 1983), and the pyridine derivatives OKY-1581 (*Feuerstein* and *Ramwell* 1981), U-63557A (*Gorman* et al. 1983), and CGS-13080 (*Ku* et al. 1983).

Dazoxiben (UK-37248-01) protected rabbits against sudden death which was induced by injection of sodium arachidonate. It prevented the increase in circulating TXB_2 levels and the pulmonary artery thrombosis usually observed following arachidonate injection (*Lefer* et al. 1981). In dogs, systemic administration of dazoxiben significantly prolonged the time required for occlusive thombus formation at sites of electrical injury in the coronary artery and simultaneously decreased venous TXB_2 concentrations (*Schumacher* and *Lucchesi* 1983). Cyclic coronary flow reductions that occur in experimentally stenosed canine coronary arteries were abolished by dazoxiben in 65% of the animals, and it significantly reduced their frequency in the other dogs. TXB_2 levels measured distal to the stenosis were reduced by dazoxiben to levels observed before the plastic constrictor was implanted (*Bush* et al. 1984).

The reports demonstrating that inhibition of TX biosynthesis was accompanied by an increased formation of 6-keto-$PGF_{1\alpha}$-like immunoreactivity (*Dale* et al. 1983; *Menys* and *Davies* 1983; *Carter* et al. 1984) might need some reevaluation, since some 6-keto-$PGF_{1\alpha}$ antibodies seem to cross-react with unidentified compounds arising after dazoxiben treatment (*Chiabrando* et al. 1984). However, *Menys* and *Davies* (1983) validated their results by radiochromatography. Administration of dazoxiben significantly reduced the ischemic response and the release of TXB_2 in the coronary sinus in patients with coronary heart disease during atrial pacing. Collagen-induced platelet aggregation ex vivo was inhibited in blood from the vena cava after dazoxiben treatment (*Thaulow* et al. 1984). An advantageous effect with respect to the inhibition of platelet aggregation has been reported in healthy volunteers after treatment with low-dose aspirin in combination with dazoxiben, as compared to either drug alone (*Cerletti* et al. 1984).

UK-38485 administered orally significantly reduced serum TXB_2 concentrations. In a multiple-dose, double-blind, placebo-controlled, crossover study, UK-38485 given twice daily reduced excretion of the major urinary TX metabolite, whereas excretion of a major 6-keto-$PGF_{1\alpha}$ metabolite showed no significant increase compared to levels in a placebo period (*Fischer* et al. 1983). Compared with dazoxiben, inhibition of TXB_2 synthesis by UK-38485 was prolonged, and the effective doses were considerably lower.

Similar to UK-37248-01, OKY-046 protected rabbits against arachidonate-induced sudden death by inhibiting TX synthesis and pulmonary thrombosis (*Edmonds* and *Lefer* 1984). In dogs with coronary artery ligation, OKY-046 prevented cardiac lactate release, the increase in TX in the great cardiac vein, and coronary venous flow reduction (*Ogawa* et al. 1983).

U-63557A is a potent inhibitor of TX synthetase in human platelets in vitro and in monkey platelets ex vivo. After systemic administration to monkeys, it effectively prevented the blockade of experimentally stenosed coronary arteries caused by platelet aggregation (*Gorman* et al. 1983).

OKY-1581, a pyridine derivative, administered in vitro or in vivo, inhibited TXA_2 generation related to clotting or aggregation induced by arachidonic acid or collagen in rabbit platelets (*Smith* and *Jubiz* 1981; *Naito* et al. 1983). Administered systemically in man, the compound reduced plasma TXB_2 levels as well as TXB_2 production in whole blood in vitro, whereas 6-keto-$PGF_{1\alpha}$ levels were simultaneously significantly increased in the serum. Platelet aggregation ex vivo induced by arachidonic acid was also significantly reduced (*Yui* et al. 1984). In dogs, OKY-1581 was found to be an effective inhibitor of coronary thrombosis induced by partial obstruction or electrical injury (*Aiken* et al. 1981; *Shea* et al. 1984a).

Oral administration of CGS-13080 in man prolonged bleeding time and inhibited TX formation (*FitzGerald* and *Oates* 1984; *MacNab* et al. 1984). Endogenous PGI_2 biosynthesis as measured by excretion of the major urinary metabolite was concomitantly doubled. Arachidonic acid-induced platelet aggregation was inhibited, whereas collagen-induced aggregation remained unaffected. The effect of 200 mg CGS-13080 was estimated to be lower than that of 20 mg aspirin (*FitzGerald* and *Oates* 1984). The half-life of about 1 h might limit possible uses of the drug. Nevertheless, during acute myocardial ischemia in cats, infusion of CGS-13080 preserved myocardial tissue, as judged by plasma and tissue creatine kinase activities and myocardial aminonitrogen concentration. In plasma, TXB_2 concentrations did not increase during myocardial ischemia, and platelet TX production ex vivo was found to be inhibited (*Burke* et al. 1983a).

2.8.4 The Use of Prostaglandin I_2 in Coronary Heart Disease

During experimentally induced acute myocardial infarction in anesthetized cats, PGI_2 infusion was shown to inhibit ischemia-induced platelet activation, to restore the infarction-induced decrease in circulating platelet count, and to suppress the ischemia-induced increase in myocardial cAMP (*Ohlendorf* et al. 1980; *Rösen* et al. 1981; *Schrör* et al. 1981a). The systemic administration of PGI_2 also prevented or moderated the increase of myocardial lactate content as well as the loss of creatine phosphokinase from ischemic myocardial tissue, and abbreviated the duration of significant ST-elevations as compared to controls (*Ogletree* et al. 1979; *Rösen* et al. 1981).

In dogs, PGI_2 infusion prevented coronary artery thrombosis initiated by electrical stimulation or the complete blockade of partially obstructed coronary arteries, which correlated with inhibition of platelet aggregation (*Aiken* et al. 1979; *Romson* et al. 1981). The concomitant use of the phosphodiesterase inhibitor aminophylline permitted a 20-fold decrease in the PGI_2 dose necessary for maintaining an equal antithrombotic efficacy (*Romson* et al. 1983b). Infarct size was reported to be smaller after acute permanent occlusion of coronary arteries in dogs treated with intravenous PGI_2 infusions, and coronary blood flow increased in these dogs during the infusion period throughout the occluded bed, but not in control dogs (*Jugdutt* et al. 1981; *Ribeiro* et al. 1981). Although PGI_2 significantly increased collateral flow, it was suggested that other cellular and metabolic effects might contribute to the protective action of PGI_2 (*Jugdutt* et al. 1981; *Ribeiro* et al. 1981).

Intravenous infusion of PGI_2 in patients with stable angina pectoris induced a decrease in coronary vascular resistance (*Bergman* et al. 1981; *Ganz* et al. 1984). Whereas the cold pressure response was unaffected by infusion of PGI_2 (*Ganz* et al. 1984), mean atrial pacing time to angina rose significantly from 142 to 241 s, while lactate production during rapid atrial pacing was concomitantly decreased (*Bergman* et al. 1981). There are conflicting results with respect to the efficacy of PGI_2 in various syndromes of myocardial ischemia. PGI_2 was found to be effective in relieving episodes of spontaneous angina at rest (*Szczeklik* et al. 1980), a finding which was not confirmed by others (*Chierchia* et al. 1982), who reported intravenous administration of PGI_2 to be without effect on the number, severity, and duration of ischemic episodes in eight out of nine patients with variant angina. In contrast to *Bergman* et al. (1981), *Szczeklik* et al. (1980) did not find that PGI_2 made any beneficial contribution in pacing-induced angina.

In patients with acute myocardial infarction, PGI_2 has been successfully employed in the recanalization of acutely obstructed coronary arteries by intracoronary administration either alone or in combination with streptokinase or urokinase (*Blaskó* et al. 1983; *Uchida* et al. 1983). Intravenous infusion of PGI_2 in patients with acute myocardial infarction has been reported to limit infarct size and to protect against infarct extension (*Wennmalm* et al. 1984).

Although the pharmacologic actions of PGI_2 are potentially valuable in the treatment of certain cardiovascular disorders, its clinical use is limited by its ineffectiveness when given orally and by its tendency to cause hypotension. Therefore, numerous attempts have been made to synthesize stable PGI_2 analogues.

In carbacyclin, the 6,9-enolether in the α-side chain of PGI_2 was replaced by a methylene group, resulting in a chemically stable but metabolically unstable PGI_2 analogue. The pharmacologic profile is similar to that of the parent compound, but it is only about 1/10 as potent (*Aiken* and *Shebuski* 1980; *Whittle* et al. 1980). A carbocyclic derivative with a modified ω-side chain, ONO-41483, was several times more active than carbacyclin in inhibiting ADP-induced ex vivo platelet aggregation in man after oral or intravenous administration (*Adaikan* et al. 1983).

Another carboprostacyclin derivative, ZK-36374, was reported to have a greater ex vivo antiaggregatory activity than PGI_2 itself, whereas the vasodepressor activity was equipotent to PGI_2 in baboons (*Adaikan* et al. 1984). Intravenous infusion of ZK-36374 in healthy volunteers significantly decreased aggregation to ADP and collagen in platelet-rich plasma or whole blood, but did not significantly affect blood pressure or heart rate (*Belch* et al. 1984). In human platelet-rich plasma, ZK-36374 was found to be between two and five times more potent than PGI_2 (*Casals-Stenzel* et al. 1983). It has been shown that ZK-36374 has beneficial effects in acute myocardial ischemia in cats, reducing ST-segment elevation and preventing the loss of myocardial creatine kinase-specific activities and of myocardial catecholamines from adrenergic nerve terminals (*Schrör* et al. 1981d, 1982; *Smith* et al. 1984). The protective effect was still significant when ZK-36374 infusion was started 4 h after coronary artery ligation and was suggested to be dissociated from the antiplatelet effects (*Schrör* et al. 1982; *Smith* et al. 1984). In primates, ZK-36374 antagonized vasopressin-induced ECG changes (*Adaikan* et al. 1984).

Recently, CG-4203, a new stable PGI_2 analogue, has also been shown to produce cardioprotective actions in rat models of cardiac hypoxia and ischemia (*Müller* et al. 1984).

Another interesting approach is the stimulation of endogenous PGI_2 biosynthesis by drugs. Nafazatrom, a pyrazolone derivative, has been shown to stimulate PGI_2 release from endothelial cells and cardiac vessel walls (*Vermylen* et al. 1979; *Klitzke* 1984). The compound exerted beneficial effects on experimental coronary thrombosis or myocardial ischemia in rats, dogs, and rabbits (*Fiedler* 1983, 1984; *Shea* et al. 1984b). Interestingly, nitrates have also been suggested to stimulate coronary vascular PGI_2 synthesis (*Schrör* et al. 1981b). In man, however, increased excretion of the major urinary PGI_2 metabolite was not observed after intravenous infusion of nitroglycerine (*D.J. FitzGerald* et al. 1984).

3 Anti-Inflammatory Drugs and the Inflammatory Component in Acute Myocardial Infarction

3.1 Myocardial Infarction and Inflammation

The pathophysiological events following acute myocardial infarction resemble in many aspects those during acute inflammation. There is cellular infiltration with platelets, leukocytes (*Weiss* et al. 1977; *Davies* et al. 1981), and other inflammatory cells (*Sommers* and *Jennings* 1964), which normally assist in phagocytosis, but have the potential of damaging nonnecrotic myocardial tissue as well. Edema (*Reimer* and *Jennings* 1979), pain, and loss of function occur (*Hillis* and *Braunwald* 1977). In addition, there is release of chemical mediators such as PGs, including PGI_2 (*Berger* et al. 1976; *Ogawa* et al. 1980; *Coker* et al. 1981; *Sakai* et al. 1982), TXA_2 (*Coker* et al. 1981; *Sakai* et al. 1982; *Tanabe* et al. 1982), mono-HETEs (*Mullane* et al. 1984), and other unidentified chemotactic factors, some of which are probably complement derived (*Hill* and *Ward* 1971; *Hartmann* et al. 1977).

Histopathological studies have confirmed the importance of the inflammatory component in acute myocardial infarction in man (*Mallory* et al. 1939; *Fishbein* et al. 1978). The acute inflammation predominating in the 1st week is followed by chronic inflammation in the 2nd week, and connective tissue proliferation and collagen deposition in the 3rd week, until healing is completed by formation of a firm, contracted scar after 6–9 weeks (*Mallory* et al. 1939; *Fishbein* et al. 1978).

Although the presence of leukocytes in infarcted tissue has long been recognized (*Mallory* et al. 1939), most of the recent studies have used indium-111-labeled leukocytes to detect cell infiltration into the myocardium (*Weiss* et al. 1977; *Davies* et al. 1981). Methodological problems with this technique prevent an early assessment of the cellular invasion. However, cell migration as early as 24 h after vascular occlusion and reperfusion of the canine left circumflex coronary artery has been demonstrated (*Romson* et al. 1982a). In a recent experimental study in dogs, polymorphonuclear, leukocyte adhesion to the endothelium and diapedesis into subendothelial layers, already apparent after a 60-min occlusion period plus a 30-min reperfusion period, were reported (*Mullane* et al. 1984). After 5 h, extensive banding of white cells within the damaged tissue and large numbers of polymorphonuclear leukocytes distributed throughout the infarct were observed.

Among other oxygenation products of arachidonic acid, rabbit and human polymorphonuclear leukocytes have been shown to synthesize 12-HETE and 5,12-DHETE (dihydroxy eicosatetraenoic acid), which was later named LTB_4 (*Borgeat* and *Samuelsson* 1979a,b; *Samuelsson* et al. 1979). The precursor of 12-HETE, 12-HPETE, is known to constrict cat coronary arteries perfused in vitro (*Trachte* et al. 1979), whereas its reduced metabolite, 12-HETE, is a potent chemotactic factor for polymorphonuclear leukocytes (*Turner* et al 1975). LTB_4 stimulates directional (chemotaxis) and random (chemokinesis) movements of polymorphonuclear leukocytes (*Ford-Hutchinson* et al. 1980; *Palmblad* et al. 1981) and induces their aggregation (*Ford-Hutchinson* et al. 1980), besides causing adhesion of leukocytes to endothelial cells (*Dahlén* et al. 1981; *Björk* et al. 1982). Although neutrophils respond to most lipoxygenase products of arachidonic acid metabolism by undergoing chemotaxis (*Goetzl* and *Pickett* 1980), the complete stimulus-secretion coupling sequence with release of lysosomal enzymes and generation of superoxide anions is launched only by LTB_4 (*Goetzl* and *Pickett* 1980). Although no LT has been described yet as occurring in infarcted myocardium, 12- and 15-HETE and another yet unidentified HETE have been detected in infarcted tissue of the dog, and the production of these lipoxygenase products of arachidonic acid metabolism has been related to the invading leukocytes (*Mullane* et al. 1984).

On the basis of these pathophysiological findings, one would infer that controlling the inflammatory response might be beneficial in acute myocardial infarction. Thus, inactivation of chemotactic C3-fragments by a cobra venom factor during myocardial infarction in the rat successfully inhibited the inflammatory response (*Hill* and *Ward* 1971). The same strategy used in dogs resulted in a protective effect against ischemic injury, as judged by ST-segment elevation and cardiac creatine kinase content (*Hartmann* et al. 1977). Depletion of neutrophils by administration of neutrophil antiserum or treatment with hydroxyurea resulted in significantly smaller infarct sizes in dogs (*Romson* et al. 1982b, 1983a;

Mullane et al. 1984). However, the most practical approach seems to be the use of anti-inflammatory drugs inhibiting leukocyte migration.

3.2 Myocardial Infarction and Corticosteroids

Anti-inflammatory steroids prevent the release of arachidonic acid (*Gryglewski* 1976; *Hong* and *Levine* 1976), thereby inhibiting the formation of not only PG and TX, but also LT and other oxygenated derivatives of arachidonic acid. It was recently proposed that anti-inflammatory steroids act by stimulating the synthesis of an inhibitor of phospholipase A_2 (*Blackwell* et al. 1980; *Hirata* et al. 1980). Corticosteroids were reported by some investigators to decrease experimental ischemic injury and infarct size (*Johnson* et al. 1953; *Libby* et al. 1973; *Maclean* et al. 1978a) and stabilize lysosomal and cellular membranes (*Spath* et al. 1974; *Welman* et al. 1979). Others found no improvement in metabolic or mechanical function (*Osher* et al. 1976; *Karmazyn* 1983) and no decrease in infarct size, but merely a delay in myocardial enzyme depletion (*Vogel* et al. 1977). Corticosteroids inhibited early inflammation and phagocytosis in rats, but they delayed the clearance of dead myocytes, resulting in mummification and impaired collagen deposition and healing (*Kloner* et al. 1978). Thus, corticosteroid treatment of rats with acute myocardial infarction resulted in thinning of the scar (*Maclean* et al. 1978a). Furthermore, in patients with acute myocardial infarction, the use of corticosteroids was associated with the formation of ventricular aneurysms and threatening ventricular rupture (*Bulkley* and *Roberts* 1974).

In another study, completed infarcts of methylprednisolone-treated patients exceeded predicted size by 72%, which was accompanied by prolonged elevations in serum levels of creatine phosphokinase and persistence of malignant ventricular dysrhythmia (*de Mello* et al. 1975; *Roberts* et al. 1976). In rats with acute myocardial infarction, 10%–20% mummified cells were seen after 48 h in groups receiving no treatment, cobra venom factor, or ibuprofen, as compared to about 72% in the group receiving anti-inflammatory steroids (*Kloner* et al. 1978; *Maclean* et al. 1978a,b). Furthermore, thinning of the remaining myocardial scar, which has been described after treatment with corticosteroids (*Maclean* et al. 1978a), was not detected after treatment with cobra venom factor, hyaluronidase, or ibuprofen (*Kloner* et al. 1978; *Maclean* et al. 1978a,b). Therefore it seems that the mummification is due to the corticosteroids rather than to the anti-inflammatory action.

3.3 Myocardial Infarction and Nonsteroidal Anti-Inflammatory Drugs

The cyclooxygenase inhibitor aspirin did not reduce infarct size (*Bonow* et al. 1981) and had little effect on leukocyte migration into an inflammatory lesion (*Higgs* et al. 1980). Pretreatment of dogs with indomethacin resulted in an increased infarct size (*Jugdutt* et al. 1979), which is probably due to the coronary vasoconstriction caused by this substance in man and dog (*Kirmser* et al. 1976; *Friedman* et al. 1981). The same drug did not affect infarct size or leukocyte infiltration into the infarcted canine myocardium when given 10 min before reperfusion (*Mullane* et al. 1984). Also in cats no protective effect of indomethacin was observed as judged by biochemical parameters (*Ogletree* and *Lefer* 1976).

Recently it has been shown that there is a differential activity of several nonsteroidal anti-inflammatory drugs on leukocyte migration and PG synthesis (*Higgs* et al. 1980), which could lead to different clinical effects according to the dose used. It was suggested that the inhibition of leukocyte accumulation in inflammation by these substances is due to an inhibitory effect on lipoxygenase with subsequently reduced synthesis of chemotactic arachidonic acid metabolites (*Higgs* et al. 1980).

Naproxen, which has been shown to inhibit leukocyte migration at low doses (*Higgs* et al. 1980) while having no effect or a slightly potentiating one at high doses (*Meacock* and *Kitchen* 1976), has effects on myocardial infarct size in dogs closely correlating with the effects on leukocytes. Thus a dose of 1 mg/kg has been found to protect the ischemic myocardium without altering hemodynamic parameters (*Smith* and *Lefer* 1981), whereas a dose of 30 mg/kg did not have any beneficial effect on infarct size (*Bolli* et al. 1981).

Ibuprofen, which inhibits leukocyte accumulation during inflammation (*Meacock* and *Kitchen* 1976; *Higgs* et al. 1980), has been reported by several groups to salvage ischemic myocardium (*Maclean* et al. 1978b; *Lefer* and *Polansky* 1979; *Jugdutt* et al. 1980; *Romson* et al. 1982c; *Wilkerson* and *Conran* 1982). While ibuprofen protected the myocardium in vivo without any apparent hemodynamic effects (*Lefer* and *Polansky* 1979; *Jugdutt* et al. 1980), a markedly decreased leukocyte infiltration has been observed in the infarcted tissue of dogs treated with this drug (*Romson* et al. 1982a).

Fluribiprofen, a halogenated derivative of ibuprofen, also revealed a protective effect in experimental myocardial infarction (*Smith* et al. 1980; *Darsee* et al. 1981) and has been shown to inhibit leukocyte migration if higher doses are employed (*Adams* et al. 1977; *Higgs* et al. 1980).

The dual inhibitor BW-755C, which inhibits the cyclooxygenase pathway as well as the lipoxygenase pathway of arachidonic acid metabolism (*Higgs* et al. 1979), completely abolished release of 12-HETE from infarcted tissue, which has been found to release ten times more 12 HETE/15-HETE than normal tissue (*Mullane* et al. 1984). In dogs, salvage of ischemic myocardium by BW-755C has been reported (*Lucchesi* et al. 1982; *Mullane* and *Moncada* 1982). A considerable decrease in polymorphonuclear leukocyte infiltration was accompanied by a significant 65% reduction in infarct size. Similar data were obtained by neutrophil depletion using hydroxyurea (*Mullane* et al. 1984). In contrast to many other studies BW-755C has been successfully administered after the ischemic period. Recently, BW-755C has also been shown to reduce the rate of tissue damage in experimental inflammation (*Higgs* et al. 1984).

Thus it seems reasonable to assume that the protective effects of certain anti-inflammatory drugs are at least partly due to their inhibitory action on the synthesis of chemotactic arachidonic acid metabolites, apart from their effects on platelets, lysosomal membranes, and various cell functions.

4 Eicosanoids in Cardiac Anaphylaxis

4.1 The Heart and the Anaphylactic Reaction

According to the classification of *Coombs* and *Gell* (1968), anaphylaxis is a type I (immediate hypersensitivity) immunopathological event mediating tissue injury. At the cellular level, anaphylaxis is caused by the interaction of the antigen with the appropriate mast cell-associated IgE antibodies in sensitized tissue. However, in some species, including the guinea pig, rat, and mouse, some subclasses of IgG antibodies which can also fix to homologous mast cells might be involved as well. The interaction of the antigen with the adjacent IgE molecules triggers biochemical events resulting in the degranulation of mast cells and the release of mediators of anaphylaxis, including histamines, 5-hydroxytryptamine, slow-reacting substance of anaphylaxis, platelet-activating factor (PAF), and other potent compounds (*Plaut* and *Lichtenstein* 1978). The anaphylactic syndrome is presumably caused by the effects of the liberated mediators. Symptoms like urticaria, nausea, diarrhea, laryngeal edema and bronchospasm, as well as arrhythmia or cardiac arrest, may occur either alone or in combination, and reveal the pathophysiological changes of organ function (*Kelly* and *Patterson* 1974). Death may occur from severe respiratory obstruction, irreversible cardiovascular collapse or both (*Kelly* and *Patterson* 1974; *Austen* 1978).

In man, anaphylactic reactions to drugs or to some foods have been accompanied by electrocardiographic changes, including all types of cardiac arrhythmias, heart blocks, bundle-branch block, and even acute myocardial infarctions (*Binder* et al. 1950; *Lilienfeld* et al. 1950; *McManus* and *Lawlor* 1950; *Foster* and *Layman* 1952; *Bernreiter* 1959; *Hanashiro* and *Weil* 1967; *Booth* and *Patterson* 1970; *Stein* and *Wecksell* 1970; *Criep* and *Woehler* 1971; *Petsas* and *Kotler* 1973). Anaphylactic cardiac arrest has been reported in a parturient after intravenous injection of propanidid (*Baraka* and *Sfeir* 1980). Cases of coronary spasm were associated with urticaria (*Foster* and *Layman* 1952; *Bristow* et al. 1982), and within minutes of a bee sting symptoms and signs of acute myocardial infarction were observed (*Hirsh* 1982).

The heart has now been recognized as a primary target organ in anaphylactic reactions (*Feigen* and *Prager* 1969; *Criep* and *Woehler* 1971; *Capurro* and *Levi* 1975; *Zavecz* and *Levi* 1977; *Levi* et al. 1982). A substantial body of clinical evidence suggests that arrhythmia and cardiac failure may occur without preceding respiratory distress, which otherwise might be the cause of secondary cardiac symptoms (*Austen* 1965; *Stein* and *Wecksell* 1970; *Austen* 1978). For rabbits it was claimed very early that fatal outcome of anaphylaxis is due to cardiac failure (*Auer* 1911). In guinea pigs, on the other hand, the dramtic course of anaphylactic shock, resulting in death by suffocation within a few minutes, has directed the attention to the lungs as a primary shock organ (*Auer* and *Lewis* 1910). However, isolated sensitized guinea pig hearts react very strongly to antigen, with a long-lasting reduction in coronary flow coupled with marked impairment of cardiac function (*Went* and *Lissâk* 1935; *Wilcox* and *Andrus* 1938; *Feigen* and *Prager* 1969; *Hahn* and *Bernauer* 1970), and electrocardiographic changes have been found that are similar to those observed in humans during anaphylaxis (*Criep* 1931; *Capurro* and *Levi* 1975; *Zavecz* and *Levi* 1977). Furthermore, upon immunologic challenge the release of histamine and other mediators such as SRS-A and other eicosanoids from hearts has been shown (*Brocklehurst* 1960; *Chakravarty* 1960a,b; *Feigen* et al. 1960; *Giotti* et al. 1966; *Levi* 1972; *Liebig* et al. 1975; *Allan* and *Levi* 1981; *Aehringhaus* et al. 1983).

4.2 Cyclooxygenase Products of Arachidonic Acid in Cardiac Anaphylaxis

Release of PG during cardiac anaphylaxis was first described by *Liebig* et al. (1975) in the isolated guinea pig heart. Ovalbumin challenge of the sensitized heart induced release of $PGF_{2\alpha}$ and PGE_2, the latter being found in trace amounts only. These data were confirmed by others using passively sensitized guinea pigs (*Levi* et al. 1976). Indomethacin was found to inhibit PG release, while release of histamine and SRS-A was simultaneously increased (*Liebig* et al. 1975; *Levi* et al. 1976). Since tachycardia and arrhythmia were even more severe than expected on the basis of the enhanced histamine release, it has been suggested that PG might modulate histamine release and the effects of released histamine during cardiac anaphylaxis (*Levi* et al. 1976).

In addition, release of TXB_2 (*Anhut* et al. 1977, 1978a,b; *Allan* and *Levi* 1981; *Aehringhaus* et al. 1983), PGD_2 (*Anhut* et al. 1978a), and 6-keto-$PGF_{1\alpha}$ (*Peskar* et al. 1979; *Allan* and *Levi* 1981; *Aehringhaus* et al. 1983) from anaphylactic guinea pig hearts has been reported. As in the guinea pig lung, TX release predominates during anaphylaxis in guinea pig hearts (*Dawson* et al. 1976; *Anhut* et al. 1977, 1978a; *Allan* and *Levi* 1981; *Aehringhaus* et al. 1983), while under basal conditions lungs (*Gryglewski* et al. 1978) and hearts (*De Deckere* et al. 1977; *Schrör* et al. 1978) release mainly PGI_2. TXA_2, the unstable precursor of TXB_2, is known to cause platelet aggregation and contraction of rabbit aorta (*Hamberg* et al. 1975). In addition, TXA_2 has been described as strongly contracting porcine coronary artery strips (*Ellis* et al. 1976; *Svensson* and *Hamberg* 1976) and canine coronary arteries (*Imai* et al. 1983), and to induce coronary vasospasm without direct negative inotropic effects in isolated, working guinea pig hearts (*Terashita* et al. 1978). These data suggest a participation of endogenous TX in anaphylactic coronary vasoconstriction. In fact, *Allan* and *Levi* (1981) reported that inhibition of TX release by 1-(2-isopropylphenyl)imidazole or high concentrations of aspirin or indomethacin ($> 10^{-5}$ mol/l) was accompanied by an inhibition of antigen-induced coronary vasoconstriction during a 10-min period after challenge. However, considering the high concentrations of the inhibitors used, nonspecific effects cannot be excluded. *Anhut* et al. (1978a), on the other hand, using imidazole as a TX synthetase inhibitor and employing a more differentiated observation schedule for coronary flow, reported a significant decrease in TXB_2 release from the hearts, but no change in anaphylactic coronary vasoconstriction was observed. These authors found an increased release of $PGF_{2\alpha}$ and PGD_2, which is supposedly due to an increased substrate availability after inhibition of TX synthetase by imidazole (*Nijkamp* et al. 1977).

In contrast to PGI_2 and PGE_2 (*Schrör* and *Krebs* 1976; *Schrör* and *Moncada* 1979; *Allan* and *Levi* 1980; *Imai* et al. 1983), $PGF_{2\alpha}$ and PGD_2 exhibit vasoconstrictor activity in coronary arteries of the guinea pig, pig, and dog (*Svensson* and *Hamberg* 1976; *Anhut* et al. 1978a; *Schrör* 1978; *Allan* and *Levi* 1980; *Imai* et al. 1983). Therefore, it could not be excluded that the increased release of vasoconstrictor PG mimicked the effect of TXA_2 during imidazole infusion. The involvement of cyclooxygenase products of arachidonic acid metabolism in the vascular response during cardiac anaphylaxis is further strengthened by the finding that indomethacin delays the onset of coronary vasoconstriction immediately after antigen injection, while the vasoconstriction later in the experiment was not influenced by the drug (*Anhut* et al. 1977).

Others (*Levi* et al. 1976), observing coronary flow for a shorter period of time, reported even complete elimination of the coronary flow reduction in anaphylactic guinea pig hearts. However, cyclooxygenase products of arachidonic acid metabolism are not only vasoactive, but they might also have some direct cardiac effects. PGE_2, PGI_2, and $PGF_{2\alpha}$ have been found to increase the sinus rate, while PGD_2 decreased it in the isolated guinea pig heart (*Allan* and *Levy* 1980). *Schrör* and *Moncada* (1979), however, did not observe effects of PGE_2 and PGI_2 on heart rate in this preparation. The decrease in ventricular contractile force produced by TXA_2, $PGF_{2\alpha}$, and PGD_2, as well as the increase caused by PGI_2, might be due to the vascular actions of these compounds in this preparation (*Terashita* et al. 1978; *Allan* and *Levi* 1980). In dog heart-lung preparations, on the other hand, PGE_1, PGE_2, and $PGF_{2\alpha}$ produced a definite positive inotropic and chronotropic effect (*Imai* et al. 1983). Thus, while it is obvious that cyclooxygenase products are released during cardiac anaphylaxis and participate in the pathophysiological events, the precise role of these arachidonic acid metabolites in the anaphylactic reaction of the heart awaits further elucidation.

4.3 Lipoxygenase Products of Arachidonic Acid in Cardiac Anaphylaxis

The term "slow-reacting substance" (SRS) was originally coined by *Feldberg* and *Kellaway* (1938) and describes a smooth muscle-contracting activity which appears in the effluent of perfused lungs of cats and guinea pigs after cobra venom treatment. The appearance of a material with a similar pharmacologic profile was reported in the effluent of sensitized guinea pig lungs following antigen challenge (*Kellaway* and *Trethewie* 1940). *Brocklehurst* (1960) suggested that the SRS found in the effluent from shocked lungs be designated "slow-reacting substance of anaphylaxis" (SRS-A) to differentiate this material from that wich is of nonimmunologic origin. The term SRS-A then generally refers to any slow-reacting substance released immunologically (*Orange* and *Austen* 1969).

Release of SRS-A from anaphylactic hearts was first shown by *Brocklehurst* (1960) and *Chakravarty* (1960b) and was later confirmed by others (*Liebig* et al. 1975; *Levi* and *Burke* 1980). The experiments by *Brocklehurst* (1960) and *Chakravarty* (1960b) already stressed the importance of vascular tissue as a source of SRS-A. This point was further supported by data demonstrating the generation of SRS-A from isolated chopped vascular tissue from pig, guinea pig, rat, rabbit, and dog blood vessels (*Piper* et al. 1983a,b; *Wittenberg* et al. 1983; *Wölbling* et al. 1983). Interestingly, the largest amounts of SRS-A were released by coronary and pulmonary arteries (*Piper* et al. 1983a,b). While the adventitia was the most active layer, the vascular smooth muscle layer also containing some adherent endothelium revealed only little SRS-A synthetizing capacity (*Piper* et al. 1983a; *Wittenberg* et al. 1983; *Wölbling* et al. 1983).

LT C_4, D_4, and E_4, which are 5-lipoxygenase-derived products of arachidonic acid metabolism (*Hammarström* 1983; *Samuelsson* 1983), have recently been demonstrated to be constituents of SRS-A (*Murphy* et al. 1979; *Lewis* et al. 1980a; *Morris* et al. 1980). The SRS released from murine mastocytoma cells was characterized as LTC_4 (*Hammarström* et al. 1979; *Murphy* et al. 1979). Human SRS consists mainly of LTD_4, but also contains some LTC_4 (*Lewis* et al. 1980b), while rat SRS consists of LTE_4 in addition to LTC_4 and LTD_4 (*Lewis* et al. 1980a,b). Due to highly active γ-glutamyltransferase, the major biological activity of guinea pig SRS is LTD_4 (*Morris* et al. 1980).

With the availability of synthetic pure LTs research in this field has very much intensified. So it was possible to develop a highly sensitive radioimmunoassay for LTC_4 (*Aehringhaus* et al. 1982). Employing this technique, release of endogenous LT, measured as LTC_4-like immunoreactivity, from isolated sensitized guinea pig hearts could be demonstrated after antigen challenge (*Aehringhaus* et al. 1983; *Wittenberg* et al. 1983), confirming former results obtained by bioassay (*Liebig* et al. 1975; *Levi* and *Burke* 1980; *Levi* et al. 1982). Using radioimmunoassay, formal identification of the LT released by the heart is not possible. However, the antibodies used recognize both the fatty acid moiety and the presence of an amino acid side chain in the LT molecule (*Aehringhaus* et al. 1982). In addition, cross-reaction of LTB_4, various PGs, and TXB_2 with the antibodies is negligible, while LTD_4, or LTE_4 might significantly contribute to the total amounts of LTC_4-like immunoreactivity measured (*Aehringhaus* et al. 1982). From a comparison of radioimmunoassay data with bioassay data a significant correlation was observed (*Wittenberg* et al. 1983). Stimulation of muscarinic receptors by infusion of methacholine increased the antigen-induced release of LTC_4-like immunoreactivity in a dose-dependent manner. Various PGs, such as the vasodilators PGI_2, PGE_1, and PGE_2 or the vasoconstrictors PGD_2, $PGF_{2\alpha}$, and 11,9-epoxymethano-PGH_2, did not affect the anaphylactic LT release, while the various PGs in the concentrations used decreased the LT-dependent anaphylactic coronary constriction (*Aehringhaus* et al. 1984; *Peskar* et al. 1984; *Wittmann* et al. 1984).

In agreement with former bioassay results, indomethacin increased the release of LTC_4-like material, probably by diversion of arachidonic acid metabolism (*Peskar* et al. 1984; *Wittmann* et al. 1984). The lipoxygenase inhibitor nordihydroguaiaretic acid (NDGA) (*Tappel* et al. 1953; *Hamberg* 1976) did not prevent the coronary flow reduction in the early phase after antigen injection, but inhibited the less pronounced long-lasting coronary flow reduction in the later phase of cardiac anaphylaxis, as well as release of LTC_4-like material (*Aehringhaus* et al. 1983; *Wittenberg* et al. 1983). The SRS-A antagonist FPL-55712 (*Augstein* et al. 1973) did not inhibit release of LTC_4-like immunoreactivity, while having effects on coronary flow in the anaphylactic guinea pig heart comparable to those of NDGA (*Aehringhaus* et al. 1983; *Ezeamuzie* and *Assem* 1983; *Wittenberg* et al. 1983).

Therefore, during cardiac anaphylaxis LT might be a relatively more important mediator in the late phase of coronary constriction, while the effects of other mediators, particularly vasoconstrictor cyclooxygenase products such as PGD_2 and $PGF_{2\alpha}$, seem to prevail in the early phase, which was shown to be inhibited by indomethacin (*Anhut* et al. 1977). Interestingly, FPL-55712 and NDGA induced an increase in basal coronary flow (*Aehringhaus* et al. 1983; *Wittenberg* et al. 1983). Thus, the question arises whether lipoxygenase products of arachidonic acid metabolism might contribute to basal coronary tone. However, it should be mentioned that both compounds at higher concentrations have effects not related to the inhibition of lipoxygenase and antagonism of SRS-A, respectively (*Krell* et al. 1981; *Welton* et al. 1981; *Scott-Miller* and *McMillan* 1983).

Partially purified SRS-A, which has been shown to have a dose-dependent negative inotropic effect (*Burke* and *Levi* 1980), prolonged and potentiated the positive chronotropic effect of exogenous histamine. Both potentiation and prolongation, as well as the negative inotropic effect, were antagonized by FPL-55712. Thus it was suggested thatSRS-A exerts a direct effect via the decrease of ventricular contraction, and indirect effects via sensitization of the anaphylactic heart to the tachyarrhythmic effects of histamine released during anaphylaxis (*Levi* et al. 1979, 1982; *Burke* and *Levi* 1980; *Levi* and *Burke* 1980). In isolated, spontaneously beating guinea pig atria, partially purified SRS-A (*Burke* and *Levi* 1980) or synthetic pure LTC_4 and LTD_4 (*Letts* and *Piper* 1981, 1982; *Terashita* et al. 1981) did not affect the rate or the force of contractions. So no direct chronotropic or inotropic action was observed. In contrast to LTC_4, LTD_4 was found to potentiate the positive chronotropic effect of histamine in this preparation (*Burke* et al. 1981, 1982a). No arrhythmia was observed in guinea pig or rat isolated hearts upon administration of exogenous LT.

In contrast to LTB_4, the sulfidopeptide LTs C_4, D_4 and E_4 are potent coronary vasoconstrictors, causing a dose-dependent decrease of coronary flow in the isolated guinea pig and rat heart (*Burke* et al. 1981, 1982a; *Letts* and *Piper* 1981, 1983; *Terashita* et al. 1981; *Piper* et al. 1982; *Ezeamuzie* and *Assem* 1983; *Letts* et al. 1983b; *Roth* and *Lefer* 1983). LTE_4 was found to be less potent than LTC_4 or LTD_4 (*Burke* et al. 1982a;

Letts and *Piper* 1983; *Letts* et al. 1983b). While *Burke* et al. (1981, 1982a) reported LTD_4 to be five times more potent than LTC_4 in the guinea pig heart, others found contrasting results (*Letts* and *Piper* 1981, 1982, 1983; *Terashita* et al. 1981; *Ezeamuzie* and *Assem* 1983). Besides inducing coronary flow reduction, both LTC_4 and LTD_4 exert a negative inotropic effect in isolated guinea pig and rat hearts (*Burke* et al. 1981, 1982a; *Letts* and *Piper* 1981, 1982, 1983; *Ezeamuzie* and *Assem* 1983; *Letts* et al. 1983b). *Burke* et al. (1982a) have shown that an angiotensin II-induced coronary flow reduction equal to an LT-induced decrease in coronary flow was accompanied by only a slightly negative inotropic effect. In potassium-depolarized guinea pig myocardium, LTC_4 and LTD_4 have been reported to inhibit calcium-dependent contractile responses, and it was suggested that the negative inotropic effect of LTs might result from a reduction in calcium ion influx through the sarcolemma (*Hattori* and *Levi* 1984).

Furthermore, in human heart muscle specimens as well as in isolated paced guinea pig atria, LTs were found to decrease contraction force, suggesting that the negative inotropic action of LTC_4 or LTD_4 was unlikely to be secondary to coronary flow reduction (*Burke* et al. 1982a). On the other hand, *Letts* and *Piper* (1981, 1982) did not observe a decrease in contractile force in isolated atria or electrically driven ventricular muscle strips after treatment with LTC_4 or LTD_4, and even a slightly positive inotropic effect of LTC_4 and LTD_4 has been reported in isolated guinea pig atria (*Terashita* et al. 1981). Thus, the mechanism of the negative inotropic effect of LTs observed by several investigators remains to be elucidated.

In guinea pig and rat hearts FPL-55712 has been shown to antagonize all acute effects of exogenous LTC_4 and LTD_4 except the prolonged phase of LTD_4-induced coronary vasoconstriction in the perfused guinea pig heart (*Burke* et al. 1982a; *Letts* and *Piper* 1982, 1983; *Piper* et al. 1982; *Letts* et al. 1983b). The cyclooxygenase inhibitor indomethacin slightly delayed the vasoconstrictor activity of LTD_4 in isolated guinea pig hearts, while the LTC_4 effect on coronary flow, contractility, and heart rate was significantly reduced, suggesting participation of an endogenous cyclooxygenase product(s) in the effects of LTC_4 on isolated guinea pig hearts (*Letts* and *Piper* 1982, 1983; *Piper* et al. 1982; *Letts* et al. 1983b). The cyclooxygenase product(s) involved have not been identified so far (*Letts* and *Piper* 1983; *Letts* et al. 1983b). TXA_2 is unlikely to be involved, since UK-37248, a TX synthesis inhibitor, did not influence coronary flow reduction by exogenous LTC_4 or LTD_4 (*Letts* et al. 1983b). In fact, release of TXB_2 by exogenous LTC_4 and LTD_4 was not observed, while, on the other hand, release of PGI_2, which is, however, a vasodilator, was detected (*Terashita* et al. 1981). In isolated rat hearts release of

PGI_2 upon LTC_4 injection has been reported as well (*Klitzke* and *Trompler* 1984). The PGI_2 released could possibly act as a negative feedback mechanism, resulting in a decreased vasoconstrictor response to LTC_4.

In sheep LTD_4-induced coronary vasoconstriction and overall impaired cardiac function were not influenced by the cyclooxygenase inhibitor ibuprofen. In this species release of TXB_2 or 6-keto-$PGF_{1\alpha}$ by exogenous LTD_4 could not be detected (*Michelassi* et al. 1982, 1983). FPL-55712 eliminated the vasoconstriction observed after intracoronary LTD_4 administration. However, significant reduction of regional wall shortening persisted, indicating that the negative inotropic effect of LTD_4 was not secondary to coronary vasoconstriction in this in vivo model (*Michelassi* et al. 1982, 1983).

Similar data were obtained from experiments in pigs, showing that LTD_4-induced coronary flow reduction (*Boyd* et al. 1983; *Letts* et al. 1983b) was inhibited by FPL-55712, while this compound by itself did not affect coronary flow, left ventricular end-diastolic pressure, or heart rate (*Boyd* et al. 1983). Similar to rat and sheep hearts, cyclooxygenase products do not seem to mediate LT effects in the pig heart, since indomethacin did not affect the coronary vasoconstrictor activity of LTD_4 in this species (*Boyd* et al. 1983; *Letts* et al. 1983b).

Both LTC_4 and LTD_4 have been shown to be potent constrictors of coronary arteries in dogs (*Imai* et al. 1983; *Letts* et al. 1983a,b; *Panzenbeck* and *Kaley* 1983; *Woodman* and *Dusting* 1983; *Ertl* et al. 1984; *Fiedler* et al. 1984), while LTE_4 did not affect coronary flow (*Woodman* and *Dusting* 1983). The LT-induced coronary flow reduction was not changed by indomethacin, suggesting a direct action not mediated by vasoconstrictor cyclooxygenase products (*Letts* et al. 1983a,b; *Panzenbeck* and *Kaley* 1983; *Woodman* and *Dusting* 1983). In mongrel dogs increasing doses of LTD_4 induced dose-dependent decreases in cardiac output and left ventricular dp/dt, which was paralleled by coronary flow reduction. It was suggested that the negative inotropic effect observed in this model might be secondary to coronary vasoconstriction (*Panzenbeck* and *Kaley* 1983). These authors also reported a frequent appearance of arrhythmias following intracoronary injection of LTD_4. In contrast, LTD_4 did not seem to affect heart rate in anesthetized greyhounds (*Letts* et al. 1983a,b). Although there was no evidence that LTD_4 induced vascular leakage or structural damage to the endothelium, hemostasis along the surface of the large coronary arteries has been observed (*Letts* et al. 1983a,b). Interestingly, this pathophysiological phenomenon has also been observed in animals which had suffered fatal anaphylactic shock (*Auer* and *Lewis* 1910; *Auer* 1911).

Despite the species differences, which occur with other effects of LTs, the sulfidopeptide LTs exhibit potent coronary vasoconstrictor actions in all animal species investigated so far. The pathophysiological role of the cardiovascular effects of LT in man remains, however, to be investigated.

5 Platelet-Activating Factor and Eicosanoids in Coronary Blood Flow

5.1 Interaction Between Platelet-Activating Factor and Eicosanoids

Synthesis of PGs requires liberation of the substrate, usually arachidonic acid, from phospholipids. As mentioned earlier (Sect. 1.2.1), unsaturated fatty acids are generally esterified at the β-glyceryl carbon and are liberated by the hydrolytic activity of phospholipase A_2 (*Blackwell* and *Flower* 1983). However, the action of the enzyme results not only in the liberation of unesterified arachidonic acid, but also gives rise to β-lysophosphatides (*Mencia-Huerta* et al. 1981), which are known to have cytotoxic and membrane-perturbing properties (*Weltzien* 1979) and are implicated in ischemia-induced lethal arrhythmias (*Sobel* and *Corr* 1979). A modified phospholipid derived from the products of phospholipase action has been described as platelet-activating factor (PAF) (*Benveniste* et al. 1977; *Mencia-Huerta* et al. 1981).

PAF, which is a mediator of inflammation, was first described as being released from sensitized basophils (*Benveniste* et al. 1972). It is also produced by a variety of cells (*Chignard* et al. 1979; *Mencia-Huerta* and *Benveniste* 1979; *Arnoux* et al. 1980), including human polymorphonuclear neutrophils (*Lotner* et al. 1980). PAF induces platelet aggregation, which has been demonstrated in several species, including man (*Benveniste* et al. 1972, 1975; *Fésüs* et al. 1977; *Cazenave* et al. 1979; *McManus* et al. 1981). In rabbit platelets the PAF-induced plateled aggregation and secretion has previously been postulated to be independent of the cyclooxygenase pathway of arachidonic acid metabolism (*Shaw* et al. 1978; *Cazenave* et al. 1979). Subsequently, however, PAF-stimulation of human (*Heffner* et al. 1983; *Korth* et al. 1984) as well as rabbit (*Shaw* et al. 1978, 1981; *McManus* et al. 1983) platelets was reported to initiate platelet PG and TX release.

It was suggested that secondary aggregation of human platelets induced by PAF might be at least in part PG endoperoxide and TX dependent (*MacIntyre* et al. 1982). In isolated rabbit lungs PAF-stimulated platelets caused pulmonary hypertension and edema, which were apparently dependent on TX release (*Heffner* et al. 1983). The time course of the pathophysiological events was closely related to the PAF-induced TX release (*Heffner* et al. 1983). In contrast, in isolated rat lungs PAF caused pulmonary vasoconstriction and edema formation independently of the presence of platelets (*Voelkel* et al. 1982). On the other hand, however, release of LTC_4 and LTD_4 which are mediators of anaphylactic reactions, by PAF has been demonstrated (*Voelkel* et al. 1982). PAF is a potent inducer of systemic anaphylactoid reactions in animals (*McManus* et al. 1980; *O'Flaherty* et al. 1981; *Stimler* et al. 1981), exhibiting spasmogenic and vascular permeability-enhancing activity (*O'Flaherty* et al. 1981; *Stimler* et al. 1981).

Interestingly, the sulfidopeptide LTs reveal a surprisingly similar pattern of action (*Hammarström* 1983; *Piper* 1983; *Samuelsson* 1983). The divalent cation ionophore A23187 is a commonly employed experimental stimulus of LT synthesis. It has been reported that the ionophore A23187 not only causes release of endogenous substrate, but also activates the enzymatic system involved in the further transformation of arachidonic acid through the lipoxygenase pathway (*Borgeat* and *Samuelsson* 1979a). It is also noteworthy that PAF has been shown to possess ionophoretic activity, including promotion of the influx of calcium ions into rabbit platelets that is time-, temperature-, and concentration-dependent (*Lee* et al. 1981).

5.2 Effects of Platelet-Activating Factor on the Coronary Circulation

In rabbits intravenous injection of PAF causes sudden death which is preceded by cessation of respiration, a marked decrease in mean arterial blood pressure, and up to 30-fold increases in plasma TXB_2 concentrations (*McManus* et al. 1983; *Lefer* et al. 1984). However, only a modest elevation of 6-keto-$PGF_{1\alpha}$ concentrations in plasma was found (*Lefer* et al 1984). Pretreatment of the animals with ibuprofen or TX synthetase inhibitors like dazoxiben, CGS-13080, or OKY-046 increased survival rates up to 100%, and it was suggested that PAF-induced sudden death in rabbits is mediated by TXA_2 release, resulting in an impaired PGI_2/TXA_2 balance (*Lefer* et al. 1984).

Similar pathophysiological data have been reported from experiments performed on dogs. PAF was found to be a potent shock-inducer after intravenous injection and the long-lasting fall in systemic blood pressure and coronary artery perfusion was paralleled by high increases in TXB_2 and smaller increases in 6-keto-$PGF_{1\alpha}$ levels in blood plasma (*Bessin* et al. 1983).

PAF has been shown to decrease myocardial contractile force and coronary blood flow in isolated guinea pig hearts (*Burke* et al. 1982b; *Benveniste* et al. 1983). Recently it has been reported that PAF induces release of immunoreactive TXB_2 and 6-keto-$PGF_{1\alpha}$ after intracoronary injection in open-chest domestic pigs, whereas release of immunoreactive LTC_4 remained unchanged (*Feuerstein* et al. 1984a). In this preparation intracoronary injection of PAF produced biphasic changes in coronary blood flow. A dose-dependent increase was followed by a decrease in coronary flow (*Feuerstein* et al. 1984a,b). While the increase in coronary blood flow remained unchanged, the decrease induced by PAF was only slightly affected by the SRS-A antagonist FPL-55712, but blocked by indomethacin. It was, therefore, suggested that PAF can produce coronary vasoconstriction sufficient to cause myocardial ischemia through a mechanism involving the production of TXA_2 rather than cysteinyl LT (*Feuerstein* et al. 1984a,b). However, it has been reported that PAF-induced cardiovascular effects are not only mediated by hemodynamic alterations related to platelet involvement, but can also be ascribed to a direct action of PAF on the cardiac muscle, since electrocardiographic alterations were found in control animals as well as in platelet-depleted rabbits (*Camussi* et al. 1984). Furthermore, during anaphylaxis in the isolated guinea pig heart, a platelet activating factor-like substance was found to be released into the coronary effluent. Intracardiac administration of PAF into nonsensitized hearts induced a dose-related decrease in left ventricular contractile force and coronary flow, as well as impaired atrioventricular conduction (*Levi* et al. 1984). Since these cardiac effects were not modified by indomethacin or FPL-55712, the authors concluded that the action of PAF in this model might be independent of arachidonic acid metabolites.

Although several reports suggest a close interaction between PAF and eicosanoids, its pathophysiological relevance in the coronary circulation awaits further exploration.

References

Abrahamsson S, Bergström S, Samuelsson B (1962) The absolute configuration of prostaglandin F_{2-1}. Proc Chem Soc:332

Adaikan PG, Lau LC, Tai MY, Karim SMM (1983) Inhibition of platelet aggregation with intravenous and oral administration of a carboprostacyclin analogue, 15-cyclopentyl-ω-pentanor-5(E)-carbacyclin (ONO 41483) in man. Prostaglandins Leukotrienes Med 10:53–64

Adaikan PG, Karim SMM, Lau LC, Tai MY, Kottegoda SR (1984) Inhibition of platelet aggregation and antagonism of vasopressin-induced ECG changes in primates by a carboprostacyclin analogue, ZK 36374. Thromb Res 33:333–340

Adams SS, Burrows CA, Skeldon N, Yates DB (1977) Inhibition of prostaglandin synthesis and leucocyte migration by flurbiprofen. Curr Med Res Opin 5:11–16

Aehringhaus U, Wölbling RH, König W, Patrono C, Peskar BM, Peskar BA (1982) Release of leukotrience C_4 from human polymorphonuclear leukocytes as determined by radioimmunoassay. FEBS Lett 146:111–114

Aehringhaus U, Peskar BA, Wittenberg HR, Wölbling RH (1983) Effect of inhibition of synthesis and receptor antagonism of SRS-A in cardiac anaphylaxis. Br J Pharmacol 80:73–80

Aehringhaus U, Dembinska-Kiéc A, Peskar BA (1984) Effects of exogenous prostaglandins on the release of leukotriene C_4-like immunoreactivity and on coronary flow in indomethacin-treated anaphylactic guinea pig hearts. Naunyn-Schmiedebergs Arch Pharmacol 326:368–374

Ahumada GG, Sobel BE, Needleman P (1980) Synthesis of prostaglandins by cultured rat heart myocytes and cardiac mesenchymal cells. J Mol Cell Cardiol 12:685–700

Aiken JW, Shebuski RJ (1980) Comparison in anaesthetized dogs of the anti-aggregatory and hemodynamic effects of prostacyclin and a chemically stable prostacyclin analog, 6a-carba-PGI_2 (carbacyclin). Prostaglandins 19:629–643
Aiken JW, Gorman RR, Shebuski RJ (1979) Prevention of blockade of partially obstructed coronary arteries with prostacyclin correlates with inhibition of platelet aggregation. Prostaglandins 17:483–494
Aiken JW, Shebuski RJ, Miller OV, Gorman RR (1981) Endogenous prostacyclin contributes to the efficacy of a thromboxane synthetase inhibitor for preventing coronary artery thrombosis. J Pharmacol Exp Ther 219:299–308
Allan G, Levi R (1980) The cardiac effects of prostaglandins and their modification by the prostaglandin antagonist N-0164. J Pharmacol Exp Ther 214:45–49
Allan G, Levi R (1981) Thromboxane and prostacyclin release during cardiac immediate hypersensitivity reactions in vitro. J Pharmacol Exp Ther 217:157–161
Anderson WH, Mohammad SF, Chuang HYK, Mason RG (1980) Heparin potentiates synthesis of thromboxane A_2 in human platelets. Adv Prostaglandin Thromboxane Res 6:287–291
Änggård E, Larsson C, Samuelsson B (1971) The distribution of 15-hydroxyprostaglandin dehydrogenase and prostaglandin Δ^{13}-reductase in tissues of the swine. Acta Physiol Scand 81:396–404
Anhut H, Bernauer W, Peskar BA (1977) Radioimmunological determination of thromboxane release in cardiac anaphylaxis. Eur J Pharmacol 44:85–88
Anhut H, Bernauer W, Peskar BA (1978a) Pharmacological modification of thromboxane and prostaglandin release in cardiac anaphylaxis. Prostaglandins 15:889–900
Anhut H, Peskar BA, Bernauer W (1978b) Release of 15-keto-13,14-dihydro-thromboxane B_2 and prostaglandin D_2 during anaphylaxis as measured by radioimmunoassay. Naunyn-Schmiedebergs Arch Pharmacol 305:247–252
Anturane Reinfarction Italian Study Group (1982) Sulphinpyrazone in post-myocardial infarction. Lancet 1:237–242
Anturane Reinfarction Trial Research Group (1978) Sulfinpyrazone in the prevention of cardiac death after myocardial infarction. The anturane reinfarction trial. N Engl J Med 298:289–295
Anturane Reinfarction Trial Research Group (1980) Sulfinpyrazone in the prevention of sudden death after myocardial infarction. N Engl J Med 302:250–256
Armstrong JM, Chapple D, Dusting GJ, Hughes R, Moncada S, Vane JR (1977) Cardiovascular actions of prostacyclin (PGI_2) in chloralose anaesthetized dogs. Br J Pharmacol 61:136P
Arnoux B, Duval D, Benveniste J (1980) Release of platelet-activating factor (PAF-acether) from alveolar macrophages by the calcium ionophore A 23187 and phagocytosis. Eur J Clin Invest 10:437–441
Aspirin Myocardial Infarction Study Research Group (1980) A randomized, controlled trial of aspirin in persons recovered from myocardial infarction. JAMA 243:661–669
Auer J (1911) Lethal cardiac anaphylaxis in the rabbit. J Exp Med 14:476–496
Auer J, Lewis PA (1910) The physiology of the immediate reaction of anaphylaxis in the guinea pig. J Exp Med 12:151–175
Augstein J, Farmer JR, Lee TB, Sheard P, Tattersall ML (1973) Selective inhibition of slow reacting substance of anaphylaxis. Nature (New Biol) 245:215–217
Austen K (1965) Systemic anaphylaxis in man. JAMA 12:116–118
Austen K (1978) The anaphylactic syndrome. In: Samter M (ed) Immunological diseases, vol 2, 3rd edn. Little Brown, Boston, pp 885–899
Bakhle YS (1983) Synthesis and catabolism of cyclo-oxygenase products. Br Med Bull 39:214–218
Baraka A, Sfeir S (1980) Anaphylactic cardiac arrest in a parturient. JAMA 243: 1745–1746
Beitz J, Förster W (1980) Influence of human low density and high density lipoprotein cholesterol on the in vitro prostaglandin I_2 synthetase activity. Biochim Biophys Acta 620:352–355

Belch JJF, Greer I, McLaren M, Saniabadi AR, Miller S, Sturrock RD, Forbes CD (1984) The effects of intravenous ZK36-374, a stable prostacyclin analogue, on human volunteers. Prostaglandins 28:67–77
Benveniste J (1974) Platelet-activating factor, a new mediator of anaphylaxis and immune complex deposition from rabbit and human basophils. Nature 249:581–582
Benveniste J, Henson PM, Cochrane CG (1972) Leukocyte-dependent histamine release from rabbit platelets: the role of IgE, basophils and a platelet-activating factor. J Exp Med 136:1356–1377
Benveniste J, Le Couedic JP, Kamoun P (1975) Aggregation of human platelets by platelet-activating factor. Lancet 1:344–345
Benveniste J, Le Couedic JP, Polonsky J, Tence M (1977) Structural analysis of purified platelet-activating factor by lipases. Nature 269:170–171
Benveniste J, Boullet C, Brink C, Labat C (1983) The actions of Paf-acether (platelet-activating factor) on guinea-pig isolated heart preparations. Br J Pharmacol 80:81–83
Berger HJ, Zaret BL, Speroff L, Cohen LS, Wolfson S (1976) Regional cardiac prostaglandin release during myocardial ischemia in anaesthetized dogs. Circ Res 38:566–571
Bergman G, Daly K, Atkinson L, Rothman M, Richardson PJ, Jackson G, Jewitt DE (1981) Prostacyclin: haemodynamic and metabolic effects in patients with coronary artery disease. Lancet 1:569–572
Bergström S, Sjövall J (1960a) The isolation of prostaglandin F from sheep prostate glands. Acta Chem Scand 14:1693–1700
Bergström S, Sjövall J (1960b) The isolation of prostaglandin E from sheep prostate glands. Acta Chem Scand 14:1701:1705
Bergström S, Ryhage R, Samuelsson B, Sjövall J (1963) Prostaglandins and related factors. The structure of prostaglandin E_1, $F_{1\alpha}$ and $F_{1\beta}$. J Biol Chem 238:3555–3564
Bergström S, Danielsson H, Samuelsson B (1964) The enzymatic formation of prostaglandin E_2 from arachidonic acid. Biochim Biophys Acta 90:207–210
Bergström S, Carlson LA, Weeks JR (1968) The prostaglandins: a family of biologically active lipids. Phrmacol Rev 20:1–48
Bernreiter M (1959) Electrocardiogram of patient in anaphylactic shock. JAMA 170: 1628–1630
Bessin P, Bonnet J, Apffel D, Soulard C, Desgroux L, Pelas I, Benveniste J (1983) Acute circulatory collapse caused by platelet-activating factor (PAF-acether) in dogs. Eur J Pharmacol 86:403–413
Best LC, Martin TJ, Russell RGG, Preston FE (1977) Prostacyclin increases cyclic AMP levels and adenylate cyclase activity in platelets. Nature 267:850–852
Binder MJ, Gunderson HJ, Cannon J, Rosove L (1950) Electrocardiographic changes associated with allergic reactions to penicillin. Am Heart J 40:940–944
Björk J, Hedqvist P, Arfors KE (1982) Increase in vascular permeability induced by leukotriene B_4 and the role of polymorphonuclear leukocytes. Inflammation 6: 189–200
Blackwell GJ, Flower RJ (1983) Inhibition of phospholipase. Br Med Bull 39:260–264
Blackwell GJ, Flower RJ, Russell-Smith N, Salmon JA, Thorogood PB, Vane JR (1978) Prostacyclin is produced in whole blood. Br J Pharmacol 64:436P
Blackwell GJ, Carnuccio R, DiRosa M, Flower RJ, Parente L, Persico P (1980) Macrocortin: a polypeptide causing the anti-phospholipase effect of glucocorticoids. Nature 287:147–149
Blaskó G, Berentey E, Harsányi A, Sas G (1983) Intracoronarily administered prostacyclin and streptokinase for treatment of myocardial infarction. Adv Prostaglandin Thromboxane Leukotrienes Res 11:385–390

Block AJ, Feinberg H, Herbacynska-Cedro K, Vane JR (1975) Anoxia-induced release of prostaglandins in rabbit isolated hearts. Circ Res 36:34–42

Bolli R, Goldstein RE, Davenport N, Epstein SE (1981) Influence of sulfinpyrazone and naproxen on infarct size in the dog. Am J Cardiol 47:841–847

Bolton HS, Chanderbhan R, Bryant RW, Bailey JM, Weglicki WB, Vahouny GV (1980) Prostaglandin synthesis by adult heart myocytes. J Mol Cell Cardiol 12:1287–1298

Bonow RO, Lipson LC, Sheehan FH, Capurro NL, Isner JM, Roberts WC, Goldstein RE, Epstein SE (1981) Lack of effect of aspirin on myocardial infarct size in the dog. Am J Cardiol 47:258–264

Booth BH, Patterson R (1970) Electrocardiographic changes during human anaphylaxis. JAMA 211:627–631

Borgeat P, Samuelsson B (1979a) Arachidonic acid metabolism in polymorphonuclear leukocytes: Effects of ionophore A23187. Proc Natl Acad Sci USA 76:2148–2152

Borgeat P, Samuelsson B (1979b) Metabolism of arachidonic acid in polymorphonuclear leukocytes: structure analysis of novel hydroxylated compounds. J Biol Chem 254:7865–7869

Borgeat P, Hamberg M, Samuelsson B (1976) Transformation of arachidonic acid and homo-γ-linolenic acid by rabbit polymorphonuclear leukocytes. Monohydroxy acids from novel lipoxygenase. J Biol Chem 251:7816–7820

Born GV (1962) Aggregation of blood platelets by adenosine diphosphate and its reversal. Nature 194:927–929

Boston Collaborative Drug Surveillance Group (1974) Regular aspirin intake and acute myocardial infarction. Br Med J 1:440–443

Boyd LM, Ezra D, Feuerstein G, Goldstein RE (1983) Effects of FPL-55712 or indomethacin on leukotriene-induced coronary constriction in the intact pig heart. Eur J Pharmacol 89:307–311

Breddin K, Loew D, Lechner K, Überla K, Walter E (1979) Secondary prevention of myocardial infarction. Comparison of acetylsalicylic acid, phenprocoumon and placebo. A multicenter two-year prospective study. Thromb Haemost 41:225–236

Bristow MR, Ginsburg R, Kantrowitz NE, Baim DS, Rosenbaum JT (1982) Coronary spasm associated with urticaria: Report of a case mimicking anaphylaxis. Clin Cardiol (Lond) 5:238–240

Brocklehurst WE (1953) Occurrence of an unidentified substance during anaphylaxic shock in cavy lung. J Physiol 120:16P–17P

Brocklehurst WE (1960) The release of histamine and formation of a slow-reacting substance (SRS-A) during anaphylactic shock. J Physiol 151:416–435

Brox JH, Nordøy A (1983) The effect of polyunsaturated fatty acids on endothelial cells and their production of prostacyclin, thromboxane and platelet inhibitory activity. Thromb Haemost 50:762–767

Brox JH, Killie JE, Gunnes S, Nordøy A (1981) The effect of cod liver oil and corn oil on platelets and vessel wall in man. Thromb Haemost 46:606–611

Bulkley BH, Roberts WC (1974) Steroid therapy during acute myocardial infarction: A cause of delayed healing and ventricular aneurysm. Am J Med 56:244–250

Bunting S, Gryglewski R, Moncada S, Vane JR (1976) Arterial walls generate from prostaglandin endoperoxides a substance (prostaglandin X) which relaxes strips of mesenteric and coeliac arteries and inhibits platelet aggregation. Prostaglandins 12: 897–913

Bunting S, Moncada S, Vane JR (1983) The prostacyclin-thromboxane A_2 balance: pathophysiological and therapeutic implications. Br Med Bull 39:271–276

Burke JA, Levi R (1980) Slow reacting substance of anaphylaxis (SRS-A): Direct and indirect cardiac effects. Fed Proc 39:389

Burke JA, Levi R, Corey EJ (1981) Cardiovascular effects of pure synthetic leukotrienes C and D. Fed Proc 40:1015

Burke JA, Levi R, Guo ZG, Corey EJ (1982a) Leukotriene C_4, D_4 and E_4: Effects on human and guinea-pig cardiac preparations in vitro. J Pharmacol Exp Ther 221: 235–241

Burke JA, Levi R, Hanahan DJ, Pinckard RN (1982b) Cardiac effects of acethyl glyceryl ether phosphorylcholine. Fed Proc 41:823
Burke SE, DiCola G, Lefer AM (1983a) Protection of ischemic cat myocardium by CGS-13080, a selective potent thromboxane A_2 synthesis inhibitor. J Cardiovasc Pharmacol 5:842–847
Burke SE, Lefer AM, Nicolaou KC, Smith GM, Smith JB (1983b) Reponsiveness of platelets and coronary arteries from different species to synthetic thromboxane and prostaglandin endoperoxide analogs. Br J Pharmacol 78:287–292
Bush LR, Campbell WB, Buja LM, Tilton GD, Willerson JT (1984) Effects of the selective thromboxane synthetase inhibitor dazoxiben on variation in cyclic blood flow in stenosed canine coronary arteries. Circulation 69:1161–1170
Butkus A, Skrinska VA, Schumacher OP (1980) Thromboxane production and platelet aggregation in diabetic subjects with clinical complications. Thromb Res 19:211–223
Camussi G, Montrucchio G, Alloatti G, Mariano F, Coda R, Tetta C, Emanuelli G (1984) Platelet-dependent and independent cardiovascular effects of platelet-activating factor (PAF) (Abstr). In: Neri Serneri GG, Masotti G, Gensini GF (eds) Platelets prostaglandins and the cardiovascular system. Boehringer, Ingelheim, Florence, p 160
Capurro N, Levi R (1975) The heart as a target organ in systemic allergic reactions. Comparison of cardiac anaphylaxis in vivo and in vitro. Circ Res 36:520–528
Carter AJ, Bevan JA, Hanley SP, Morgan WE, Turner DR (1984) A comparison of human pulmonary arterial and venous prostacyclin and thromboxane synthesis – effect of a thromboxane synthetase inhibitor. Thromb Haemost 51:257–260
Carvalho ACA, Colman RW, Lees RS (1974) Platelet function in hyperlipoproteinemia. N Engl J Med 290:434–438
Casals-Stenzel J, Buse M, Losert W (1983) Comparison of the vasodepressor action of ZK 36374, a stable prostaglandin derivative, PGI_2 and PGE_1, with their effect on platelet aggregation and bleeding time in rats. Prostaglandins Leukotrienes Med 10: 197–212
Cazenave JP, Benveniste J, Mustard JF (1979) Aggregation of rabbit platelets by platelet-activating factor is independent of the release reaction and the arachidonate pathway and inhibited by membrane active drugs. Lab Invest 41:275–285
Cerletti C, Rajtar G, Bertelé V, de Gaetano G (1984) Inhibition of arachidonate-induced human platelet aggregation by a single low oral dose of aspirin in combination with a thromboxane synthetase inhibitor. Thromb Haemost 52:215
Chakravarty N (1960a) The mechanism of histamine release in anaphylactic reaction in guinea pig and rat. Acta Physiol Scand 48:146–166
Chakravarty N (1960b) The occurrence of a lipid-soluble smooth muscle-stimulating principle ("SRS") in anaphylactic reaction. Acta Physiol Scand 48:167–177
Chapple DJ, Dusting GJ, Hughes R, Vane JR (1978) A vagal reflex contributes to the hypotensive effect of prostacyclin in anaesthetized dogs. J Physiol (Lond) 281: 43P–44P
Chapple DJ, Dusting GJ, Hughes R, Vane JR (1980) Some direct and reflex cardiovascular actions of prostacyclin (PGI_2) and prostaglandin E_2 in anaesthetized dogs. Br J Pharmacol 68:437–447
Chiabrando C, Castagnoli MN, Noseda A, Fanelli R, Rajtar G, Cerletti C, de Gaetano G (1984) Comparison of radioimmunoassay and high-resolution gas chromatography mass spectrometry for the quantitative determination of serum thromboxane B_2 and 6-keto-$PGF_{1\alpha}$ after pharmacological blockade of thromboxane synthetase. Prostaglandins Leukotrienes Med 16:79–88
Chiavarelli M, Moncada S, Mullane KM (1982) Prostacyclin can either increase or decrease heart rate depending on the basal state. Br J Pharmacol 75:243–249
Chien KR, Han A, Sen A, Buja LM, Willerson JT (1984) Accumulation of unesterified arachidonic acid in ischemic canine myocardium. Relationship to a phosphatidylcholine deacylation – reacylation cycle and the depletion of membrane phospholipids. Circ Res 54:313–322

Chierchia S, Patrono C, Crea F, Ciabattoni G, De Caterina R, Cinotti GA, Distante A, Maseri A (1982) Effects of intravenous prostacyclin in variant angina. Circulation 65:470–477

Chignard M, Le Couedic JP, Tence M, Vargaftig BB, Benveniste J (1979) The role of platelet-activating factor in platelet aggregation. Nature 279:799–800

Christ-Hazelhof E, Nugteren DH, Van Dorp DA (1976) Conversion of prostaglandin endoperoxides by glutathione-s-transferases and serum albumins. Biochim Biophys Acta 450:450–461

Christofinis GJ, Moncada S, Bunting S, Vane JR (1979) Prostacyclin release of rabbit aorta and human umbilical vein endothelial cells after prolonged subculture. In: Vane JR, Bergström S (eds) Prostacyclin. Raven, New York, pp 77–84

Coker SJ, Parrat JR, Ledingham I McA, Zeitlin IJ (1981) Early release of thromboxane and 6-keto-$PGF_{1\alpha}$ from the ischemic canine myocardium; relation to early post-infarction arrhythmias. Nature 291:323–324

Colli S, Lombroso M, Maderna P, Tremoli E, Nicosia S (1983) Effects of PGI_2 on platelet aggregation and adenylate cyclase activity in human type IIA hypercholesterolemia. Biochem Pharmacol 32:1989–1993

Coombs RRA, Gell PGH (1968) Classification of allergic reactions responsible for clinical hypersensitivity and disease. In: Gell PGH, Coombs RRA (eds) Clinical aspects of immunology, 2nd edn. Blackwell, Oxford, pp 575–596

Corey EJ, Niwa N, Falck JR, Mioskowski C, Arai Y, Marfat A (1980) Recent studies on the chemical synthesis of eicosanoids. Adv Prostaglandin Thromboxane Res 6: 19–25

Coronary Drug Project Research Group (1976) Aspirin in coronary heart disease. J Chron Dis 29:625–642

Cortellaro M, Boschetti C, Antoniazzi V, Moreo G, Repetto S, Verna E, Boscarini M, Limido A, Binaghi G, Polli EE (1983) Transcoronary platelet thromboxane A_2 formation without platelet trapping in patients with coronary stenosis – effect of sulphinpyrazone treatment. Thromb Haemost 50:857–859

Criep LH (1931) Electrocardiographic studies of the effect of anaphylaxis in the cardiac mechanism. Arch Intern Med 48:1098–1109

Criep LH, Woehler TR (1971) The heart in human anaphylaxis. Ann Allergy 29:399–409

Culp BR, Lands WEM, Lucchesi BR, Pitt B, Romson J (1980) The effect of dietary supplementation of fish oil on experimental myocardial infarction. Prostaglandins 20:1021–1030

Dahlén SE, Björk J, Hedqvist P, Arfors KE, Hammarström S, Lindgren JA, Samuelsson B (1981) Leukotrienes promote plasma leakage and leukocyte adhesion in postcapillary venules: In vivo effects with relevance to the acute inflammatory response. Proc Natl Acad Sci USA 78:3887–3891

Dale J, Thaulow E, Myhre E, Parry J (1983) The effect of a thromboxane synthetase inhibitor dazoxiben and acetylsalicylic acid on platelet function and prostaglandin metabolism. Thromb Haemost 50:703–706

D'Angelo V, Villa S, Mysliwiec M, Donati MB, de Gaetano B (1978) Defective fibrinolytic and prostacyclin-like activity in human atheromatous plaques. Thromb Haemost 39:535–536

Darsee JR, Kloner RA, Braunwald E (1981) Demonstration of lateral and epicardial border zone salvage by flurbiprofen using an in vivo method for assessing myocardium at risk. Circulation 63:29-35

Davi G, Custro N, Novo S, Mattina S, Strano A (1983) The effect of two low doses of aspririn on whole blood thromboxane and prostacyclin generation in healthy subjects. Thromb Haemost 50:669–670

Davies RA, Thakur ML, Berger HJ, Wackers PJTH, Gottschalk A, Zaret BL (1981) Imaging the inflammatory response to acute myocardial infarction in man using indium-111-labeled autologous platelets. Circulation 63:826–832

Davis K, Ginsburg R, Bristow M, Harrison DC (1980) Biphasic action of prostacyclin in the human coronary artery. Clin Res 28:3A

Dawson W, Boot JR, Cockerill AF, Mallen DNB, Osborne DJ (1976) Release of novel prostaglandins and thromboxanes after immunological challenge of guinea-pig lung. Nature 262:699–702

De Deckere EAM (1979) Effects of prostaglandins on coronary flow rate and left ventricular work in isolated rat heart. Eur J Pharmacol 58:211–213

De Deckere EAM, Nugteren DH, Ten Hoor F (1977) Prostacyclin is the major prostaglandin released from the isolated perfused rabbit and rat heart. Nature 268:160–163

Dembinska-Kiéc A, Gryglewska T, Zmuda A, Gryglewski RJ (1977) The generation of prostacyclin by arteries and by the coronary vascular bed is reduced in experimental atherosclerosis in rabbits. Prostaglandins 14:1025–1034

Dembinska-Kieć A, Rücker W, Schönhöfer PS (1979) PGI_2 enhanced cAMP content in bovine coronary arteries in the presence of isobutylmethylxanthine. Naunyn Schmiedebergs Arch Pharmacol 308:107–110

De Mello VR, Roberts R, Sobel BE (1975) Deleterious effects of methylprednisolone in patients with evolving myocardial infarction. Clin Res 23:179A

Dreyfuss F, Zahavi J (1973) Adenosine diphosphate induced platelet aggregation in myocardial infarction and ischemic heart disease. Atherosclerosis 17:107–120

Dusting GJ, Angus JA (1984) Interaction of epoprostenol (PGI_2) with vasoconstrictors on diameter of large coronary arteries of the dog. J Cardiovasc Pharmacol 6: 20–27

Dusting GJ, Moncada S, Vane JR (1977a) Prostacyclin is a weak contractor of coronary arteries of the pig. Eur J Pharmacol 45:301–304

Dusting GJ, Moncada S, Vane JR (1977b) Prostacyclin (PGX) is the endogenous metabolite responsible for relaxation of coronary arteries induced by arachidonic acid. Prostaglandins 13:3–15

Dusting GJ, Chapple DJ, Hughes R, Moncada S, Vane JR (1978) Prostacyclin (PGI_2) induces coronary vasodilatation in anaesthetized dogs. Cardiovasc Res 12:720–730

Dyerberg J, Bang HO (1979) Haemostatic function and platelet polyunsaturated fatty acids in Eskimos. Lancet 2:433–435

Dyerberg J, Jørgensen KA (1980) The effect of arachidonic- and eicosapentaenoic acid on the synthesis of prostacyclin-like material in human umbilical vasculature. Artery 8:12–17

Dyerberg J, Bang HO, Hjørne N (1975) Fatty acid composition of the plasma lipids in Greenland Eskimos. Am J Clin Nutr 28:958–966

Dyerberg J, Bang HO, Stoffersen E, Moncada S, Vane JR (1978) Eicosapentaenoic acid and prevention of thrombosis and atherosclerosis. Lancet 2:117–119

Dyerberg J, Jørgensen KA, Arnfred T (1981) Human umbilical blood vessel converts all cis-5,8-11,14,17 eicosapentaenoic acid to prostaglandin I_3. Prostaglandins 22: 857–862

Editorial (1980) Aspirin after myocardial infarction. Lancet 1:1172–1173

Edlund A, Berglund B, Kaijser L, Patrono C, Sollevi A, van Dorne D, Wennmalm A (1983) Release of adenosine and prostacyclin from ischemic human hearts. Adv Prostaglandin Thromboxane Leukotriene Res 11:365–370

Edmonds LC, Lefer AM (1984) Protective actions of a new thromboxane synthetase inhibitor in arachidonate induced sudden death. Life Sci 35:1763–1768

Egan RW, Paxton J, Kuehl FA Jr (1976) Mechanism of irreversible self-deactivation of prostaglandin synthetase. J Biol Chem 251:7329–7335

Ellis EF, Oelz O, Roberts LJ II, Payne NA, Sweetman BJ, Nies AS, Oates JA (1976) Coronary arterial smooth muscle contraction by a substance released from platelets: Evidence that it is thromboxane A_2. Science 193:1135–1137

Elwood PC, Sweetnam PM (1979) Aspirin and secondary mortality after myocardial infarction. Lancet 2:1313–1317

Elwood PC, Cochrane AL, Burr ML, Sweetnam PM, Williams G, Welsby E, Hughes SJ, Renton R (1974) A randomized controlled trial of acetyl salicylic acid in the secondary prevention of mortality from myocardial infarction. Br Med J 1:436–440

EPSIM Research Group (1982) A controlled comparison of aspirin and oral anticoagulants in prevention of death after myocardial infarction. N Engl J Med 307: 701–708

Ertl G, Fiedler V, Abram TS, Kochsiek K (1984) Coronary effects of leukotriene C_4 and D_4 at normal and reduced coronary perfusion (Abstr). 2nd International Symposium on Prostaglandins in the Cardiovascular System, Nürnberg-Fürth, May 9–11

Ezeamuzie IC, Assem ESK (1983) Effects of leukotrienes C_4 and D_4 on guinea-pig heart and the participation of SRS-A in the manifestation of guinea-pig cardiac anaphylaxis. Agents Actions 13:182–187

Feigen GA, Prager DJ (1969) Experimental cardiac anaphylaxis. Physiologic, pharmacologic and biochemical aspects of immune reactions in the isolated heart. Am J Cardiol 24:474–491

Feigen GA, Vaughan Williams EM, Peterson JK, Nielsen CB (1960) Histamine release and intracellular potentials during anaphylaxis in the isolated heart. Circ Res 8: 713–723

Feldberg W, Kellaway CH (1938) Liberation of histamine and formation of lysocithin-like substances by cobra venom. J Physiol (Lond) 94:187–226

Ferreira SH, Moncada S, Vane JR (1971) Indomethacin and aspirin abolish prostaglandin release from the spleen. Nature (New Biol) 231:237–239

Fésüs L, Csaba B, Muszbek L (1977) Platelet-activating factor, the trigger of hemostatic alterations in rat anaphylaxis. Clin Exp Immunol 27:512–515

Feuerstein G, Ezra D, Hayes E, Ramwell PW, Goldstein RE (1984a) Effects of platelet activating factor on coronary circulation, cardiac function and coronary vein level of 6-keto-$PGF_{1\alpha}$, TXB_2 and leukotriene C_4-like immunoreactivity of the intact domestic pig heart (Abstr). In: Neri Serneri GG, Masotti G, Gensini GF (eds) Platelets prostaglandins and the cardiovascular system. Boehringer-Ingelheim, Florence, p 159

Feuerstein G, Boyd LM, Ezra D, Goldstein RE (1984b) Effect of platelet-activating factor on coronary circulation of the domestic pig. Am J Physiol 246:466–471

Feuerstein N, Ramwell PW (1981) OKY-1581, a potential selective thromboxane synthetase inhibitor. Eur J Pharmacol 69:533–534

Fiedler VB (1983) Reduction of myocardial infarction and dysrhythmic activity by nafazatrom in the conscious rat. Eur J Pharmacol 88:263–267

Fiedler VB (1984) Reduction of acute myocardial ischemia in rabbit hearts by nafazatrom. J Cardiovasc Pharmacol 6:318–324

Fiedler VB, Mardin M, Abram TS (1984) Nifedipine on cardiovascular leukotriene D_4 actions in the anaesthetized dog. Eur J Pharmacol 104:159–164

Fischer S, Weber PC (1983) Thromboxane A_3 (TXA_3) is formed in human platelets after dietary eicosapentaenoic acid (C20:5ω3). Biochem Biophys Res Commun 116:1091–1099

Fischer S, Weber PC (1984) Prostaglandin I_3 is formed in vivo in man after dietary eicosapentaenoic acid. Nature 307:165–168

Fischer S, Struppler M, Böhlig B, Bernutz C, Weber W, Weber PC (1983) The influence of selective thromboxane synthetase inhibition with a novel imidazole derivative, UK-38 485, on prostanoid formation in man. Circulation 68:821–826

Fishbein MC, Maclean D, Maroko PR (1978) The histopathologic evaluation of myocardial infarction. Chest 73:843–849

FitzGerald DJ, Roy L, Robertson RM, FitzGerald GA (1984) The effect of organic nitrates on prostacyclin biosynthesis and platelet function in humans. Circulation 70:297–302

FitzGerald GA, Oates JA (1984) Selective and nonselective inhibition of thromboxane formation. Clin Pharmacol Ther 35:633–640

FitzGerald GA, Sherry S (1982) Pharmacology and pharmacokinetics of platelet-active drugs under current clinical investigation. In: Oates JA (ed) Prostaglandins and the cardiovascular system. Raven, New York, pp 107–172
FitzGerald GA, Friedman LA, Miyamori I, Grady JO, Lewis PJ (1979) A double blind placebo controlled crossover study of prostacyclin in man. Life Sci 25:665–672
FitzGerald GA, Oates JA, Hawiger J, Maas RL, Roberts LJ, Lawson JA, Brash AR (1983a) Endogenous biosynthesis of prostacyclin and thromboxane and platelet function during chronic administration of aspirin in man. J Clin Invest 71:676–688
FitzGerald GA, Pedersen AK, Patrono C (1983b) Analysis of prostacyclin and thromboxane biosynthesis in cardiovascular disease. Circulation 67:1174–1177
FitzGerald GA, Smith B, Pedersen AK, Brash AR (1984) Increased prostacyclin biosynthesis in patients with severe atherosclerosis and platelet activation. N Engl J Med 310:1065–1068
Fitzpatrick FA, Gorman RR (1978) A comparison of imidazole and 9,11-azoprosta-5,13-dienoic acid; two selective thromboxane synthetase inhibitors. Biochim Biophys Acta 539:162–173
Flower RJ (1974) Drugs which inhibit prostaglandin biosynthesis. Phrarmacol Rev 26: 33–67
Flower RJ, Blackwell GJ (1976) The importance of phospholipase A_2 in prostaglandin biosynthesis. Biochem Pharmacol 25:285–291
Ford-Hutchinson AW, Bray MA, Doig MV, Shipley ME, Smith MJH (1980) Leukotriene B, a potent chemokinetic and aggregating substance released from polymorphonuclear leukocytes. Nature 286:264–265
Foster RF, Layman JD (1952) Generalized urticaria with electrocardiographic changes simulating myocardial infarction. JAMA 148:203–205
Friedman PL, Brown EJ, Gunther S, Alexander RW, Barry WH, Mudge GH Jr, Grossman W (1981) Coronary vasoconstrictor effect of indomethacin in patients with coronary-artery disease. N Engl J Med 305:1171–1175
Ganz P, Gaspar J, Colucci WS, Barry WH, Mudge GH, Alexander RW (1984) Effects of prostacyclin on coronary hemodynamics at rest and in response to cold pressor testing in patients with angina pectoris. Am J Cardiol 53:1500–1504
Gerrard JM, Stuart MJ, Rao GHR, Steffes MW, Mauer SM, Brown DM, White JG (1980) Alteration in the balance of prostaglandin and thromboxane synthesis in diabetic rats. J Lab Clin Med 95:950–958
Ginsburg R, Bristow MR, Harrison DC, Stinson EB (1980) Studies with isolated human coronary arteries. Some general observations, potential mediators of spasm, role of calcium antagonists. Chest 78:180–186
Giotti A, Guidotti A, Mannaioni PF, Zilletti L (1966) The influence of adrenotropic drugs and noradrenaline on the histamine release in cardiac anaphylaxis in vitro. J Physiol (Lond) 184:924–941
Goetzl EJ, Pickett WC (1980) The human PMN leukocyte chemotactic activity of complex hydroxyeicosatetraenoic acids (HETEs). J Immunol 125:1789–1791
Goldblatt MW (1935) Properties of human seminal plasma. J Physiol (Lond) 84:208–218
Gorman RR, Bunting S, Miller OV (1977a) Modulation of human platelet adenylate cyclase by prostacyclin (PGX). Prostaglandins 13:377–388
Gorman RR, Fitzpatrick FA, Miller OV (1977b) A selective thromboxane synthetase inhibitor blocks the cAMP lowering activity of PGH_2. Biochem Biophys Res Commun 79:305–313
Gorman RR, Johnson RA, Spilman CH, Aiken JW (1983) Inhibition of platelet thromboxane A_2 synthase activity by sodium 5-(3′-pyridinyl methyl)benzofuran-2-carboxylate. Prostaglandins 26:325–342
Green LH, Seroppian E, Handin RI (1980) Platelet activation during exercise-induced myocardial ischemia. N Engl J Med 302:193–197
Greenwald JE, Bianchine JR, Wong LK (1979) The production of the arachidonate metabolite HETE in vascular tissue. Nature 281:588–589

Gryglewski RJ (1976) Steroid hormones, anti-inflammatory steroids and prostaglandins. Pharmacol Res Commun 8:337–348
Gryglewski R, Vane JR (1972) The release of prostaglandins and rabbit aorta contracting substance (RCS) from rabbit spleen and its antagonism by anti-inflammatory drugs. Br J Pharmacol 45:37–47
Gryglewski RJ, Bunting S, Moncada S, Flower RJ, Vane JR (1976) Arterial walls are protected against deposition of platelet thrombi by a substance (Prostaglandin X) which they make from prostaglandin endoperoxides. Prostaglandins 12:685–713
Gryglewski RJ, Korbut R, Ocetkiewicz A (1978) Generation of prostacyclin by lungs in vivo and its release into the arterial circulation. Nature 273:765–767
Gryglewski RJ, Salmon JA, Ubatuba FB, Weatherly BC, Moncada S, Vane JR (1979) Effects of all cis-5,8,11,14,17 eicosapentaenoic acid and PGH_3 on platelet aggregation. Prostaglandins 18:453–478
Guyton JR, Willerson JT (1977) Peripheral venous platelet aggregates in patients with unstable angina pectoris and acute myocardial infarction. Angiology 28:695–701
Hahn F, Bernauer W (1970) Studies on heart anaphylaxis. III. Effect of antigen and histamine on perfused guinea-pig heart. Arch Int Pharmacodyn 184:129–157
Halushka PV, Rogers RC, Loadholt CB, Colwell JA (1981) Increased platelet thromboxane synthesis in diabetes mellitus. J Lab Clin Med 97:87–96
Hamazaki T, Hirai A, Terano T, Sajiki J, Kondo S, Fujita T, Tamura Y, Kumagai A (1982) Effects of orally administered ethyl ester of eicosapentaenoic acid (EPA; C20:5, ω-3) on PGI_2-like substance production by rat aorta. Prostaglandins 23: 557–567
Hamberg M (1976) On the formation of thromboxane B_2 and 12-hydroxy-5,8,10,14-eicosatetraenoic acid (12ho-20:4) in tissues from the guinea pig. Biochim Biophys Acta 431:651–654
Hamberg M, Fredholm BB (1976) Isomerization of prostaglandin H_2 into prostaglandin D_2 in the presence of serum albumin. Biochim Biophys Acta 431:189–193
Hamberg M, Samuelsson B (1974) Prostaglandin endoperoxides. Novel transformations of arachidonic acid in human platelets. Proc Natl Acad Sci USA 71:3400–3404
Hamberg M, Svensson J, Samuelsson B (1974a) Prostaglandin endoperoxides. A new concept concerning the mode of action and release of prostaglandins. Proc Natl Acad Sci USA 71:3824–3828
Hamberg M, Svensson J, Wakabayashi T, Samuelsson B (1974b) Isolation and structure of two prostaglandin endoperoxides that cause platelet aggregation. Proc Natl Acad Sci USA 71:345–349
Hamberg M, Svensson J, Samuelsson B (1975) Thromboxanes: a new group of biologically active compounds derived from prostaglandin endoperoxides. Proc Natl Acad Sci USA 72:2994–2998
Hammarström S (1983) Leukotrienes. Annu Rev Biochem 52:355–377
Hammarström S, Falardeau P (1977) Resolution of prostaglandin endoperoxide synthase and thromboxane synthase of human platelets. Proc Natl Acad Sci USA 74: 3691–3695
Hammarström S, Murphy RC, Samuelsson B, Clark DA, Mioskowski C, Corey EJ (1979) Structure of leukotriene C. Identification of the amino acid part. Biochem Biophys Res Commun 91:1266–1272
Hanashiro PK, Weil MH (1967) Anaphylactic shock in man. Report of two cases with detailed hemodynamic and metabolic studies. Arch Intern Med 119:129–140
Harrison HE, Reece AH, Johnson M (1978) Decreased vascular prostacyclin in experimental diabetes. Life Sci 23:351–356
Hartmann JR, Robinson JA, Gunnar RM (1977) Chemotactic activity in the coronary sinus after experimental myocardial infarction: Effects of pharmacologic interventions on ischemic injury. Am J Cardiol 40:550–555
Hassam AG, Crawford MA (1976) The differential incorporation of labelled linoleic, γ-linolenic, dihomo-γ-linolenic and arachidonic acids into the developing rat brain. J Neurochem 27:967–968

Hattori Y, Levi R (1984) Negative inotropic effect of leukotrienes: leukotriene C_4 and D_4 inhibit calcium-dependent contractile responses in potassium-depolarized guinea-pig myocardium. J Pharmacol Exp Ther 230:646–651

Heffner JE, Shoemaker SA, Canham EM, Patel M, McMurtry IF, Morris HG, Repine JE (1983) Acethyl glyceryl ether phosphorylcholine-stimulated human platelets cause pulmonary hypertension and edema in isolated rabbit lungs. J Clin Invest 71: 351–357

Hemler M, Lands WEM, Smith WL (1976) Purification of the cyclooxygenase that forms prostaglandins. Demonstration of two forms of iron in the holoenzyme. J Biol Chem 251:5575–5579

Herbaczynska-Cedro K, Moncada S, Mullane KM, Vane JR (1982) Effects of thromboxane A_2 on the coronary circulation in vivo (Abstr). 5th International Conference on Prostaglandins, Florence, May 18–21, p 277

Herman AG, Claeys M, Moncada S, Vane JR (1979) Biosynthesis of prostacyclin (PGI_2) and 12L-hydroxy-5,8,10,14 eicosatetraenoic acid (HETE) by pericardium, pleura, peritoneum and aorta of the rabbit. Prostaglandins 18:439–452

Higgs GA, Vane JR (1983) Inhibition of cyclooxygenase and lipoxygenase. Br Med Bull 39:265–270

Higgs GA, Bunting S, Moncada S, Vane JR (1976) Polymorphonuclear leukocytes produce thromboxane A_2-like activity during phagocytosis. Prostaglandins 12: 749–757

Higgs GA, Moncada S, Vane JR (1977) Prostacyclin (PGI_2) inhibits the formation of platelet thrombi induced by adenosine diphosphate (ADP) in vivo. Br J Pharmacol 61:137P

Higgs GA, Flower RJ, Vane JR (1979) A new approach to anti-inflammatory drugs. Biochem Pharmacol 28:1959–1961

Higgs GA, Eakins KE, Mugridge KG, Moncada S, Vane JR (1980) The effects of nonsteroid anti-inflammatory drugs on leukocyte migration in carrageenin-induced inflammation. Eur J Pharmacol 66:81–86

Higgs GA, Mugridge KG, Moncada S, Vane JR (1984) Inhibition of tissue damage by the arachidonate lipoxygenase inhibitor BW755C. Proc Natl Acad Sci USA 81: 2890–2892

Hill JH, Ward PA (1971) The phlogistic role of C3 leukotactic fragments in myocardial infarcts of rats. J Exp Med 133:885–900

Hillis LD, Braunwald E (1977) Myocardial ischemia. N Engl J Med 296:971–978

Hillis LD, Braunwald E (1978) Coronary-artery spasm. N Engl J Med 299:695–702

Hintze TH, Kaley G (1984) Ventricular receptors activated following myocardial prostaglandin synthesis initiate reflex hypotension, reduction in heart rate, and redistribution of cardiac output in the dog. Circ Res 54:239–247

Hintze T, Kaley G, Martin EG, Messina EJ (1978) PGI_2 induces bradycardia in the dog. Prostaglandins 15:712

Hintze TH, Martin EG, Messina EJ, Kaley G (1979) Prostacyclin (PGI_2) elicits reflex bradycardia in dogs: evidence for vagal mediation. Proc Soc Exp Biol Med 162: 96–100

Hintze TH, Panzenbeck MJ, Messina EJ, Kaley G (1981) Prostacyclin (PGI_2) lowers heart rate in the conscious dog. Cardiovasc Res 15:538–546

Hintze TH, Kichuk MR, Stern H, Harrison J, Kaley G (1984) Prostacyclin dilates large coronary arteries in conscious dogs (Abstr). In: Neri Serneri GG, Masotti G, Gensini GF (eds) Platelets prostaglandins and the cardiovascular system. Florence, p 96

Hirata F, Schiffmann E, Venkatasubramanian K, Salomon D, Axelrod J (1980) A phospholipase A_2 inhibitory protein in rabbit neutrophils induced by glucocorticoids. Proc Natl Acad Sci USA 77:2533–2536

Hirsh SA (1982) Acute allergic reaction with coronary vasospasm. Am Heart J 103:928

Hirsh PD, Hillis LD, Campbell WB, Firth BG, Willerson JT (1981) Release of prostaglandins and thromboxane into the coronary circulation in patients with ischemic heart disease. N Engl J Med 304:685–691

Hirsh PD, Firth BG, Campbell WB, Willerson JT, Hillis LD (1982) Influence of blood sampling site and technique on thromboxane concentrations in patients with ischemic heart disease. Am Heart J 104:234–237

Hong SCL, Levine L (1976) Inhibition of arachidonic acid release from cells as the biochemical action of anti-inflammatory corticoids. Proc Natl Acad Sci USA 73: 1730–1734

Hornstra G, Christ-Hazelhof E, Haddeman E, ten Hoor F, Nugteren DH (1981) Fish oil feeding lowers thromboxane and prostacyclin production by rat platelets and aorta and does not result in the formation of prostaglandin I_3. Prostaglandins 21:727–738

Hsueh W, Needleman P (1978) Sites of lipase activation and prostaglandin synthesis in isolated, perfused rabbit hearts and hydronephrotic kidneys. Prostaglandins 16: 661–681

Imai S, Takeda K, Nakagawa Y, Nakazawa M, Shimamoto N, Katano Y (1983) Effects of prostaglandins and related compounds on the canine coronary artery and arteriole. Adv Prostaglandin Thromboxane Leukotriene Res 11:401–405

Ingerman-Wojenski C, Silver MJ, Smith JB, Macarak E (1981) Bovine endothelial cells in culture procedure thromboxane as well as prostacyclin. J Clin Invest 67:1292–1296

Irvine RF (1982) How is the level of free arachidonic acid controlled in mammalian cells? Biochem J 204:3–16

Isakson PC, Raz A, Denny SE, Wyche A, Needleman P (1977) Hormonal stimulation of arachidonate release from isolated perfused organs. Relationship to prostaglandin biosynthesis. Prostaglandins 14:853–871

Ito T, Ogawa K, Enomoto I, Hashimoto H, Kai I, Satake T (1980) Cormparison of the effects of PGI_2 and PGE_1 on coronary and systemic hemodynamics and coronary arterial cyclic nucleotide levels in dogs. Adv Prostaglandin Thromboxane Res 7: 641–646

Jakubowski JA, Ardlie NG (1979) Evidence for the mechanism by which eicosapentaenoic acid inhibits human platelet aggregation and secretion-implications for the prevention of vascular disease. Thromb Res 16:205–217

Johnson AS, Scheinberg SR, Gerisch RA, Saltzstein HC (1953) Effect of cortisone on the size of experimentally produced infarcts. Circulation 7:224–228

Johnson RA, Morton DR, Kinner JH, Gorman RR, McGuire JC, Sun FF, Wittacker N, Bunting S, Salmon J, Moncada S, Vane JR (1976) The chemical structure of prostaglandin X (prostacyclin). Prostaglandins 12:915–928

Jørgensen KA, Steffersen E, Dyerberg J (1979) Stability of prostacyclin in plasma. Lancet 1:1352

Jouve R, Rolland PH, Delboy C, Mercier C (1984) Thromboxane B_2, 6-keto-$PGF_{1\alpha}$, PGE_2, $PGF_{2\alpha}$ and PGA_1 plasma levels in arteriosclerosis obliterans: Relationship to clinical manifestation, risk factors, and arterial pathoanatomy. Am Heart J 107: 45–52

Jugdutt BI, Hutchins GM, Bulkley BH, Pitt B, Becker LC (1979) Effect of indomethacin on collateral blood flow and infarct size in the conscious dog. Circulation 4:734–743

Jugdutt BI, Hutchins GM, Bulkley BH, Becker LC (1980) Salvage of ischemic myocardium by ibuprofen during infarction in the conscious dog. Am J Cardiol 46: 74–82

Jugdutt BI, Hutchins GM, Bulkley GH, Becker LC (1981) Dissimilar effects of prostacyclin, prostaglandin E_1 and prostaglandin E_2 on myocardial infarct size after coronary occlusion in conscious dogs. Circ Res 49:685–700

Kaley G, Hintze TH, Messina EJ (1980) Role of prostaglandins in ventricular reflexes. Adv Prostaglandin Thromboxane Res 7:601–608

Karim SMM, Sandler M, Williams ED (1967) Distribution of prostaglandins in human tissues. Br J Pharmacol Chemother 31:340–344

Karim SMM, Hillier K, Devlin J (1968) Distribution of prostaglandins E_1, E_2, $F_{1\alpha}$ and $F_{2\alpha}$ in some animal tissues. J Pharm Pharmacol 20:749–753

Karmazyn M (1983) Reduction of enzyme release from perfused ischemic hearts by steroidal and non-steroidal prostaglandin synthesis inhibitors. Prostaglandins Leukotrienes Med 11:299–315

Karmazyn M, Horrobin DF, Manku MS, Cunnane SC, Karmali RA, Ally AI, Morgan RO, Nicolaou KC, Barnette WE (1978) Effects of prostacyclin on perfusion pressure, electrical activity, rate and force of contraction in isolated rat and rabbit hearts. Life Sci 22:2079–2086

Kellaway CH, Trethewie ER (1940) The liberation of a slow-reacting smooth-muscle stimulating substance in anaphylaxis. Q J Exp Physiol 30:121–145

Kelly JF, Patterson P (1974) Anaphylaxis. Course, mechanism, and treatment. JAMA 227:1431–1436

Kirmser R, Berger HJ, Cohen LS, Wolfson S (1976) Effect of indomethacin, a prostaglandin inhibitor on epicardial ST elevation and myocardial flow after coronary occlusion. Circulation (Suppl 2) 53/54:II–194

Klitzke AK (1984) Nafazatrom (BAYg6575), a potent stimulator of prostacyclin release from cardiac and renal vessel wall. Arch Int Pharmacodyn 271:220–228

Klitzke AK, Trompler AT (1984) Leukotriene C_4 stimulates prostacyclin release from the cardiac vessel wall (Abstr). In: Neri Serneri GG, Masotti G, Gensini GF (eds) Platelets prostaglandins and the cardiovascular system. Bohringer-Ingelheim, Florence, p 43

Kloner RA, Fishbein MC, Lew H, Maroko PR, Braunwald E (1978) Mummification of the infarcted myocardium by high dose corticosteroids. Circulation 57:56–63

Körner CF, Hausmann G, Gemsa D, Resch K (1984) Rate of prostaglandin synthesis is not controlled by phospholipase A activity but by reincorporation of released fatty acids into phospholipids. Agents Actions 15:28–30

Korth R, Riess H, Brehm G, Hiller E (1984) Influence of unsaturated platelet-activating factor on aggregation, serotonin release and thomboxane synthesis of human thrombocytes (Abst). In: Neri Serneri GG, Masotti G, Gensini GF (eds) Platelets prostaglandins and the cardiovascular system. Boehringer-Ingelheim, Florence, p 157

Kraemer RJ, Phernetton TM, Folts JD (1976) Prostaglandin-like substances in coronary venous blood following myocardial ischemia. J Pharmacol Exp Ther 199: 611–619

Krell RD, Osborn R, Falcone K, Vickery L (1981) Enhancement of isolated airway response to bronchoconstrictive agonists by the slow-reacting substance of anaphylaxis FPL 55712. Prostaglandins 22:423–432

Kröner EE, Peskar BA, Fischer H, Ferber E (1981) Control of arachidonic acid accumulation in bone marrow-derived macrophages by acetyltransferases. J Biol Chem 256:3690–3697

Ku EC, McPherson SE, Signor C, Chertock H, Cash WD (1983) Characterization of imidazo(1,5-a)pyridine-5-hexanoic acid (CGS-13080) as a selective thromboxane synthetase inhibitor using in vitro and in vivo biochemical models. Biochem Biophys Res Commun 112:899–906

Kukovetz WR, Holzmann S, Wurm A, Pöch G (1979) Prostacyclin increases cAMP in coronary arteries. J Cyclic Nucleotide Res 5:469–475

Kulmacz RJ, Lands WEM (1983) Characteristics of prostaglandin H synthetase. Adv Prostaglandin Thromboxane Leukotriene Res 11:93–97

Kunze H, Vogt W (1971) Significance of phospholipase A for prostaglandin formation. Ann NY Acad Sci 180:123–125

Kurzrock R, Lieb CC (1930) Biochemical studies of human semen. II. The action of semen on the human uterus. Proc Soc Exp Biol Med 28:268–272

Lands WEM (1979) The biosynthesis and metabolism of prostaglandins. Annu Rev Physiol 41:633–652

Lands WEM, Samuelsson B (1968) Phospholipid precursors of prostaglandins. Biochim Biophys Acta 164:426–429

Larrue J, Leroux C, Daret D, Bricaud H (1982) Decreased prostaglandin production in cultured smooth muscle cells from atherosclerotic rabbit aorta. Biochim Biophys Acta 710:257–263

Lee T, Malone B, Blank ML, Snyder F (1981) 1-alkyl-2-acethyl-sn-glycero-3-phosphocholine (platelet-activating factor) stimulates calium influx in rabbit platelets. Biochem Biophys Res Commun 102:1262–1268

Lefer AM, Polansky EW (1979) Beneficial effects of ibuprofen in acute myocardial ischemia. Cardiology 64:265–274

Lefer AM, Ogletree ML, Smith JB, Silver MJ, Nicolaou KC, Barnette WE, Gasic GP (1978) Prostacyclin: A potentially valuable agent for preserving myocardial tissue in acute myocardial ischemia. Science 200:52–54

Lefer AM, Okamatsu S, Smith EF III, Smith JB (1981) Beneficial effect of a new thromboxane synthetase inhibitor in arachidonate-induced sudden death. Thromb Res 23:265–273

Lefer AM, Müller HF, Smith JB (1984) Pathophysiological mechanisms of sudden death induced by platelet activating factor. Br J Pharmacol 83:125–130

Letts LG, Piper JP (1981) The effects of leukotrienes (LT) D_4 and C_4 on the guinea-pig isolated heart. J Physiol (Lond) 317:94P–95P

Letts LG, Piper PJ (1982) The actions of leukotrienes (LT) C_4 and D_4 on the guinea-pig isolated heart. Br J Pharmacol 76:169–176

Letts LG, Piper PJ (1983) Cardiac actions of leukotrienes B_4, C_4, D_4 and E_4 in guinea pig and rat in vitro. Adv Prostaglandin Thromboxane Leukotriene Res 11:391–395

Letts LG, Newman DL, Greenwald SE, Piper PJ (1983a) Effects of intra-coronary administration of leukotriene D_4 in the anaesthetized dog. Prostaglandins 26: 563–572

Letts LG, Piper PJ, Newman DL (1983b) Leukotrienes and their action on the coronary circulation. In: Piper PJ (ed) Leukotrienes and other lipoxygenase products. Research Studies Press, Chichester, pp 94–107

Levi R (1972) Effects of exogenous and immunologically released histamine on the isolated heart: a quantitative comparison. J Pharmacol Exp Ther 182:227–238

Levi R, Burke JA (1980) Cardiac anaphylaxis: SRS-A potentiates and extends the effects of released histamine. Eur J Pharmacol 62:41–49

Levi R, Allan G, Zavecz JH (1976) Prostaglandins and cardiac anaphylaxis. Life Sci 18:1255–1264

Levi R, Burke JA, Holland BA (1979) Slow reacting substance of anaphylaxis (SRS-A): Possible role in cardiac hypersensitivity reactions. Fed Proc 38:261

Levi R, Burke JA, Corey EJ (1982) SRS-A, leukotrienes, and immediate hypersensitivity reactions of the heart. In: Samuelsson B, Paoletti R (eds) Leukotrienes and other lipoxygenase products. Raven, New York, pp 215–222

Levi R, Burke JA, Guo ZG, Hattori Y, Hoppens CM, McManus LM, Hanahan DJ, Pinckard RN (1984) Acetyl glyceryl ether phosphorylcholine (AGEPC). A putative mediator of cardiac anaphylaxis in the guinea pig. Circ Res 54:117–124

Levine PH (1973) On acute effect of cigarette smoking on platelet function. Circulation 48:619–623

Levy JV (1978) Contractile responses to prostacyclin (PGI_2) of isolated human saphenous and rat venous tissue. Prostaglandins 16:93–97

Lewis HD, Davis JW, Archibal DG, Steinke WE, Smitherman TC, Doherty JE, Schnaper HW, Le Winter MM, Linares E, Pouget JM, Sabharwal SC, Chesler E, DeMots H (1983) Protective effects of aspirin against acute myocardial infarction and death in men with unstable angina. Results of a Veterans Administration Cooperative Study. N Engl J Med 309:396–403

Lewis RA, Drazen JM, Austen KF, Clark DA, Corey EJ (1980a) Identification of the C(6)-S-conjugate of leukotriene A with cysteine as a naturally occurring slow reacting substance of anaphylaxis (SRS-A). Importance of the 11-cis-geometry for biological activity. Biochem Biophys Res Commun 96:271–277

Lewis RA, Austen KF, Drazen MJ, Clark DA, Marfat A, Corey EJ (1980b) Slow reacting substances of anaphylaxis: Identification of leukotrienes C-1 and D from human and rat sources. Proc Natl Acad Sci USA 77:3710–3714
Lewy RI, Smith JB, Silver MJ, Saia J, Walinsky P, Wiener L (1979a) Detection of thromboxane B_2 in peripheral blood of patients with Prinzmetal's angina. Prostaglandin Med 5:243–248
Lewy RI, Wiener L, Smith JB, Walinsky P, Silver MJ, Saia J (1979b) Comparison of plasma concentrations of thromboxane B_2 in Prinzmetal's variant angina and classical angina pectoris. Clin Cardiol 2:404–406
Lewy RI, Wiener L, Walinsky P, Lefer AM, Silver MJ, Smith JB (1980) Thromboxane release during pacing-induced angina pectoris: possible vasoconstrictor influence on the coronary vasculature. Circulation 61:1165–1171
Libby P, Maroko PR, Bloor CM, Sobel BE, Braunwald E (1973) Reduction of experimental myocardial infarct size by corticosteroid administration. J Clin Invest 52: 599–607
Liebig R, Bernauer W, Peskar BA (1975) Prostaglandin, slow-reacting substance, and histamine release from anaphylactic guinea pig hearts, and its pharmacological modification. Naunyn Schmiedebergs Arch Pharmacol 289:65–76
Lilienfeld A, Hochstein E, Weiss W (1950) Acute myocarditis with bundle branch block due to sulfonamide sensitivity. Circulation 1:1060–1064
Lotner GZ, Lynch JM, Betz SJ, Henson PM (1980) Human neutrophil derived platelet-activating factor. J Immunol 124:676–684
Lucchesi BR, Jolly SR, Bailie MB, Abrams GD (1982) Protection of ischemic myocardium by BW755C. Fed Proc 41:8576
Luchi RJ, Chahine RA (1981) Coronary artery spasm, coronary artery thrombosis and myocardial infarction. Ann Intern Med 95:502–505
MacIntyre DE, Pearson JD, Gordon JL (1978) Localization and stimulation of prostacyclin production in vascular cells. Nature 271:549–551
MacIntyre DE, Shaw AM, Pollock WK, Marks G, Westwick J (1982) The role of endogenous arachidonate metabolites in phospholipid-induced human platelet activation (Abstr). 5th International Conference on Prostaglandins, Florence, May 18–21, p 714
Maclean D, Fishbein MC, Braunwald E (1978a) Long-term preservation of ischemic myocardium after experimental coronary artery occlusion. J Clin Invest 61:541–551
Maclean D, Fishbein MC, Blum RI, Braunwald E, Maroko PR (1978b) Long-term preservation of ischemic myocardium by ibuprofen after experimental coronary artery occlusion. Am J Cardiol 41:394
MacNab MW, Foltz EL, Graves BS, Rinehart RK, Tripp SL, Feliciano NR, Sen S (1984) The effects of a new thromboxane synthetase inhibitor, CGS-13080, in man. J Clin Pharmacol 24:76–83
Mallory GK, White PD, Salcedo-Salgar J (1939) The speed of healing of myocardial infarction. A study of the pathologic anatomy in seventy-two cases. Am Heart J 18: 647–671
Marooka S, Kobayashi M, Shimamoto T (1977) Experimental ischemic heart disease induced by thromboxane A_2 in rabbits. Jpn Circ J 41:1373–1379
Maseri A, L'Abbate A, Chierchia S, Parodi O, Severi S, Biagini A, Distante A, Marzilli M, Ballestra AM (1979) Significance of spasm in the pathogenesis of ischemic heart disease. Am J Cardiol 44:788–792
McManus JF, Lawlor JJ (1950) Myocardial infarction following the administration of tetanus antitoxin. N Engl J Med 242:17–19
McManus LM, Hanahan DJ, Demopoulos CA, Pinckard RN (1980) Pathobiology of the intravenous infusion of acetyl glyceryl ether phosphorylcholine (AGEPC), a synthetic platelet-activating factor (PAF), in the rabbit. J Immunol 124:2919–2924
McManus LM, Hanahan DJ, Pinckard RN (1981) Human platelet stimulation by acetyl glyceryl ether phosphorylcholine. J Clin Invest 67:903–906

McManus LM, Fitzpatrick FA, Hanahan DJ, Pinckard RN (1983) Thromboxane B_2 release following acetyl glyceryl ether phosphorylcholine infusion in the rabbit. Immunopharmacology 5:197–207

Meacock SCR, Kitchen EA (1976) Some effects of non-steroidal anti-inflammatory drugs on leukocyte migration. Agents Actions 6:320–325

Metha J, Mehta P, Burger C, Pepine JC (1978) Platelet aggregation studies in coronary artery disease. Part 4. Effect of aspirin. Atherosclerosis 31:169–175

Mehta J, Mehta P, Conti CR (1980a) Platelet function studies in coronary heart disease. IX. Increased platelet prostaglandin generation and abnormal platelet sensitivity to prostacyclin and endoperoxide analog in angina pectoris. Am J Cardiol 46:943–947

Mehta J, Mehta P, Pepine CJ, Conti CR (1980b) Platelet function studies in coronary artery disease. VII. Effect of aspirin and tachycardia stress on aortic and coronary venous blood. Am J Cardiol 45:945–951

Mehta J, Mehta P, Zipper R, Horalek C (1981a) Thromboxane/prostacyclin equilibrium at rest and during exercise in man and its relationship to myocardial ischemia. Clin Res 29:223A

Mehta J, Nichols WW, Mehta P, Pepine C, Conti CR (1981b) Effects of prostacyclin on systemic and coronary hemodynamics in the dog. Am Heart J 102:835–840

Mehta J, Mehta P, Feldman RL (1982) Severe intracoronary thromboxane release preceding acute coronary artery occlusion. Prostaglandins Leukotrienes Med 8: 599–605

Mehta J, Mehta P, Feldman RL, Hovalek C (1984a) Thromboxane release in coronary artery disease: spontaneous versus pacing-induced angina. Am Heart J 107:286–292

Mehta J, Nichols WW, Goldman R (1984b) Prostacyclin release following endoperoxide analogue infusion in the intact dog. Am J Physiol 246:R205–R210

Mencia-Huerta JM, Benveniste J (1979) Platelet-activating factor and macrophages. I. Evidence for the release from rat and mouse peritoneal macrophages and not from mastocytes. Eur J Immunol 9:409–415

Mencia-Huerta JM, Ninio E, Landes A, Godfroid JJ, Benveniste J (1981) Enzymatic assembly of platelet-activating factor (PAF-acether). Fed Proc 40:1015

Menys VC, Davies JA (1983) Selective inhibition of thromboxane synthetase with dazoxiben – basis of its inhibitory effect on platelet adhesion. Thromb Haemost 49:96–101

Michelassi F, Landa L, Hill RD, Lowenstein E, Watkins WD, Petkau AJ, Zapol WM (1982) Leukotriene D_4: A potent coronary artery vasoconstrictor associated with impaired ventricular contraction. Science 217:841–843

Michelassi F, Landa L, Hill RD, Huttemeier P, Lowenstein E, Zapol WM, Watkins WD (1983) Leukotriene D_4: A potent sheep coronary vasoconstrictor not blocked by ibuprofen. Adv Prostaglandin Thromboxane Leukotriene Res 11:397–399

Miyamoto T, Ogino N, Yamamoto S, Hayaishi O (1976) Purification of prostaglandin endoperoxide synthetase from bovine vesicular gland microsomes. J Biol Chem 251:2629–2636

Moncada S (1982) Biological importance of prostacyclin. Br J Pharmacol 76:3–31

Moncada S, Vane JR (1978) Unstable metabolites of arachidonic acid and their role in haemostasis and thrombosis. Br Med Bull 34:129–135

Moncada S, Vane JR (1979) Pharmacology and endogenous roles of prostaglandin endoperoxides, thromboxane A_2, and prostacyclin. Pharmacol Rev 30:293–333

Moncada S, Gryglewski R, Bunting S, Vane JR (1976a) An enzyme isolated from arteries transforms prostaglandin endoperoxides to an unstable substance that inhibits platelet aggregation. Nature 263:663–665

Moncada S, Gryglewski RJ, Bunting S, Vane JR (1976b) A lipid peroxide inhibits the enzyme in blood vessel microsomes that generates from prostaglandin endoperoxides the substance (prostaglandin X) which prevents platelet aggregation. Prostaglandins 12:715–737

Moncada S, Bunting S, Mullane K, Thorogood P, Vane JR, Raz A, Needleman P (1977a) Imidazole: A selective inhibitor of thromboxane synthetase. Prostaglandins 13:611–618

Moncada S, Herman AG, Higgs EA, Vane JR (1977b) Differential formation of prostacyclin (PGX or PGI_2) by layers of the arterial wall. An explanation for the anti-thrombotic properties of vascular endothelium. Thromb Res 11:323–344

Morley J, Bray MA, Jones RW, Nugteren DH, van Dorp DA (1979) Prostaglandin and thromboxane production by human and guinea-pig macrophages and leukocytes. Prostaglandins 17:729–736

Morris HR, Taylor GW, Piper PJ, Tippins JR (1980) Structure of slow-reacting substance of anaphylaxis from guinea pig lung. Nature 285:104–106

Mullane KM, Moncada S (1982) The salvage of ischemic myocardium by BW755C in anaesthetized dogs. Prostaglandins 24:255–266

Mullane KM, Bradley G, Moncada S (1982) The interaction of platelet derived mediators on isolated canine coronary arteries. Eur J Pharmacol 84:115–118

Mullane KM, Read N, Salmon JA, Moncada S (1984) Role of leukocytes in acute myocardial infarction in anaesthetized dogs: Relationship to myocardial salvage by anti-inflammatory drugs. J Pharmacol Exp Ther 228:510–522

Müller B, Schneider J, Hennies HH, Flohé L (1984) Cardioprotective action of the new stable epoprostenol analogue CG 4203 in rat models of cardiac hypoxia and ischemia. Drug Res 34:1506–1509

Murphy RC, Hammarström S, Samuelsson B (1979) Leukotriene C: a slow reacting substance from murine mastocytoma cells. Proc Natl Acad Sci USA 76:4275–4279

Mustard JF, Kinlough-Rathbone RL, Packham MA (1983) Aspirin in the treatment of cardiovascular disease: A Review. Am J Med 74:43–49

Nadler JL, Yamamoto J, Zipser R, Horton R (1984) Nicotine inhibits prostacyclin and stimulates thromboxane in chronic smokers (Abstr). Kyoto Conference on Prostaglandins, Kyoto, November 25–28, p 231

Naito J, Komatsu H, Ujiie A, Hamano S, Kubota T, Tsuboshima M (1983) Effects of thromboxane synthetase inhibitors on aggregation of rabbit platelets. Eur J Pharmacol 91:41–48

Narumiya S, Salmon JA, Cottee FH, Weartherley BC, Flower RJ (1981) Arachidonic acid 15-lipoxygenase from rabbit peritoneal polymorphonuclear leukocytes. Partial purification and properties. J Biol Chem 256:9583–9592

Needleman P (1978) Characterization of the reaction sequence involved in phospholipid labeling and deacylation and prostaglandin synthesis and actions. J Allergy Clin Immunol 62:96–102

Needleman P, Key SC, Isakson PC, Kulkarni PS (1975) Relationship between oxygen tension, coronary vasodilatation and prostaglandin biosynthesis in the isolated heart. Prostaglandins 9:123–134

Needleman P, Moncada S, Bunting S, Vane JR, Hamberg M, Samuelsson B (1976) Identification of an enzyme in platelet microsomes which generates thromboxane A_2 from prostaglandin endoperoxides. Nature 261:558–560

Needleman P, Kulkarni PS, Raz A (1977) Coronary tone modulation: Formation and actions of prostaglandins, endoperoxides and thromboxanes. Science 195:409–412

Needleman P, Raz A, Minkes MS, Ferendelli JA, Sprecher H (1979) Triene prostaglandins: prostacyclin and thromboxane biosynthesis and unique biological properties. Proc Natl Acad Sci USA 76:944–948

Neri Serneri GG (1984) Prostaglandins and ischemic heart disease (Abstr). In: Neri Serneri GG, Masotti G, Gensini GF (eds) Platelets, prostaglandins and the cardiovascular system. Boehringer-Ingelheim, Florence, pp 20–21

Neri-Serneri GG, Gensini GF, Abbate R, Mugnaini C, Favilla S, Brunelli C, Chierchia S, Parodi O (1981a) Increased fibrinopeptide A formation and thromboxane A_2 production in patients with ischemic heart disease: relationships to coronary pathoanatomy, risk factors, and clinical manifestations. Am Heart J 101:185–194

Neri Serneri GG, Masotti G, Gensini GF, Abbate R, Poggesi L, Galanti G, Favilla S (1981b) Prostacyclin, thromboxane, and ischemic heart disease. In: Johnsson Hegyeli R (ed) Prostaglandins and cardiovascular disease. Raven, New York, pp 139–157

Neri Serneri GG, Masotti G, Poggesi L, Galanti G, Morettini A, Scarti L (1982) Reduced prostacyclin production in patients with different manifestations of ischemic heart disease. Am J Cardiol 49:1146–1151

Neri Serneri GG, Abbate R, Gensini GF, Panetta A, Casolo GC, Carini M (1983) TXA_2 production by human arteries and veins. Prostaglandins 25:753–766

Neri Serneri GG, Gensini GF, Abbate R, Prisco D, Rogasi PG, Casolo GC, Di Donato M, Dabizzi MP, Fantini F (1984a) Abnormal cardiopulmonary synthesis of thromboxane A_2 in patients with spontaneous angina (Abstr). In: Neri Serneri GG, Masotti G, Gensini GF (eds) Platelets prostaglandins and the cardiovascular system. Bohringer-Ingelheim, Florence, pp 109–111

Neri Serneri GG, Gensini GF, Masotti G, Abbate R, Morettini A, Poggesi L, Fortini A (1984b) Role of prostacyclin and thromboxane A_2 in ischemic heart disease. Adv Exp Med Biol 164:175–185

Nijkamp FP, Moncada S, White HL, Vane JR (1977) Diversion of prostaglandin endoperoxide metabolism by selective inhibition of thromboxane A_2 biosynthesis in lung, spleen or platelets. Eur J Pharmacol 44:179–186

Nowak J, Kaijser L, Wennmalm A (1980) Cardiac synthesis of prostaglandins from arachidonic acid in man. Prostaglandins Med 4:205–214

Nugteren DH (1975) Arachidonate lipoxygenase in blood platelets. Biochim Biophys Acta 380:299–307

Nugteren DH, Hazelhof E (1973) Isolation and properties of intermediates in prostaglandin biosynthesis. Biochim Biophys Acta 326:448–461

Nugteren DH, van Dorp DA, Bergström S, Hamberg M, Samuelsson B (1966) Absolute configuration of the prostaglandins. Nature 212:38–39

O'Flaherty JT, Lees CJ, Stimler NP (1981) Anaphylatoxic polymorphonuclear neutrophil (PMN)-stimulating actions of platelet activating factor (PAF). Fed Proc 40: 1015p

Ogawa K, Ito T, Enomoto I, Hashimoto H, Kai I, Satake T (1980) Increase of coronary flow and levels of PGE_1 and $PGF_{2\alpha}$ from the ischemic area of experimental myocardial infarction. Adv Prostaglandin Thromboxane Res 7:665–669

Ogawa K, Sakai K, Ito T, Watanabe J, Satake T (1983) Effects of selective thromboxane synthetase inhibitor and indomethacin on prostacyclin and thromboxane A_2 from ischemic canine heart. Adv Prostaglandin Thromboxane Leukotriene Res 11: 371–376

Ogino N, Miyamoto T, Yamamoto S, Hayaishi O (1977) Prostaglandin endoperoxide E isomerase from bovine vesicular gland microsomes, a glutathione-requiring enzyme. J Biol Chem 252:890–895

Ogletree ML, Lefer AM (1976) Influences of nonsteroidal anti-inflammatory agents on myocardial ischemia in the cat. J Pharmacol Exp Ther 197:582–593

Ogletree ML, Smith JB, Lefer AM (1978) Actions of prostaglandins on isolated perfused cat coronary arteries. Am J Physiol 235:H400–H406

Ogletree ML, Lefer AM, Smith JB, Nicolaou KC (1979) Studies on the protective effect of prostacyclin in acute myocardial ischemia. Eur J Pharmacol 56:95–103

O'Grady J, Warrington S, Moti MJ, Bunting S, Flower R, Fowle ASE, Higgs EA, Moncada S (1980) Effects of intravenous infusion of prostacyclin (PGI_2) in man. Prostaglandins 19:319–332

Ohki S, Ogino N, Yamamoto S, Hayashi O (1979) Prostaglandin hydroperoxidase, an integral part of prostaglandin endoperoxide synthetase from bovine vesicular gland microsomes. J Biol Chem 254:829–836

Ohlendorf R, Perzborn E, Schrör K (1980) Prevention of infarction-induced decrease in circulating platelet count by prostacyclin. Thromb Res 19:447–453

Oliva PB (1983) The role of coronary artery spasm in acute myocardial infarction. Cardiovasc Clin 14:45-58
Orange RP, Austen KF (1969) Slow reacting substance of anaphylaxis. Adv Immunol 10:105–144
Osher J, Lang TW, Meerbaum S, Hashimoto K, Farcot JC, Corday E (1976) Methylprednisolone treatment in acute myocardial infarction. Effect on regional and global myocardial function. Am J Cardiol 37:564–571
Palmblad J, Malmsten CL, Udén A, Rådmark O, Engstedt L, Samuelsson B (1981) Leukotriene B_4 is a potent and stereospecific stimulator of neutrophil chemotaxis and adherence. Blood 58:658–661
Panzenbeck MJ, Kaley G (1983) Leukotriene D_4 reduces coronary blood flow in the anaesthetized dog. Prostaglandins 25:661–670
Patrignani P, Filabozzi P, Patrono C (1982) Selective cumulative inhibition of platelet thromboxane production by low-dose aspirin in healthy subjects. J Clin Invest 69: 1366–1372
Petsas AA, Kotler MN (1973) Electrocardiographic changes associated with penicillin anaphylaxis. Chest 64:66–69
Persantine-Aspirin Reinfarction Study Research Group (1980) Persantine and aspirin in coronary heart disease. Circulation 62:449–461
Peskar BA, Steffens Ch, Peskar BM (1979) Radioimmunoassay of 6-keto-prostaglandin $F_{1\alpha}$ in biological material. In: Albertini A, Da Prada M, Peskar BA (eds) Radioimmunoassay of drugs and hormones in cardiovascular medicine. Elsevier/North-Holland, Amsterdam, pp 239–250
Peskar BA, Aehringhaus U, Weinerowski P, Wittmann G (1984) Pharmacological modification of leukotriene release and coronary constrictor effect in cardiac anaphylaxis (Abstr). In: Neri Serneri GG, Masotti G, Gensini GF (eds) Platelets prostaglandins and the cardiovascular system. Boehringer-Ingelheim, Florence, p C1
Pifer DD, Cagen LM, Chesney CM (1981) Stability of prostaglandin I_2 in human blood. Prostaglandins 21:165–175
Piper PJ (1983) Pharmacology of leukotrienes. Br Med Bull 39:255–259
Piper PJ, Vane J (1971) The release of prostaglandins from lung and other tissues. Ann NY Acad Sci 180:363–383
Piper PJ, Letts LG, Samhoun MN, Tippins JR, Palmer MA (1982) Actions of leukotrienes on vascular, airway and gastrointestinal smooth muscle. In: Samuelsson B, Paoletti R (eds) Leukotrienes and other lipoxygenase products. Raven, New York, pp 169–181
Piper PJ, Letts LG, Galton SA (1983a) Generation of leukotriene like substance from porcine vascular and other tissues. Prostaglandins 25:591–599
Piper PJ, Letts LG, Tippins JR, Barnett K (1983b) Generation of a substance with the biological properties of a leukotriene from porcine vascular tissue. In: Piper PJ (ed) Leukotrienes and other lipoxygenase products. Research Studies Press, Chichester, pp 299–306
Plaut M, Lichtenstein LM (1978) Histamine, 5-hydroxytryptamine SRS-A: discussion of type I hypersensitivity (anaphylaxis). In: Vane JR, Ferreira SH (eds) Inflammation. Springer, Berlin Heidelberg New York, pp 345–373 (Handbook of Pharmacology, vol 50/1)
Ramwell PW, Leovey EMK, Sintetos AL (1977) Regulation of the arachidonic acid cascade. Biol Reprod 16:70–87
Reimer KA, Jennings RB (1979) The changing anatomic reference base of evolving myocardial infarction. Underestimation of myocardial collateral blood flow and overestimation of experimental anatomic infarct size due to tissue edema, haemorrhage and acute inflammation. Circulation 60:866–876
Ribeiro LGT, Brandon TA, Hopkins DG, Reduto LA, Taylor AA, Miller RR (1981) Prostacyclin in experimental myocardial ischemia effects on hemodynamics, regional myocardial blood flow, infarct size and mortality. Am J Cardiol 47:835–840

Roberts R, DeMello V, Sobel BE (1976) Deleterious effects of methylprednisolone in patients with myocardial infarction. Circulation (Suppl 1) 53:I204–I206
Robertson RM, Robertson D, Friesinger GC, Timmons S, Hawiger J (1980) Platelet aggregates in peripheral and coronary sinus blood in patients with spontaneous coronary artery spasm. Lancet 2:829–831
Robertson RM, Robertson D, Roberts LJ, Maas RL, FitzGerald GA, Friesinger GC, Oates JA (1981) Thromboxane A_2 in vasotonic angina pectoris. Evidence for direct measurements and inhibitor trials. N Engl J Med 304:998–1003
Romson JL, Haack DW, Abrams GD, Lucchesi BR (1981) Prevention of occlusive coronary artery thrombosis by prostacyclin infusion in the dog. Circulation 64: 906–914
Romson JL, Hook B, Rigot V, Swanson D, Lucchesi B (1982a) Effect of ibuprofen on the accumulation of 111-In leukocytes and platelets in infarcted dog myocardium. Fed Proc 41:8752
Romson JL, Hook B, Kunkel S, Abrams G, Lucchesi BR (1982b) Reduction of myocardial infarct size by neutrophil depletion in the dog. Circulation (Suppl 2) 66:337
Romson JL, Bush LR, Jolly SR, Lucchesi BR (1982c) Cardioprotective effects of ibuprofen in experimental regional and global myocardial ischemia. J Cardiovasc Pharmacol 4:187–196
Romson JL, Hook BG, Kunkel SL, Abrams GD, Schork MA, Lucchesi BR (1983a) Reduction of the extent of ischemic myocardial injury by neutrophil depletion in the dog. Circulation 67:1016–1023
Romson JL, Hook BG, Lucchesi BR (1983b) Potentiation of the antithrombotic effect of prostacyclin by simultaneous administration of aminophylline in a canine model of coronary artery thrombosis. J Pharmacol Exp Ther 227:288–294
Rösen R, Rösen P, Ohlendorf R, Schrör K (1981) Prostacyclin prevents ischemia-induced increase of lactate and cyclic AMP in ischemic myocardium. Eur J Pharmacol 69:489–491
Rösen P, Rösen R, Hohl C, Reinauer H, Klaus W (1984) Reduced transcoronary exchange and prostaglandin synthesis in diabetic rat heart. Am J Physiol 247:H563–H569
Roth DM, Lefer AM (1983) Studies on the mechanism of leukotriene induced coronary artery constriction. Prostaglandins 26:573–581
Roy L, Knapp HR, Robertson RM, FitzGerald GA (1983) Endogenous prostacyclin (PGI_2) biosynthesis is stimulated by cardiac catheterization (CC) and angiography (A) in man. Circulation (Suppl 3) 68:III–104
Sakai K, Ito T, Ogawa K (1982) Role of endogenous prostacyclin and thromboxane A_2 in the ischemic canine heart. J Cardiovasc Pharmacol 4:129–135
Salzman EW (1977) Interrelation of prostaglandin endoperoxide (prostaglandin G_2) and cyclic 3′5′-adenosine monophosphate in human blood platelets. Biochim Biophys Acta 499:48–60
Salzman PM, Salmon JA, Moncada S (1980) Prostacyclin and thromboxane A_2 synthesis by rabbit pulmonary artery. J Pharmacol Exp Ther 215:240–247
Samuelsson B (1983) Leukotrienes: Mediators of immediate hypersensitivity reactions and inflammation. Science 220:568–575
Samuelsson B, Granström E, Green K, Hamberg M, Hammarström S (1975) Prostaglandins. Annu Rev Biochem 44:669–695
Samuelsson B, Goldyne M, Granström E, Hamberg M, Hammarström S, Malmsten C (1978) Prostaglandins and thromboxanes. Annu Rev Biochem 47:997–1029
Samuelsson B, Borgeat P, Hammarström S, Murphy RC (1979) Introduction of a nomenclature: Leukotrienes. Prostaglandins 17:785–787
Sanders TAB, Naismith DJ, Haines AP, Vickers M (1980) Cod-liver oil, platelet fatty acids, and bleeding time. Lancet 1:1189
Scherhag R, Kramer HJ, Düsing R (1982) Dietary administration of eicosapentaenoic acid and linolenic acid increases arterial blood pressure and suppresses vascular prostacyclin synthesis in the rat. Prostaglandins 23:369–382

Schmidt B, Flesch I, Ecker B, Hovestadt I, Ferber E (1984) Membrane phospholipid changes during macrophage activation. Agents Actions 15:21–27
Scholtholt J, Birringer H, Fiedler VB, Schölkens B (1981) Effect of prostacyclin (PGI_2) and adenosine (ASN) on total and regional blood flow of isolated, collateralized dog hearts. Basic Res Cardiol 76:313–327
Schrör K (1978) Prostaglandin D_2 (PGD_2) – a potent coronary vasoconstrictor agent in the guinea pig isolated heart. Naunyn Schmiedebergs Arch Pharmacol 302:61–62
Schrör K, Krebs R (1976) On the action of PGE_2 on coronary vessels. A comparative study with adenosine. Naunyn Schmiedebergs Arch Pharmacol 293:R27
Schrör K, Moncada S (1979) Effects of prostacyclin on coronary circulation, heart rate and myocardial contractile force in isolated hearts of guinea pig and rabbit – comparison with prostaglandin E_2. Prostaglandins 17:367–373
Schrör K, Rösen P (1979) Prostacyclin (PGI_2) decreases the cyclic AMP level in coronary arteries. Naunyn Schmiedebergs Arch Pharmacol 306:101–103
Schrör K, Moncada S, Ubatuba FB, Vane JR (1978) Transformation of arachidonic acid and prostaglandin endoperoxides by the guinea pig heart. Formation of RCS and prostacyclin. Eur J Pharmacol 47:103–114
Schrör K, Link HB, Rösen R, Klaus W, Rösen P (1980a) Prostacyclin-induced coronary vasodilatation. Interactions with adenosine, cyclic AMP, and energy charge in the rat heart in vitro. Eur J Pharmacol 64:341–348
Schrör K, Rösen P, Link HB, Rösen P (1980b) Physiological and biochemical parameters of prostacyclin action on the heart and the coronary vasculature. Adv Prostaglandin Thromboxane Res 7:625–630
Schrör K, Addicks K, Darius H, Ohlendorf R, Rösen P (1981a) PGI_2 inhibits ischemia-induced platelet activation and prevents myocardial damage by inhibition of catecholamine release from adrenergic nerve terminals. Evidence for cAMP as common denominator. Thromb Res 21:175–180
Schrör K, Grodzinska L, Darius H (1981b) Stimulation of coronary vascular prostacyclin and inhibition of human thromboxane A_2 after low-dose-nitroglycerin. Thromb Res 23:59–67
Schrör K, Köhler P, Müller M, Peskar BA, Rösen P (1981c) Prostacyclin-thromboxane interactions in the platelet-perfused in vitro heart. Am J Physiol 241:H18–H25
Schrör K, Ohlendorf R, Darius H (1981d) Beneficial effects of a new carbacyclin derivative, ZK 36374, in acute myocardial ischemia. J Pharmacol Exp Ther 219: 243–249
Schrör K, Darius H, Ohlendorf R, Matzky R, Klaus W (1982) Dissociation of antiplatelet effects from myocardial cytoprotective activity during acute myocardial ischemia in cats by a new carbacyclin derivative (ZK 36374). J Cardiovasc Pharmacol 4:554–561
Schumacher WA, Lucchesi BR (1983) Effect of thromboxane synthetase inhibitor UK-37,248 (dazoxiben) upon platelet aggregation, coronary artery thrombosis and vascular reactivity. J Pharmacol Exp Ther 227:790–796
Schwartz CJ (1982) Thrombosis in the pathogenesis of sudden cardiac death and myocardial infarction. In: Oates JA (ed) Prostaglandins and the cardiovascular system. Raven, New York, pp 1–14
Scott-Miller AM, McMillan RM (1983) Aggregation and degranulation in human neutrophils: Do lipoxygenase metabolites play a role? In: Piper PJ (ed) Leukotrienes and other lipoxygenase products. Research Studies Press, Chichester, pp 255–256
Shaw JO, Printz MP, Hirabayashi K, Henson PM (1978) Role of prostaglandin synthesis in rabbit platelet activation induced by basophil-derived platelet-activating factor. J Immunol 121:1939–1945
Shaw JO, Klusick SJ, Hanahan DJ (1981) Activation of rabbit platelet phospholipase and thromboxane synthesis by 1-O-hexadecyl/octadecyl-2-acetyl-sn-glyceryl-3-phosphorylcholine (platelet activating factor). Biochim Biophys Acta 663:222–229

Shea MJ, Driscoll EM, Romson JL, Pitt B, Lucchesi BR (1984a) Effect of OKY-1581, a thromboxane synthetase inhibitor, on coronary thrombosis in the conscious dog. Eur J Pharmacol 105:285–291
Shea MJ, Driscoll EM, Romson JL, Pitt B, Lucchesi BR (1984b) The beneficial effects of nafazatrom (BAYg6575) on experimental coronary thrombosis. Am Heart J 107:629–637
Sherry S (1982) Effect of prostaglandin-mediated platelet-suppressant drugs on acute cardiovascular catastrophes. In: Oates JA (ed) Prostaglandins and the cardiovascular system. Raven, New York, pp 173–210
Sherry S (1983) Role of platelet-active drugs in coronary artery disease. Cardiovasc Clin 14:173–189
Shimamoto T (1978) Stroke and heart attack induced experimentally by thromboxane A_2 in rabbits and effect of EG 626, a thromboxane A_2 antagonist. In: Wissler RW et al. (eds) International Symposium: State of prevention and therapy in human artherosclerosis and in animal models. Westdeutscher Verlag, Colognes, pp 139–152
Siess W, Roth P, Scherer B, Kurzmann I, Böhlig B, Weber PC (1980) Platelet-membrane fatty acids, platelet aggregation, and thromboxane formation during a mackerel diet. Lancet 1:441–444
Silberbauer K, Schernthaner G, Sinzinger H, Piza-Katzer H, Winter M (1979) Decreased vascular prostacyclin in juvenile-onset diabetes. N Engl J Med 300:366–367
Sinzinger H, Feigl W, Silberbauer K (1979) Prostacyclin generation in atherosclerotic arteries. Lancet 2:469
Sinzinger H, Scherntaner G, Kaliman J (1981) Sensitivity of platelets to prostaglandins in coronary heart disease and angina pectoris. Prostaglandins 22:773–781
Sinzinger H, O'Grady J, Fitscha P, Kaliman J (1984) Extremely-low-dose aspirin (one milligram per day) renders human platelets more sensitive to antiaggregation prostaglandins. N Engl J Med 311:1052
Smirnov IE, Mentz PR, Markov CM (1980) Effects of prostacyclin on coronary blood flow and cardiac activity in normotensive and spontaneously hypertensive rats. Adv Prostaglandin Thromboxane Res 7:631–633
Smith EF III, Lefer AM (1981) Stabilization of cardiac lysosomal and cellular membranes in protection of ischemic myocardium due to coronary occlusion: Efficacy of the nonsteroidal anti-inflammatory agent, naproxen. Am Heart J 101:394–402
Smith EF III, Carrow BA, Lefer AM (1980) Effects of flurbiprofen on myocardial cell damage during acute myocardial ischemia. Res Commun Chem Pathol Pharmacol 28:413–433
Smith EF III, Lefer AM, Nicolaou KC (1981a) Mechanisms of coronary vasoconstriction induced by carbocylic thromboxane A_2. Am J Physiol 240:H493–497
Smith EF III, Lefer AM, Aharony D, Smith JB, Magolda RL, Claremon D, Nicolaou KC (1981b) Carbocyclic thromboxane A_2: Aggravation of myocardial ischemia by a new synthetic thromboxane A_2 analog. Prostaglandins 21:443–456
Smith EF III, Gallenkämper W, Beckmann R, Thomsen T, Mannesmann G, Schrör K (1984) Early and late administration of a PGI_2-analogue, ZK36374 (iloprost): effects on myocardial preservation, collateral blood flow and infarct size. Cardiovasc Res 18:163–173
Smith JB, Jubiz W (1981) OKY-1581: a selective inhibitor of thromboxane synthesis in vivo and in vitro. Prostaglandins 22:353–363
Smith JB, Willis AL (1971) Aspirin selectively inhibits prostaglandin production in human platelets. Nature (New Biol) 231:235–237
Smith WL, Lands WEM (1972) Oxygenation of polyunsaturated fatty acids during prostaglandin biosynthesis by sheep vesicular gland. Biochemistry 11:3276–3285
Smitherman TC, Milam M, Woo J, Willerson JT, Frenkel EP (1981) Elevated beta thromboglobulin in peripheral venous blood of patiens with acute myocardial ischemia: direct evidence for enhanced platelet reactivity in vivo. Am J Cardiol 48:395–402

Sobel BE, Corr PB (1979) Biochemical mechanisms potentially responsible for lethal arrhythmias induced by ischemia: the lysolipid hypothesis. Adv Cardiol 26:76–85
Sommers HM, Jennings RB (1964) Experimental acute myocardial infarction. Histologic and histochemical studies of early myocardial infarcts induced by temporary or permanent occlusion of a coronary artery. Lab Invest 13:1491–1503
Spath JA, Lane DL, Lefer AM (1974) Protective action of methylprednisolone on the myocardium during experimental myocardial ischemia in the cat. Circ Res 35:44–51
Spector AA, Kaduce TL, Figard PH, Norton KC, Haak JC, Czervionke RL (1983) Eicosapentaenoic acid and prostacyclin production by cultured human endothelial cells. J Lipid Res 24:1595–1604
Stein I, Wecksell I (1970) Cardiac disease accompanying allergic drug reactions. J Allergy 45:48–54
Stimler NP, Bloor CM, Hugli TE, Wykle RL, McCall CE, O'Flaherty JT (1981) Anaphylactic actions of platelet-activating factor. Am J Pathol 105:64–69
Struijck CB, Beerthuis RL, Pabon HJJ, Van Dorp DA (1966) specificity of the enzyme conversion of polyunsaturated fatty acids into prostaglandins. Recl Trav Chim Pays Bas 85:1233–1250
Subbiah MTR, Deitemeyer D (1980) Altered synthesis of prostaglandins in platelet and aorta from spontaneously diabetic Wistar rats. Biochem Med 23:231–235
Sun FF (1977) Biosynthesis of thromboxanes in human platelets. I. Characterization and assay of thromboxane synthetase. Biochem Biophys Res Commun 74:1432–1440
Svensson J, Hamberg M (1976) Thromboxane A_2 and prostaglandin H_2: Potent stimulators of the swine coronary artery. Prostaglandins 12:943–950
Svensson J, Hamberg M, Samuelsson B (1975) Prostaglandin endoperoxides IX. Characterization of rabbit aorta contracting substance (RCS) from guinea pig lung and human platelets. Acta Physiol Scand 94:222–228
Szczeklik A, Gryglewski RJ (1980) Low density lipoproteins (LDL) are carriers for lipid peroxides and invalidate prostacyclin (PGI_2) biosynthesis in arteries. Artery 7:488–495
Szczeklik A, Gryglewski RJ, Musial J, Grodzinska L, Serwónska M, Marcinkiewicz E (1978a) Thromboxane generation and platelet aggregation in survivals of myocardial infarction. Thromb Haemost 40:66–74
Szczeklik A, Gryglewski RJ, Nizânkowski R, Musial J, Pieton R, Mruk J (1978b) Circulatory and antiplatelet effects of intravenous prostacyclin in healthy man. Pharmacol Res Commun 10:545–556
Szczeklik A, Szczeklik J, Nizankowski R, Gluszko P (1980) Prostacyclin for acute coronary insufficiency. Artery 8:7–11
Szczeklik A, Gryglewski RJ, Domagala B, Zmuda A, Hartwich J, Wozny E, Grzywacz M, Madej J, Gryglewska T (1981) Serum lipoproteins, lipid peroxides and prostacyclin biosynthesis in patients with coronary heart disease. Prostaglandins 12: 795–807
Tada M, Kuzuya T, Inoue M, Kodama K, Mishima M, Yamada M, Inui M, Abe H (1981) Elevation of thromboxane B_2 levels in patients with classic and variant angina pectoris. Circulation 64:1107–1115
Tanabe M, Terashita ZI, Fujiwara S, Shimamoto N, Gota N, Nishikawa K, Hirata M (1982) Coronary circulatory failure and thromboxane A_2 release during coronary occlusion and reperfusion in anaesthetized dogs. Cardiovasc Res 16:99–106
Tappel AL, Lundberg WO, Boyer PD (1953) Effect of temperature and antioxidants upon the lipoxidase-catalyzed oxidation of sodium linoleate. Arch Biochem Biophys 42:293–304
Tateson JE, Moncada S, Vane JR (1977) Effect of prostacyclin (PGX) on cyclic AMP concentrations in human platelets. Prostaglandins 13:389–397
Taylor GW, Morris HR (1983) Lipoxygenase pathways. Br Med Bull 39:219–222

Terano T, Salmon JA, Moncada S (1984a) Biosynthesis and biological activity of leukotriene B_5. Prostaglandins 27:217–232

Terano T, Salmon JA, Moncada S (1984b) Effect of orally administered eicosapentaenoic acid (EPA) on the formation of leukotriene B_4 and leukotriene B_5 by rat leukocytes. Biochem Pharmacol 33:3071–3076

Terashita ZI, Fukui H, Nishikawa K, Hirata M, Kikuchi S (1978) Coronary vasospastic action of thromboxane A_2 in isolated, working guinea pig hearts. Eur J Pharmacol 53:49–56

Terashita ZI, Fukui H, Hirata M, Terao S, Ohkawa S, Nishikawa K, Kikuchi S (1981) Coronary vasoconstriction and PGI_2 release by leukotrienes in isolated guinea pig hearts. Eur J Pharmacol 73:357–361

Thaulow E (1983) Platelet function in blood from the coronary sinus in patients with arteriosclerotic heart disease. Thromb Haemost 50:629–632

Thaulow E, Dale J, Myhre E (1984) Effects of a selective thromboxane synthetase inhibitor, dazoxiben, and of acetylsalicylic acid on myocardial ischemia in patients with coronary artery disease. Am J Cardiol 53:1255–1258

Toda N (1984) Responses of human, monkey and dog coronary arteries in vitro to carbocyclic thromboxane A_2 and vasodilators. Br J Pharmacol 83:399–408

Trachte GJ, Lefer AM, Aharony D, Smith JB (1979) Potent constriction of cat coronary arteries by hydroperoxides of arachidonic acid and its blockade by anti-inflammatory agents. Prostaglandins 18:909–914

Tremoli E, Folco GC, Agradi E, Galli C (1979) Platelet thromboxanes and serum cholesterol. Lancet 1:107–108

Tremoli E, Maderna P, Collis S, Morazzoni G, Sirtori M, Sirtori CR (1984) Increased platelet sensitivity and thromboxane B_2 formation in type II hyperlipoproteinaemic patients. Eur J Clin Invest 14:329–333

Turner SR, Tainer JA, Lynn WS (1975) Biogenesis of chemotactic molecules by the arachidonate lipoxygenase system of platelets. Nature 257:680–681

Tuvemo T, Strandberg K, Hamberg M, Samuelsson B (1976) Maintenance of the tone of the human umbilical artery by prostaglandin and thromboxane formation. Adv Prostaglandin Thromboxane Res 1:425–428

Tyler HM, Saxton CAPD, Parry MJ (1981) Administration to man of UK-37,248-01, a selective inhibitor of thromboxane synthetase. Lancet 1:629–632

Ubatuba FB, Moncada S, Vane JR (1979) The effect of prostacyclin (PGI_2) on platelet behaviour, thrombus formation in vivo and bleeding time. Thromb Haemost 41: 425–435

Uchida Y, Hanai T, Hasegawa K, Kawamura K, Oshima T (1983) Recanalization of obstructed coronary artery by intracoronary administration of prostacyclin in patients with acute myocardial infarction. Adv Prostaglandin Thromboxane Leukotriene Res 11:377–383

Van der Giessen WJ, Serruys PW, Stoel I, Hugenholtz PG, de Leeuw PW, van Vliet HHDM, Deckmyn H, Vermylen J (1983) Acute effect of cigarette smoking on cardiac prostaglandin synthesis and platelet behaviour in patients with coronary heart disease. Adv Prostaglandin Thromboxane Leukotriene Res 11:359–364

Van der Ouderaa FJ, Buytenhek M, Nugteren DH, Van Dorp DA (1977) Purification and characterization of prostaglandin endoperoxide synthetase from sheep vesicular glands. Biochim Biophys Acta 487:315–331

Van Dorp DA (1966) The biosynthesis of prostaglandins. Mem Soc Endocrinol 14: 39–47

Van Dorp DA, Beerthuis RK, Nugteren DH, Vonkeman H (1964) The biosynthesis of prostaglandins. Biochim Biophys Acta 90:204–207

Vane JR (1971) Inhibition of prostaglandin synthesis as a mechanism of action for aspirin-like drugs. Nature (New Biol) 231:232–235

Vermylen J, Chamone DAF, Verstraete M (1979) Stimulation of prostacyclin release from vessel wall by BAYg6575, an antithrombotic compound. Lancet 1:518–520

Vesterqvist O, Gréen K (1984) Urinary excretion of 2,3-dinor-thromboxane B_2 in man under normal conditions, following drugs and during some pathological conditions. Prostaglandins 27:627–644
Voelkel NF, Worthen S, Reeves JT, Henson PM, Murphy RC (1982) Nonimmunological production of leukotrienes induced by platelet-activating factor. Science 218: 286–288
Vogel WM, Zannoni VG, Abrams GD, Lucchesi BR (1977) Inability of methylprednisolone sodium succinate to decrease infarct size or preserve enzyme activity measured 24 hours after coronary occlusion in the dog. Circulation 55:588–595
Vogt W (1978) Role of phospholipase A_2 in prostaglandin formation. Adv Prostaglandin Thromboxane Res 3:89–95
Von Euler US (1934) Zur Kenntnis der pharmakologischen Wirkungen von Nativsekretion und Extrakten männlicher akzessorischer Geschlechtsdrüsen. Arch Exp Pathol Pharmakol 175:78–84
Von Euler US (1935) Über die spezifische blutdurcksenkende Substanz des menschlichen Prostata- und Samenblasensekretes. Klin Wochenschr 33:1182–1183
Von Euler US (1937) On the specific vaso-dilating and plain muscle stimulating substances from accessory genital glands in man and certain animals (prostaglandin and vesiglandin). J Physiol (Lond) 88:213–234
Vonkeman H, van Dorp DA (1968) The action of prostaglandin synthetase on 2-arachidonyl-lecithin. Biochim Biophys Acta 164:430–432
Wallis J, Moses JW, Borer JS, Weksler B, Goldberg HL, Fisher J, Kase M, Tack-Goldman K, Carter J, Calle S (1982) Coronary blood flow in coronary artery disease: heparin-induced potentiation caused by prostaglandin release. Circulation (Suppl 2) 66:II–263
Wang HH, Kulkarni PS, Eakins KE (1980) Effect of prostaglandins and thromboxane A_2 on the coronary circulation of adult dogs and puppies. Eur J Pharmacol 66: 31–41
Warrington SJ, O'Grady J (1980) Cardiovascular effects of prostacyclin in man. Adv Prostaglandin Thromboxane Res 7:619–624
Weiss ES, Ahmed SA, Thakur ML, Welch MJ, Coleman RE, Sobel BE (1977) Imaging of the inflammatory response in ischemic canine myocardium with 111-indium-labelled leukocytes. Am J Cardiol 40:195–199
Weksler BB, Marcus AJ, Jaffe EA (1977) Synthesis of prostaglandin I (prostacyclin) by cultured human and bovine endothelial cells. Proc Natl Acad Sci USA 74: 3922–3926
Welman E, Selwyn AP, Fox KM (1979) Lysosomal and cytosolic enzyme release in acute myocardial infarction: effects of methylprednisolone. Circulation 59:730–733
Welton AF, Hope WC, Tobias LD, Hamilton JG (1981) Inhibition of antigen induced histamine release and thromboxane synthetase by FPL 55712. A specific SRS-A antagonist? Biochem Pharmacol 30:1378–1382
Weltzien HU (1979) Cytolytic and membrane-perturbing properties of lysophosphatidylcholine. Biochim Biophys Acta 559:259–287
Wennmalm Å (1979) Prostaglandin-mediated inhibition of noradrenaline release. VI. On the intra-cardiac source of prostaglandins released from isolated rabbit hearts. Acta Physiol Scand 105:254–256
Wennmalm Å (1980) Nicotine inhibits hypoxia– and arachidonate-induced release of prostacyclin-like activity in rabbit hearts. Br J Pharmacol 69:545–549
Wennmalm Å (1982) Interaction of nicotine and prostaglandins in the cardiovascular system. Prostaglandins 23:139–144
Wennmalm Å (1984) Prostacyclin (PGI_2) mediates the coronary vasodilator effect of adenosine (Abstr). In: Neri Serneri GG, Masotti G, Gensini GF (eds) Platelets prostaglandins and the cardiovascular system. Boehringer-Ingelheim, Florence, p 200

Wennmalm Å, Henriksson P, Ehag O (1984) Limitation of myocardial infarct size in humans with prostacyclin (Abstr). In: Neri Serneri GG, Masotti G, Gensini GF (eds) Platelets prostaglandins and the cardiovascular system. Boehringer-Ingelheim, Florence, p 166

Went S, Lissák K (1935) Histamin- und Proteinwirkung am normalen und sensibilisierten Meerschweinchenherz. Naunyn Schmiedebergs Arch Exp Pathol Pharmakol 179:609–615

Whittle BJR, Moncada S (1983) Pharmacological interactions between prostacyclin and thromboxanes. Br Med Bull 39:232–238

Whittle BJR, Moncada S, Vane JR (1978) Comparison of the effects of prostacyclin (PGI_2) prostaglandin E_1 and D_2 on platelet aggregation in different species. Prostaglandins 16:373–388

Whittle BJR, Moncada S, Whiting F, Vane JR (1980) Carbacyclin – a potent stable prostacyclin analogue for the inhibition of platelet aggregation. Prostaglandins 19: 605–627

Wilcox HB, Andrus EC (1938) Anaphylaxis in the isolated heart. J Exp Med 67:169–180

Wilkerson RD, Conran PB (1982) Ibuprofen reduces infarct size and ventricular arrhythmias after coronary occlusion (CAO). Fed Proc 41:1737

Wittenberg HR, Wölbling RH, Aehringhaus U, Patrono C, Peskar BM, Peskar BA (1983) Release of slow-reacting substance of anaphylaxis from guinea pig heart and vascular tissue. In: Piper PJ (ed) Leukotrienes and other lipoxygenase products. Research Studies Press, Chichester, pp 121–127

Wittmann G, Weinerowski P, Aehringhaus U, Peskar BA (1984) Effect of muscarinic receptor stimulation and of prostaglandins (PG) on release of leukotriene C_4 (LTC_4)-like immunoreactivity in cardiac anaphylaxis. Naunyn Schmiedebergs Arch Pharmacol (Suppl) 325:R36

Wlodawer P, Hammarström S (1978) Thromboxane synthetase from bovine lung – solubilization and partial purification. Biochem Biophys Res Commun 80:525–532

Wölbling RH, Aehringhaus U, Peskar BM, Peskar BA (1983) Release of slow reacting substance of anaphylaxis from layers of guinea pig aorta. Prostaglandins 25:823–828

Woodman OL, Dusting GJ (1983) Coronary vasoconstriction induced by leukotrienes in the anaesthetized dog. Eur J Pharmacol 86:125–128

Wynalda MA, Fitzpatrick FA (1980) Albumins stabilize prostaglandin I_2. Prostaglandins 20:853–861

Yoshimoto T, Yamamoto S, Okuma M, Hayaishi O (1977) Solubilization and resolution of thromboxane synthesizing system from microsomes of bovine blood platelets. J Biol Chem 252:5871–5874

Yui Y, Hattori R, Takatsu Y, Nakajima H, Wakabayashi A, Kawai C, Kayama N, Hiraku S, Inagawa T, Tsubojima M, Naito J (1984) Intravenous infusion of a selective inhibitor of thromboxane A_2 synthetase in man: influence on thromboxane B_2, and 6-keto-prostaglandin $F_{1\alpha}$ levels and platelet aggregation. Circulation 70:599–605

Zavecz JH, Levi R (1977) Separation of primary and secondary cardiovascular events in systemic anaphylaxis. Circ Res 40:15–19

Ziboh VA, Maruta H, Lord J, Cagle WD, Lucky W (1979) Increased biosynthesis of thromboxane A_2 in diabetic platelets. Eur J Clin Invest 9:223–228

Rev. Physiol. Biochem. Pharmacol., Vol. 104

Methodological Aspects of Purification and Reconstitution of Transport Proteins from Mammalian Plasma Membranes

HERMANN KOEPSELL

Contents

1 Introduction

In the last 20 years knowledge about transport systems in plasma membranes has accumulated as a result of the development of (a) sophisticated techniques for in vivo measurements of net and tracer fluxes, of membrane potentials, and of membrane permeabilities and of (b) methods to isolate plasma membrane vesicles, in which flux measurements could be performed. For a molecular understanding of transport the isolation of transport proteins and their functional incorporation into lipid bilayers

Max-Planck-Institut für Biophysik, Kennedyallee 70, 6000 Frankfurt (Main) 70, Federal Republic of Germany is necessary. The ultimate identification of a transporter, which may be composed of different subunits, can only be achieved after purification and reconstitution.

The ease with which transport proteins can be purified and reconstituted depends on several factors. These are (a) the complexity of the transport system, which may consist of one or more subunits, (b) the concentration of the transporter in the membrane, (c) whether the transporter catalyzes an enzymatic reaction by which it can be identified during solubilization and purification, (d) the existence of specific inhibitors and/or activators, (e) the lipid requirement of the transporter, and (f) the ease of measuring transport in proteoliposomes.

Successful reconstitution experiments were performed with transport proteins from mammalian plasma membranes which catalyze an enzymatic reaction, namely (Na^++K^+)-ATPase from kidney or from shark rectal gland and Ca^{2+}-ATPase from erythrocytes. Purification of these transporters was possible since the enzymatic activity could be followed during purification and since either the transporter was present in a high concentration in the membrane, as in the case of (Na^++K^+)-ATPase, or affinity purification was successfully employed, as in the case of the Ca^{2+}-ATPase. After the purification of the transporters, reconstitution was achieved.

For purification and reconstitution of facilitated diffusion systems such as the D-glucose transporter from erythrocytes and the anion exchange protein from erythrocytes, additional difficulties had to be overcome. Thus for the D-glucose transporter from the erythrocyte membrane, which was not identified or purified, reconstitution was optimized with solubilized membrane extracts. By employing D-glucose uptake measured in proteoliposomes to identify the transporter during protein fractionation experiments, purification of the D-glucose transporter was achieved. Reconstitution of the anion exchange protein from erythrocytes, on the other hand, was easier since this transporter was identified by a specific inhibitor and was partially purified by selective extraction with Triton X-100 (*Rothstein* et al. 1975). In this case, however, functional reconstitution was difficult since this transport protein denaturated easily during solubilization, purification, and reconstitution.

Reconstitution and purification of Na^+ cotransport systems from renal or intestinal brush border membranes proved to be difficult since these transport proteins are present in very small concentrations, i.e., each constitutes less than 1% of the total membrane proteins in these membranes. Furthermore, since uptake is driven by an Na^+ gradient, the proteoliposomes had to be impermeable against Na^+ so that the gradient was present for a sufficiently long time. Formation of large proteoliposomes appeared to be more useful than that of small ones as their smaller surface to volume ratio means that the Na^+ gradient persists over a longer period. With the exceptions of Na^+ D-glucose cotransporter, which can be specifically inhibited by phlorizin, and the $Na^+/2Cl^-/K^+$ cotransporter, which is inhibited by furosemide (*Geck* et al. 1980; *Greger* and *Schlatter* 1981), specific high affinity inhibitors of the other cotransport systems are not available.

It is difficult to purify functional hydrophobic transport proteins since different purification and reconstitution methods may be optimal for different transporters and since the quantification of functional transporter molecules by uptake experiments is problematic. However, with knowledge of the appropriate methods, promising purification and reconstitution experiments can be done and it is a matter of time and persistence before the different transporters are purified and reconstituted. Taking some transport proteins from mammalian plasma membranes as examples, I will focus on methodological aspects in solubilization, purification, and reconstitution. Although no general theory concerning effective reconstitution and/or purification can be established, some rules for purification and reconstitution experiments can be derived. I would like to demonstrate problems and pitfalls which may arise during purification and reconstitution. If one is aware of these problems and if experimental conditions are controlled carefully, reconstitution experiments are highly reproducible even when one does not at the time know how the experimental conditions influence purification and reconstitution; they have nothing to do with magic, as has been inferred recently (*Semenza* et al. 1984)!

In this review I will first summarize some general statements concerning the structure and properties of plasma membranes and the events during solubilization and purification of hydrophobic membrane proteins. The different procedures which can be employed for reconstitution of several transport systems from mammalian plasma membranes will be described. The benefits and drawbacks of the different reconstitution methods will be discussed. It will be shown that our knowledge concerning transport proteins has been significantly increased by reconstitution experiments.

2 Remarks on Structure and Properties of Plasma Membranes

The structure and function of plasma membranes have been discussed in several reviews (*Stoeckenius* and *Engelman* 1969; *Singer* and *Nicolson* 1972; *Guidotti* 1972; *Gitler* 1972; *Singer* 1974; *Montal* 1976). Since an understanding of the organization and properties of intact membranes is necessary to perform solubilization, purification, and reconstitution experiments, some basic facts will be compiled. Plasma membranes are considered as fluid mosaic structures which are composed of a phospholipid bilayer in which hydrophobic proteins are embedded. Hydrophobic parts of the proteins may span the membrane once or several times or may be embedded in the lipid bilayer. The membrane proteins may be in contact with the intra- and/or extracellular compartments (*Singer* 1977). The bilayer is mainly formed by the phospholipids phosphatidylcholine, phosphatidylethanolamine, phosphatidylserine, and sphingomyelin. The phospholipids contain fatty acids of different chain length and with different degrees of saturation. Different membranes contain varying amounts of cholesterol, which appears to be embedded between the fatty acids of the phospholipids (*Seelig* and *Seelig* 1974; *Demel* and *de Kruyff* 1976). The cytoplasmic and extracellular membrane monolayer may be composed of different types of phospholipid. In erythrocytes phosphatidylcholine and sphingomyelin are mainly localized in the external monolayer, whereas phosphatidylethanolamine and phosphatidylserine are mainly found in the internal monolayer of the plasma membrane (*Verkleij* et al. 1973; *Zwaal* et al. 1973; *Rothman* and *Lenard* 1977). In the monolayers, cholesterol and the different phospholipids are not supposed to be uniformly distributed (*Stier* and *Sackmann* 1973; *Demel* and *de Kruyff* 1976). There is evidence that proteins embedded in the membrane are surrounded by specific lipids which may somehow be immobilized (*Lichtenberg* et al. 1979; *Brotherus* et al. 1980; *Scott* and *Coe* 1983; *Lentz* et al. 1985). Transport proteins may have specific binding sites for phospholipids or cholesterol (*Klappauf* and *Schubert* 1979). The phospholipids in the membrane may be present in an ordered "quasi crystalline" or in a "liquid crystalline" state. The phase transition of these states occurs at a specific temperature (T_c) which is dependent on the cholesterol content of the membrane, on the phospholipid head groups, on the chain length and degree of saturation of the fatty acids, on the pH, on the presence of cations and anions, and on the presence of proteins (*Ladbrooke* and *Chapman* 1969; *Träuble* and *Eibl* 1974; *Gomez-Fernandez* et al. 1980; *Scott* and *Coe* 1983). Since some types of phospholipids may aggregate below their T_c, domains of different phospholipids may be present in the membrane (*Papahadjopoulos* et al. 1973; *Sandermann* 1978; *Lichtenberg* et al. 1979). If the temperature is altered, the physical state of lipid domains and their microviscosity will change. A lateral movement of protein molecules may occur and the functional activity of transport molecules may alter (*McMurchie* and *Raison* 1979; *Yuli* et al. 1981; *Lookman* et al. 1982; *Kapitza* et al. 1984). Transport proteins in the membrane may be present as monomers, as dimers, or as polymers (e.g., *Ikemoto* 1982; *Pappert* and *Schubert* 1983). Since the different degrees of association of the monomers may be dependent on the presence of specific lipids in their close environment, the functional state of membrane proteins is in a complex way affected by temperature. At a given temperature the passive membrane permeability, which depends mainly on the physical state of the lipid domains in the membrane, is dependent on the concentrations of different phospholipids, cholesterol, and membrane proteins (*Finkelstein* and *Cass* 1967; *Ladbrooke* et al. 1968; *de Gier* et al. 1968, 1969; *Papahadjopoulos* 1971; *Kimelberg* and *Papahadjopoulos* 1971; *de Kruyff* et al. 1972, 1973; *Inoue* 1974; *Grunze* and *Deuticke* 1974; *Simon* et al. 1982). Furthermore, the passive membrane permeability is supposed to be dependent on the membrane potential (*O'Shea* et al. 1984).

3 Solubilization of Membrane Proteins

In order to purify transport proteins they have to be removed from the plasma membrane. This may be done mechanically by sonication (*Stasny* and *Crane* 1964; *Morgan* and *Hanahan* 1966; *Rosenberg* and *McIntosh* 1968), by extraction of the proteins with organic solvents (*Sandermann* and *Strominger* 1974; *Altendorf* et al. 1977; *Amanuma* et al. 1977; *Sigrist* et al. 1977; *Sandermann* et al. 1978), by chemical modification of proteins (see, e.g., *Howlett* and *Wardrop* 1978), by chaotropic reagents, by alkaline and/or EDTA treatment, by fractional digestion, by solubilization with acetic acid (*Schubert* and *Domning* 1978), by treatment with urea (*Kusaka* and *Kanai* 1978), and by solubilization with detergents. Here only the solubilization with detergents will be discussed in detail since this procedure has in many cases been applied successfully during the purification of active transport proteins. Properties of detergents, their function during membrane solubilization, and the behavior of solubilized membrane proteins have been described in several reviews (*Razin* 1972; *Helenius* and *Simons* 1975; *Tanford* and *Reynolds* 1976; *Helenius* et al. 1979; *Lichtenberg* et al. 1983). It appeared that the conditions needed to solubilize functionally active transport proteins have to be found empirically for each transport system. However, some information concerning different detergents and their action on proteins and lipids is now available which will enable the investigator to perform promising experiments. Some basic facts concerning detergents are therefore summarized below, with special emphasis on the description of detergents which have been introduced only recently.

3.1 General Properties of Detergents

Lipids and detergents are molecules which have hydrophilic and hydrophobic parts and are called amphiphiles. Lipids are insoluble amphiphiles whereas the detergents used for membrane solubilization are soluble amphiphiles. They have a more hydrophilic character than the insoluble amphiphiles and their monomer solubility is therefore several orders of magnitude higher. The insoluble amphiphiles are subdivided into nonswelling and swelling compounds. Both groups of insoluble amphiphiles form stable monolayers at the air–water interface. However, only the swelling amphiphiles can form lamellar liquid crystals whereas the nonswelling amphiphiles are not present in the bulk aqueous solution. Cholesterol, for example, is an insoluble nonswelling amphiphile whereas phos-

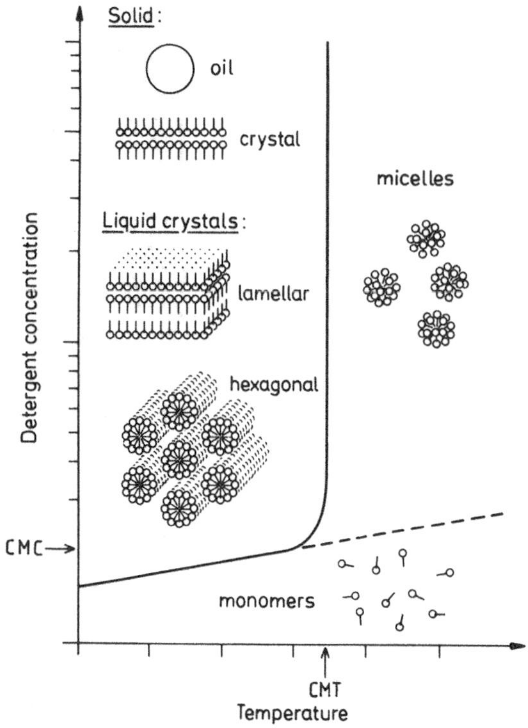

Fig. 1. Hypothetical phase diagram of soluble amphiphiles. Soluble amphiphiles with or without lyotropic mesomorphism may be present as monomers and micelles and in the solid state. Soluble amphiphiles with lyotropic mesomorphism may also be present as liquid crystals. In the graph the critical micellar concentration (*CMC*) and the critical micellar temperature (*CMT*) are indicated

pholipids are insoluble swelling amphiphiles. The soluble amphiphiles, which form unstable monolayers at the air–water interface, are divided into two subgroups: soluble amphiphiles with lyotropic mesomorphism (which can form cubic, hexagonal, or lamellar liquid crystals), and amphiphiles without lyotropic mesomorphism which do not form these crystals (see Fig. 1). Most of the nonionic, anionic, and cationic detergents are soluble amphiphiles with lyotropic mesomorphism, whereas the bile salts are soluble amphiphiles without lyotropic mesomorphism. If a detergent is added to water the detergent forms a monolayer at the air–water interface which is in equilibrium with monomers or with monomers and liquid crystals in the bulk aqueous phase. When the monomer concentration in the water reaches a critical level and the temperature is sufficiently high, detergent monomers associate to micelles in which the hydrophobic parts of the detergent molecules are associated and the hydrophilic parts are directed to the outside water phase. The detergent concentration at which micelles are formed is called the critical micellar concentration (CMC). This concentration is characteristic for the respective detergent and is dependent on temperature, ionic strength, presence of alcohols, and pH (*Shinoda* 1954; *Reynolds* and *Tanford* 1970; *Hayase* and *Hayano* 1978a,b). The formation of micelles requires a critical temperature (CMT) which may be concentration dependent (*Carey* and *Small* 1972). Below CMT, soluble amphiphiles with lyotropic mesomorphism form liquid crystals if their concentration is sufficiently high. Above CMT and CMC all detergents (soluble amphiphiles with and without lyotropic mesomorphism) form micelles. The number of detergent molecules per micelle (aggregation number), and thus the micellar weight, is dependent on the type of detergent, the detergent concentration, the ionic strength, and the temperature (*Mazer* et al. 1976). Micellar solutions of nonionic detergents become turbid if the temperature is increased above a certain level (cloud point). Above this temperature a detergent-rich phase may separate from the solution (*Bordier* 1981).

3.2 Events During Solubilization of Membrane Proteins

Although different detergents interact specifically with membrane proteins (*Kirkpatrick* et al. 1974), some general principles of the detergent solubilization of membrane proteins can be outlined. In a first step detergents bind to membrane proteins and partition between the lipid bilayer and the aqueous phase (*Becker* et al. 1975; *Jackson* et al. 1982). The incorporation of detergents into the lipid bilayer may have a "fluidizing" effect on the membrane and may perturb structure and function of transport proteins (*Richards* 1978; *Andersen* et al. 1983). The binding of detergents to proteins may occur at charged protein residues if ionic detergents are employed (see, e.g., *Burkhard* and *Stolzenberg* 1972), or to hydrophobic protein groups if ionic or nonionic detergents are used (*Tanford* 1972; *Helenius* and *Söderlund* 1973; *Simons* et al. 1973; *Makino* et al. 1973; *Crane* and *Lampen* 1974; *Riepl* and *Vidaver* 1978). Detergent binding to membrane proteins, which can be observed if membranes are incubated with low detergent concentrations, may lead to a removal of specific proteins, which have been only partially embedded in the membrane core (*Jørgensen* 1974; *Kirkpatrick* et al. 1974; *Pearlstein* and *Seaver* 1976; *Riepl* and *Vidaver* 1978; *Vasseur* et al. 1978). After incubation of the membrane with higher detergent concentrations the membranes are disintegrated and detergent–protein–lipid micelles are formed. The size of these micelles and their composition of proteins, lipid, and detergent are dependent on the employed protein/lipid/detergent ratios and on the temperature (*Simons* et al. 1973; *Becker* et al. 1975). At high lipid to protein and/or detergent to protein ratios, detergent–protein–lipid micelles are formed which contain single protein molecules. At high detergent to lipid ratios, lipids are separated from the proteins and pure protein–detergent micelles are generated (*Engelman* et al. 1967; *Pitt-Rivers* and *Impiombato* 1968; *Sardet* et al. 1976). If high detergent concentrations are employed, protein aggregates or oligomers may be dissociated (*Sanderman* 1978; *McEnery* et al. 1984; *Wagner* et al. 1984), and transport activity may be inactivated if the oligomeric structure is essential for the transport function (e.g., *Liang* and *Winter* 1977).

3.3 Specific Properties of Detergents

A large number of different detergents have been synthesized and characterized (e.g., *Kirkpatrick* and *Sandberg* 1973; *Helenius* and *Simons* 1975; *Stubbs* et al. 1976; *Helenius* et al. 1979; *Hjelmeland* 1980; *Womack* et al. 1983). Experience concerning solubilization, purification, and reconstitution of membrane transport proteins has been obtained with some of these detergents. I will discuss the properties of a number of them, focussing on newly introduced ones which appear to be specially appropriate for purification and reconstitution experiments.

The bile salts cholate and deoxycholate are soluble amphiphiles without lyotropic mesomorphism. The bean-shaped bile salt molecules have a polar and an apolar side. Above CMC they form aggregates of two to eight bile salt molecules, which are in contact at their apolar sides (*Small* et al. 1969; *Carey* and *Small* 1972). The size of the micelles, which is small compared with nonionic detergents, is dependent on parameters such as counterion concentration, temperature, and pH. In the case of deoxycholate, larger aggregates of micelles can be formed in the presence of high cation concentrations (*Carey* and *Small* 1972). Since deoxycholate and cholate have high critical micellar concentrations of about 3 and 14 m*M* respectively, these detergents can be removed easily by dialysis. During the interaction of low concentrations of bile salts with phospholipid membranes, single bile salt molecules bind to hydrophobic proteins (*Crane* and *Lampen* 1974) and invade the membrane (*Lichtenberg* et al. 1979). By increasing the bile salt concentration the bilayer is transformed into mixed micelles of detergent, lipid, and protein. According to a model introduced by *Small* and co-workers (*Small* et al. 1969), these micelles consist of bilayer sheets surrounded by bile salt molecules, the hydrophilic sides of which are in contact with the surrounding buffer. The size of these mixed micelles decreases if the detergent to lipid or protein ratio is increased. By employing an appropriate detergent concentration, the size of the solubilized membrane domains can be adjusted to be large enough to keep the function of oligomeric proteins intact and to be small enough to separate membrane domains which contain different proteins. In the case of deoxycholate complexes of proteins and phospholipids are removed from the membranes (*Kirkpatrick* et al. 1974). It was found that cholesterol remains in the membrane until more than 50% of protein and phospholipids are solubilized. Since bile salts may dissociate membrane proteins but do not have such strong denaturating properties as other ionic detergents, they may be applied successfully for the solubilization of transport proteins.

The nonionic polyethylene detergent Triton X-100 has been employed frequently for the solubilization of membrane proteins. The nonionic detergents are, in general, less denaturating than ionic surfactants (*McFarland* and *Inesi* 1971). However, in some cases bile acids may preserve functional activity of membrane proteins better (e.g., *Greenberg* and *Tsong* 1984). Compared with the bile salts, Triton X-100 has a low CMC of 0.24 m*M* (*Helenius* and *Söderlund* 1973) and can, like other alkyl and alkyl-aryl polyoxyethylene ether detergents (e.g., Brij 36T, Lubrol PX and WX), not be removed quickly by dialysis. These detergents may bind with high affinity to hydrophobic membrane proteins. The molecular weight of Triton X-100 micelles has been estimated to be between 63 000 and 150 000 (*Mankowich* 1954; *Kushner* and *Hubbard* 1954; *Dwiggins* et al. 1960; *Kuriyama* 1962). This is much higher than the molecular weight of bile salt micelles, which is below 20 000 (*Helenius* and *Simons* 1972). If phosphatidylcholine membranes are treated with small amounts of Triton X-100 (molar ratio of detergent to phospholipid < 1), detergent molecules are incorporated into the membrane. By increasing the detergent/phospholipid ratio the phospholipid membrane is fractionated into large bilayer pieces. At high detergent/phospholipid ratios the membrane is decomposed into mixed micelles. The size of these mixed micelles is similar to that of pure Triton X-100 micelles if the molar Triton to phospholipid ratio is 10 or more (*Dennis* 1974).

About ten years ago the nonionic detergent octylglucoside has been introduced for membrane solubilization experiments (*Baron* and *Thompson* 1975). Octylglucoside, which has a high CMC of 25 m*M* (*Shinoda* et al. 1961), can be easily removed by dialysis. After solubilization with octylglucoside the biological activity of several membrane proteins is better preserved than after solubilization with Triton X-100 or with other nonionic detergents (*Stubbs* et al. 1976; *Baron* and *Thompson* 1975; *Baldwin* et al. 1982). Due to the presence of the D-glucose moiety, octylglucoside should be specially suited to solubilize native D-glucose transport proteins since octylglucoside may protect the D-glucose binding site. This should be specifically valid for the Na^+ D-glucose cotransport protein from brush-border membranes of kidney or intestine since octylglucoside has been shown to be transported by the Na^+-dependent D-glucose uptake system (*Ramaswamy* et al. 1976). In fact, octylglucoside has been applied successfully in the solubilization and purification of the facilitated diffusion system for D-glucose in erythrocytes (*Baldwin* et al. 1982) and for the solubilization of the Na^+ D-glucose cotransporter from brush-border membranes of renal proximal tubules (*Lin* et al. 1979; *Im* et al. 1982; *Koepsell* et al. 1984). During solubilization octylglucoside is partitioned into the membrane up to a limiting concentration. At higher octylglucoside concentrations membrane fragments and mixed detergent–protein–lipid micelles are obtained. If excess of octylglucoside is present, protein–detergent and lipid–detergent micelles are formed (*Stubbs* and *Litman*

1978). *Stubbs* and *Litman* found that in protein–detergent micelles, which were obtained after solubilization of rhodopsin, about 270 mol octylglucoside were found associated with 1 mol protein. This value agrees well with the solubilized rhodopsin–Triton X-100 complex (*Osborne* et al. 1974; *Sardet* et al. 1976).

A serious problem which arises generally during solubilizing and purification experiments is the tendency of hydrophobic membrane proteins to form complexes after solubilization (*McFarland* and *Inesi* 1971; *Tanford* and *Reynolds* 1976). In dissociating such complexes, ionic detergents are more efficient than nonionic detergents (e.g., *Robinson* and *Capaldi* 1977) but in many cases they lead to protein denaturation. Furthermore, they cannot be employed when purification is done by ion exchange chromatography and by isoelectric focusing. Zwitterionic detergents, which should prevent protein aggregation, are considered a promising alternative. A group of zwitterionic detergents which has been introduced some years ago for membrane solubilization experiments are the N-alkyl-*N,N*-dimethyl-3-ammoniopropane sulfonates (alkyl-sulfobetains) (*Gonenne* and *Ernst* 1978). Experiments which demonstrate that cholesterol is required to reconstitute Na^+ D-glucose cotransport in proteoliposomes (*Ducis* and *Koepsell* 1983) suggest that the 1,2-cyclopentenophenanthrene structure may be important to preserve biological activity of hydrophobic proteins during solubilization. Thus, zwitterionic detergents containing this structural element are promising compounds for membrane solubilization. Recently such a compound, 3-[-(3-cholamidopropyl)-dimethylammonio]-1-propane sulfonate (CHAPS), has been synthesized (*Hjelmeland* 1980). This detergent has a CMC of about 9 m*M* and an aggregation number of about 10. Like octylglucoside it is able to solubilize membranes below the CMC. Concentrations of CHAPS, which are high enough to solubilize membranes, do not inactivate the enzymes esterase and sulfatase, the prolactin receptor, the opiate receptor, or the (Na^++K^+)-ATPase (*Simonds* et al. 1980; *Liscia* et al. 1982; *Gavish* et al. 1983; *Womack* et al. 1983; *Yoda* et al. 1984). Finally, the nonionic detergents octaethylene glycol mono-*n*-dodecyl ether and lauroyl sucrose should be mentioned. Octaethylene glycol mono-*n*-dodecyl ether has been successfully applied for the solubilization of the functional (Na^++K^+)-ATP-ase and the anion exchange protein from erythrocytes (*Cornelius* and *Skou* 1984; *Boodhoo* and *Reithmeier* 1984); lauroyl sucrose was able to solubilize the functional β-adrenergic receptor together with components of the adenylate cyclase (*Osipow* et al. 1956; *Hekman* et al. 1984).

3.4 Preservation of Biological Activity of Membrane Proteins

Preserving the biological activity during solubilization of transport proteins often proved to be difficult and demanded time-consuming trial and error experiments. A detergent has to be found which is able to remove the protein from the membrane and to separate it from associated proteins without being irreversibly inactivated. This can be achieved if an appropriate detergent and suitable experimental conditions are found for the solubilization. A number of factors which may be crucial in preventing protein inactivation may be listed: buffer osmolarity, pH, cation and anion concentrations, presence of polyols, of reducing agents, or of chelating agents, presence of substrates, detergent/protein/lipid ratio, time of incubation, and temperature (e.g., *Stipani* and *Palmieri* 1983; *Koepsell* et al. 1983; *Hjelmeland* and *Chrambach* 1984). Initially, solubilization should be tried with the lowest detergent concentration capable of dissociating the membrane in the presence of high protein and lipid concentrations. A pH value around 7, high substrate concentrations, and the presence of glycerol or sucrose and lipids are favorable. If transport can be retained after membrane dissociation with high protein to detergent and lipid to detergent ratios, the detergent concentration should be increased to find out whether endogenous lipids and the associated proteins can be removed from the transporter without irreversible inactivation. Under these conditions the addition of specific lipids often becomes necessary, and detergents may now prove to be unsuitable since they may lead to irreversible inactivation of the transport proteins.

Transport proteins which are denatured during solubilization and/or during removal of their associated lipids may, in some cases, be renatured. Thus, it has been reported that some transport proteins which lose their activity if associated lipids are removed can be reactivated by the addition of lipids or of detergents or by reconstitution into proteoliposomes (*Fenster* and *Copenhaver* 1967; *MacLennan* 1970; *Kimelberg* and *Papahadjopoulos* 1972, 1974; *Hong* and *Hubbell* 1973; *Bruni* et al. 1975; *Knowles* et al. 1976; *Dean* and *Tanford* 1977; *Krämer* and *Klingenberg* 1977, 1980; *Goodman* et al. 1978; *Simons* et al. 1978; *Sandermann* 1978; *Ottolenghi* 1979; *Dean* and *Gray* 1982; *Pick* and *Bassilian* 1982). Hopefully, the renaturation of some transport proteins after treatment with organic solvents or after treatment with the denaturing detergent SDS, as has been described for soluble proteins (*Weber* and *Kuter* 1971; *Lacks* and *Springhorn* 1980; *Clarke* 1981), will be possible in the future. Some optimism may be justified by reports in which renaturation of intrinsic membrane proteins was described. For example, *MacLennan* et al. reported in 1976 that Ca^{2+}-ATPase activity and Ca^{2+} transport from sarcoplasmic reticulum could be recovered after dilution of the SDS-treated protein with phospholipids. The renaturation was favored by the presence of sucrose, Mg^{2+}, ATP, and EGTA. In 1981 *Huang* and co-workers succeeded in renaturing delipidated bacteriorhodopsin, which was denatured by formic acid or anhydrous trifluoroacedic acid, by treatment with SDS and cholate, and by addition of retinal. Moreover, in 1984 *Tzartos* and *Changeux* succeeded in renaturing α-bungarotoxin binding and antibody binding to the α-subunit of the acetylcholine receptor, which had been purified by SDS gel electrophoresis. Finally, renaturation of a membrane-bound phosphatase from *Acholeplasma laidawii* (*Hjertén* 1983) and of meningococcal outer membrane proteins were reported (*Mandrell* and *Zollinger* 1984).

4 Methods of Purifying Transport Proteins

A partial purification of transport molecules may be achieved by procedures during which loosely bound membrane proteins are removed from the membrane. These methods may be powerful if the transporters are present in high concentrations. They include treatment of the membranes with alkaline EDTA (*Kahlenberg* 1976; *Hatefi* and *Hanstein* 1969), with protein-modifying reagents, or with low detergent concentrations (*Liljas* et al. 1974; *Jørgensen* 1974; *Klip* et al. 1979). Generally, however, these procedures are not supposed to achieve complete purification of the transporters. Sometimes removal of membrane proteins is attempted by limited proteolysis of membrane proteins with proteases (*Klip* et al. 1979; *Malathi* and *Preiser* 1983; *Kano-Kameyama* and *Hoshi* 1983). This strategy must be considered with great caution. After the use of proteases the cleaved transport proteins may remain functionally active if the transport proteins remain in their lipid environment and the protein fragments do not dissociate (*Jung* et al. 1973; *Rohn* 1983). This may even be the case if the membranes are solubilized, the transport proteins are present in detergent–protein–lipid micelles, and the solubilized proteins are reconstituted (*Anner* et al. 1984b). Only during the polypeptide analysis by SDS gel electrophoresis may the polypeptide fragments of the transporter proteins be dissociated; they will, together with other splitting products, appear in the low molecular weight region of the gels.

A novel way of partially purifying transport proteins makes use of the observations that solubilized membrane proteins may selectively aggregate with each other or with lipids and may form insoluble or soluble complexes (*Razin* et al. 1969; *Kahane* and *Razin* 1971; *Warren* et al. 1974b; *Simons* et al. 1978; *Koepsell* et al. 1983). During reconstitution, membrane proteins may be selectively associated and/or incorporated into proteoliposomes (*Utsumi* et al. 1982; *Koepsell* et al. 1983). The selective incorporation of transport proteins into liposomes is dependent on the lipid composition of the liposomes and/or on the presence of other proteins (*Kagawa* and *Racker* 1971; *Koepsell* et al. 1983). During preparation of proteoliposomes different types and sizes of liposomes containing different proteins may often be generated. The liposomes may be separated by gel chromatography or by density gradient centrifugation. If experimental conditions are found such that the density of the active liposomes can be significantly increased by the transported substrate, a "transport specificity fractionation" can be performed (*Goldin* and *Rhoden* 1978; *Papazian* et al. 1979). These methods may lead to a substantial purification and the transport activity of labile proteins may be preserved. Drawbacks of this procedure are that (a) the effects are not predictable, (b) complete purification may not be possible, (c) the recovery of transport proteins is low, and (d) the transport proteins are associated with excess lipid.

For final purification, conditions have to be found under which the solubilized transport proteins are separated from associated proteins and/or lipids without denaturation. Therefore the detergent to protein and/or lipid ratio should be adjusted such that each micelle contains only one type of protein. If the transport proteins are only present in small concentrations, sufficiently high concentrations of essential lipids may not be present in the solubilized material. Thus, the essential lipids must be identified and added during the solubilization and purification experiments. Solubilized transport proteins may be separated by conventional methods, such as precipitation procedures, density gradient centrifugation, aqueous phase partitioning, hydrophobic chromatography, gel filtration, ion exchange chromatography, adsorption chromatography, affinity chromatography with concanavalin A, or chromatography on gel-linked SH-group reagents, or by specific procedures such as affinity chromatography with specific ligands or with antibodies.

If precipitation methods are used, lack of essential lipids may not be a problem since proteins are normally precipitated together with their associated lipids. Precipitation can, for example, be achieved with ammonium sulfate or by altering the ionic composition or the osmolarity of the solubilization buffer (*Koepsell* et al. 1983). To achieve protein separation by density gradient centrifugation in the presence of detergent, the samples have to be centrifuged at high speed for many hours (*Warren* et al. 1974b). In this procedure nonspecific effects on gel materials are avoided, endogenous lipids will not be removed completely, and the proteins are not diluted extensively. Since concentrated protein solutions have to be applied to the gradient, complete dissociation of the membrane proteins may be a problem. Aqueous phase partitioning is a powerful method of separating solubilized membrane proteins. It has been applied successfully to the purification of phospholipase A 1 and of L-lactate dehydrogenase from *Escherichia coli* (*Albertsson* 1973; *Futai* and *Kimura* 1977). A drawback of this method is that the experimental conditions for separation have to be optimized empirically for the membrane protein to be separated. Chromatography on gels which carry hydrophobic groups is an effective method of separating water-soluble proteins due to their different degrees of hydrophobicity (*Hofstee* 1973; *Rosengren* et al. 1975; *Jennissen* and *Heilmeyer* 1975; *Schöpp* et al. 1976). It may also be used for the separation of membrane proteins if it is performed in the presence of some detergent (*Simmonds* and *Yon* 1976; *Homcy* et al. 1978; *Cresswell* 1979; *Carson* and *Konigsberg* 1981). Since only limited experience with the hydrophobic chromatography of membrane proteins is available as yet, the efficiency of this method is unclear. Gel filtration experiments which have to be performed in the presence of detergent may work if the detergent–protein–lipid micelles are sufficiently small that the molecular weight of the incorporated proteins increases the total micellar weights significantly. Since the weight differences of the micelles are relatively small, long columns have to employed. In this method, nonspecific adsorption of the proteins to the gel material may occur and the separation obtained may be incomplete (*Ne'eman* et al. 1972; *Acevedo* et al. 1981). In several cases, ion exchange chromatography or chromatofocusing has been used for the purification of hydrophobic membrane proteins (*Yu* and *Steck* 1975; *Kasahara* and *Hinkle* 1977a,b; *Kahlenberg* and *Zala* 1977; *Lin* et al. 1984). The separation with these methods has to be performed in the presence of nonionic detergents. Using hydroxyapatite columns, adsorption chromatography has been performed with some success (*Fröman* 1982; *Stipani* and *Palmieri* 1983). Furthermore, separation of glycoproteins by concanavalin A columns or by gel-linked SH-group reagents may lead to some enrichment of transport proteins (*Hatefi* and *Hanstein* 1969; *Pearlstein* 1977; *Lotan* and *Nicolson* 1979; *Fröman* et al. 1981; *Touraille* et al. 1981; *de Pinto* et al. 1982; *Schraw* and *Regen* 1983). Normally, the final purification of a transport protein is not achieved by any of these conventional methods. If, however, some of these methods are combined, good purification may be obtained.

Affinity chromatography with immobilized specific ligands is a powerful purification method by which proteins which are only present in small concentrations can be enriched and purified by a one-step procedure. This method can, however, only be applied if a high affinity ligand to the transporter is known and if this ligand can be immobilized without much loss of affinity. Success can only be obtained if the micelles contain single proteins and if high affinity binding of the solubilized protein and accessibility of the ligand are preserved in the presence of detergent (*Robinson* et al. 1980). Affinity purification has been shown to be highly effective for the purification of membrane-bound receptor proteins (for review see

Wilchek et al. 1984). Affinity purification of transport proteins from plasma membranes has been claimed for the Na^+ D-glucose cotransporter of kidney brush-border membranes (*Lin* et al. 1981) and for the (Ca^{2+})-ATPase from erythrocytes (*Niggli* et al. 1979). In the latter case purification by a calmodulin affinity column has been achieved, whereas in the former case a significant purification of the Na^+ D-glucose transporter could not be demonstrated.

It has been shown that monospecific polyclonal or monoclonal antibodies can be used to purify membrane proteins, even if they are present in small concentrations (e.g., receptor proteins, *Wilchek* et al. 1984; *van Rendswoude* and *Kempf* 1984). Thus one can be confident that it will also be possible to purify transporters from plasma membranes with antibodies if one succeeds in raising and identifying monoclonal, high affinity antibodies against transport proteins.

5 Reconstitution Procedures

To identify functional transport proteins during their purification, they have to be reconstituted into artificial lipid membranes. For this, the reconstitution into proteoliposomes is the chosen method since the reconstitution of active transport proteins into proteoliposomes can be performed much more easily than the reconstitution into planar lipid membranes. Several procedures to reconstitute transprt proteins into proteoliposomes have been developed and applied with varying success (for reviews see *Racker* 1977; *Racker* et al. 1979; *Hokin* 1981; *Eytan* 1982). Lipids which were associated with the transport proteins or were added during reconstitution may significantly affect reconstitution. Since the structure of most transport proteins is not known, and since the mechanism of protein reconstitution into liposomes is poorly understood, different reconstitution procedures have to be tried for each transport protein. Moreover, it was found that different reconstitution procedures may be differentially effective at different purification steps of transport proteins (*Koepsell* et al. 1984). Below I will summarize the reconstitution procedures which have been applied successfully to transport proteins from plasma membranes.

5.1 Reconstitution by Detergent Removal

Formation of proteoliposomes can be achieved if detergent is removed from mixtures of solubilized proteins and lipids. Detergent may be removed

(1) by dialysis, (2) by gel filtration or by adsorption of the detergent, or (3) by dilution.

1. The formation of proteoliposomes by detergent removal was the first method developed for reconstitution of membrane proteins (*Kagawa* and *Racker* 1971; *Racker* 1972; *Kagawa* et al. 1973; *Racker* and *Kandrach* 1973). This method, which was originally performed with cholate and is therefore often referred to as "cholate dialysis procedure," proved to be successful for the reconstitution of many different transport proteins from mammalian plasma membranes (*Hilden* et al. 1974; *Goldin* and *Tong* 1974; *Hilden* and *Hokin* 1975; *Sweadner* and *Goldin* 1975; *Anner* et al. 1977; *Goldin* 1977; *Edwards* 1977; *Baldwin* et al. 1982; *Ross* and *McConnell* 1977, 1978; *Darmon* et al. 1983; *Gietzen* et al. 1980a; *Niggli* et al. 1981a, 1982). To obtain a successful reconstitution by this procedure the employed detergent must be easily dialyzable (CMC must be high) and the protein must retain its biological activity during the dialysis. For a successful reconstitution the formed proteoliposomes must have an appropriate size, their membrane permeability must be adequate, the lipsosomal membranes must contain lipids which are essential for the incorporated transport proteins, and the transporters must be incorporated properly. The size of the liposomes and the activity of the transport proteins after reconstitution are dependent on the type of detergent employed, on the speed of detergent removal, on the pH during detergent removal, on the types of lipid used for reconstitution, and on the properties and concentrations of proteins (*Kagawa* and *Racker* 1971; *Kagawa* et al. 1973; *Zumbuehl* and *Weder* 1981; *King* and *Marchbanks* 1982; *Anner* et al. 1984a). To form proteoliposomes of homogeneous size, the starting material should contain a homogeneous population of mixed detergent–protein–lipid micelles and the rate of detergent removal should be controlled. This can be achieved by a continuously perfused dialysis chamber (*Milsmann* et al. 1978; *Zumbuehl* and *Weder* 1981). Proteins which are reconstituted by dialysis may be oriented randomly or uniformly in the lipid bilayer (*Racker* and *Kandrach* 1973; *Knowles* and *Racker* 1975; *Goldin* 1977; *Helenius* et al. 1977, 1981; *Karlish* and *Stein* 1982a). The orientation of the transporter molecules seems to be affected by the properties of the transporter and by the protein to lipid ratio employed during reconstitution (*Cornelius* and *Skou* 1984). If the experimental conditions are adjusted such that lipid vesicles are formed before the membrane proteins start to be incorporated, an oriented incorporation may be favored (*Helenius* et al. 1981). In proteoliposomes of different sizes different proteins may be distributed nonrandomly.

2. If the detergent is removed by gel filtration, liposomes of homogeneous size can be formed (*Brunner* et al. 1976; *Mimms* et al. 1981). From mixtures of phospholipids with either cholate or octylglucoside, liposomes

with diameters of respectively about 300 Å (*Brunner* et al. 1976, 1978; *Baldwin* et al. 1981; *Kelleher* et al. 1983) or about 2000 Å were formed (*Mimms* et al. 1981). In the presence of membrane proteins the formed proteoliposomes are larger and more variable in size (*Wacker* et al. 1976; *Koepsell* et al. 1980). In proteoliposomes of different size, different membrane proteins may be incorporated (*Spiess* et al. 1981). The detergent can also be removed by adsorption to specific gel materials such as Bio Bead SM-2 (*Chiesi* et al. 1978; *Baldwin* et al. 1979, 1981; *Wolosin* 1980; *Köhne* et al. 1981; *van Hoogevest* et al. 1983). It has been observed that the ionic strength of the running buffer affects the formation of the vesicles (*Chiesi* et al. 1978). *Kagawa* and collaborators (1973) found that enzymes for oxidative phosphorylation from mitochondria, which can be functionally reconstituted by dialysis, were inactive when the detergent was removed by gel filtration. Functional reconstitution was, however, achieved if during gel filtration phospholipids and some detergent were added to the running buffer (*Eytan* and *Racker* 1977). Thus for the incorporation of the transport proteins, which occurs at a critical detergent concentration, detergent removal should not be too rapid. Functional reconstitution of transport proteins by this method is possible if the experimental conditions are optimized (*Martonosi* 1968; *Baldwin* et al. 1981). This procedure is also well suited for the incorporation of enzymes or binding proteins into liposomes (*Wacker* et al. 1976; *Koepsell* et al. 1980).

3. A third method of removing the detergent from solubilized protein–lipid mixtures is to reduce the detergent concentration by dilution. By this procedure excellent results were obtained, especially if a mixture of cholate and octylglucoside was employed as detergent (*Racker* et al. 1979). Cytochrome oxidase, the ubiquinone-cytochrome c oxidoreductase from beef heart, bacteriorhodopsin, the H^+-ATPase from chloroplasts, the H^+-ATPase from mitochondria, the Ca^{2+}-ATPase from sarcoplasmic reticulum, and the (Na^++K^+)-ATPase from plasma membranes have been reconstituted by this method (*Martonosi* 1968; *Racker* 1972; *Racker* et al. 1975, 1979; *Rich* and *Clarke* 1982; *Jackson* and *Litman* 1985). In several cases, transport activity higher than with any other reconstitution procedure was achieved. Since the dilution procedure is very simple and optimal conditions can be found easily in serial experiments, this procedure should be tried if reconstitution of novel transport systems is attempted. An important advantage of this procedure is that long contact times of the proteins with the detergent are avoided. The properties of the formed proteoliposomes are dependent on the degree of dilution and on the detergent which has been employed for the protein solubilization. Thus, it has, for example, been found that relatively large proteoliposomes with diameters up to 3000 Å can be generated if octylglucoside is used as detergent (*Racker* et al. 1979).

5.2 Reconstitution by Sonication

Since transport proteins may be inactivated during treatment with detergents or may only tolerate detergents which cannot be removed easily, and since trace amounts of detergents may increase the passive permeability of proteoliposomes (*Bangham* and *Lea* 1978), reconstitution by sonication has been developed. In the sonication procedure, firstly lipids dissolved in chloroform or ether are dried under a stream of nitrogen and are hydrated by addition of buffer. Detergent-free proteins are then added and the proteoliposomes are formed by sonication (*Racker* 1973; *Racker* and *Eytan* 1973; *Racker* and *Fisher* 1975). This reconstitution procedure can be modified by sonicating the lipids, first for a lengthy period and then briefly together with the proteins, or by adding organic solvents during the sonication (*Racker* et al. 1979). By this method the H^+-ATPase from mitochondria, cytochrome oxidase, bacteriorhodopsin (*Racker* 1973), the (Na^++K^+)-ATPase (*Racker* and *Fisher* 1975), the D-glucose transporter from erythrocytes (*Kasahara* and *Hinkle* 1976; *Zala* and *Kahlenberg* 1976; *Baldwin* et al. 1980), and the Na^+ D-glucose cotransport protein from kidney and intestine (*Crane* et al. 1976a,b; *Fairclough* et al. 1979) have been reconstituted. Drawbacks of this procedure are that the experimental conditions during sonication are difficult to control, that proteins may be inactivated during sonication, and that it is difficult to produce homogeneous proteoliposomes with diameters larger than 1000 Å. These factors may explain why the preparations obtained by sonication often show smaller transport rates than other reconstitution procedures.

5.3 Reconstitution by Freezing and Thawing

Proteoliposomes formed by detergent removal or sonication normally have diameters smaller than 500 Å and their trapping volume is around 1 μl/mg phospholipid. Due to their small size the transport rate of substrates into these vesicles decreases rapidly during the measurements. This may be a severe limitation for transport assays, in particular if transport systems are analyzed which are driven by cotransported ions.

Several methods of preparing large liposomes have been described. They can be formed by fusion of sonicated liposomes (*Miller* and *Racker* 1976), from negatively charged phospholipids (*Papahadjopoulos* et al. 1975), or by solvent evaporation (*Deamer* and *Bangham* 1976; *Schieren* et al. 1978; *Szoka* and *Papahadjopoulos* 1978). Reconstitution of bacteriorhodopsin was made possible by employing a modification of the solvent evaporation method. By this method, inactivation of this protein by the organic solvent could be prevented (*Darszon* et al. 1979; *Rigaud* et al. 1983). The methods described above are, however, of limited value for the reconstitution of transport proteins since they employ organic solvents which often inactivate the transport proteins.

Kasahara and *Hinkle* (1977a,b) found that the D-glucose transporter from erythrocytes can be reconstituted into large proteoliposomes if a mixture of sonicated lipids and protein is frozen and thawed, followed by a short sonication. In recent years, such types of freeze-thaw reconstitution procedure have been applied successfully for the reconstitution of the (Na^++K^+)-ATPase (*Hokin* and *Dixon* 1979; *Karlish* and *Pick* 1981), of the D-glucose transporter from fat cells (*Shanahan* and *Czech* 1977b; *Suzuki* and *Kono* 1980; *Robinson* et al. 1982), of several Na^+ cotransport proteins (*Koepsell* et al. 1983, 1984; *Ducis* and *Koepsell* 1983), and of other transport proteins (e.g., *Pick* and *Racker* 1979; *Unden* and *Kröger* 1982). The first step in this procedure is to form lipid vesicles. Therefore, lipids solved in organic solvents are dried at the wall of a test tube. By addition of buffer and by shaking, the lipids are hydrated and multilamellar liposomes are formed. These multilamellar liposomes may be sonicated and/or fractionated by differential centrifugation or by gel chromatography (*Huang* 1969; *Reynolds* et al. 1983). In a second step, protein is added and the mixture is frozen and thawed. A short sonication or a heat-up step thereafter, to a temperature above the phase transition of the system, seems to be important to reduce the passive permeability of the proteoliposomes and to complete the reconstitution (*Pick* 1981; *Evans* and *Kwok* 1982; *Robinson* et al. 1982; *Ducis* and *Koepsell* 1983). The proteoliposomes formed by freeze-thawing have a large specific trapping volume (often about 10 μl/mg phospholipid) and large diameters (mean diameter often larger than 0.2 μm) (*Pick* 1981; *Robinson* et al. 1982; *Ducis* and *Koepsell* 1983). Their size can be increased further if they are dialyzed against hypo-osmotic buffers (*Oku* and *MacDonald* 1983). The formation of large proteoliposomes is highly dependent on the ionic strength of the buffer and on the cations present (*Pick* 1981; *Oku* and *MacDonald* 1983; *MacDonald* and *MacDonald* 1983). Furthermore, the formation of proteoliposomes is dependent on the lipid concentration and is inhibited by glycerol and sucrose (*Pick* 1981; *Oku* and *MacDonald* 1983). Reconstitution by this procedure does not work if only neutral phospholipids are present. Presence of cholesterol increases the trapping capacity of the proteoliposomes and decreases their passive permeability (*Papahadjopoulos* et al. 1971; *Pick* 1981; *Ducis* and *Koepsell* 1983). Reconstitution by the freeze-thaw procedure is supposed to lead to a random orientation of the transport proteins, which may be independent of the size of the proteoliposomes (*Sase* et al. 1982).

5.4 Incorporation Procedure

In the reconstitution procedures described above, the incorporation of transport proteins into the lipid membrane takes place during the formation of the proteoliposomes. This is different in principle from the in vivo insertion of transport proteins into preexisting cell membranes (*Singer* 1977; *Rothman* and *Lenard* 1977). These reconstitution procedures often lead to a random orientation of the transporters which are oriented unifomly in cell membranes. Studies of the lipid dependency of reconstitution with these methods will not give information on lipid dependency of the protein insertion in vivo. In contrast to the other reconstitution procedures, the incorporation procedure may mimic steps of the in vivo insertion. In the incorporation procedure, firstly monolamellar lipid vesicles are formed and then transport proteins are added which, under appropriate conditions, may insert spontaneously into the liposomal membrane (*Kagawa* and *Racker* 1971; *Eytan* et al. 1975, 1976; *Christiansen* and *Carlsen* 1983). This spontaneous incorporation may be accelerated if detergent is present in low concentrations (*Eytan* et al. 1975; *Christiansen* and *Carlsen* 1983). The spontaneous incorporation is dependent on the buffer, the Ca^{2+} or Mg^{2+} concentration, the lipid composition of the liposomes, the presence of other proteins in the liposomes, and the presence of lipids associated with the transport protein (*Kagawa* and *Racker* 1971; *Eytan* et al. 1976). When employing this incorporation procedure the incorporated proteins are normally unidirectionally oriented. The transport proteins may be incorporated into liposomes of a defined size which may be formed by different methods. A possible effect of vesicular size on the spontaneous incorporation has not been explored systematically. Compared with other reconstitution procedures the efficiency of incorporation of transport proteins is supposed to be low (*Eytan* et al. 1975).

6 Factors Which Influence Transport Activities in Proteoliposomes

Many factors influence the transport activity of reconstituted transport proteins, including the concentration of active transport proteins, their conformation, their insertion and orientation in the membrane, the self-aggregation of transport molecules, their association with other proteins, specific interactions of transport proteins with lipids (*Yeagle* 1983), and the viscosity of their microenvironment. Furthermore, the measured transport activity may be influenced by the size of the proteoliposomes and by the thickness of the lipid bilayer. Moreover, measured transport rates may be influences by the passive permeability of the proteoliposomes and by the interaction of the transport proteins with other proteins in the membrane. The conformation and aggregation of the transport proteins, as well as the passive permeability of the proteoliposomes, are dependent on the temperature, the lipid composition of the membrane, the symmetric or asymmetric lipid distribution in the two membrane halves, the concentration of detergents and alcohols in the membrane, the membrane potential, the ionic strength of the surrounding solution, and the presence of other membrane proteins (*Ladbrooke* et al. 1968; *Demel* et al. 1972; *De Kruyff* et al. 1972, 1974; *Michaelson* et al. 1973; *Rottem* et al. 1973; *Shimshick* and *McConnell* 1973; *Verkleij* et al. 1974; *Charnock* and *Bashford* 1975; *Jilka* et al. 1975; *Duppel* and *Dahl* 1976; *Shinitzky* and *Inbar* 1976; *Rothman* and *Lenard* 1977; *Aronson* 1978; *Toggenburger* et al. 1978; *Corda* et al. 1982; *Yeagle* et al. 1982; *van der Steen* et al. 1982; *Da Cruz* et al. 1983). All these parameters are highly dependent on the employed reconstitution procedure and may be affected by minor experimental modifications during reconstitution (*Racker* and *Eytan* 1973; *Gal* et al. 1983; *Koepsell* et al. 1984).

7 Lipid Dependency of Transport After Reconstitution

From experiments on the lipid dependency of membrane-bound enzymes, difficulties in testing the lipid requirement of membrane proteins have become apparent (*Levey* 1971; *Grover* et al. 1975; *Gazzotti* and *Peterson* 1977). The best approach to this question may be the careful removal of lipids, which normally leads to the inactivation of the enzymes, and the reactivation of the enzymatic activity by addition of defined lipids. Different procedures for lipid removal should be tried since the method of lipid removal may influence the data. It has been shown that membrane proteins may have a general requirement for a hydrophobic environment and a selective requirement for specific lipids (*Sandermann* 1978). If transport proteins catalyze an enzymatic reaction or bind a specific high affinity ligand, their lipid dependency can be investigated without the transport proteins being functionally incorporated in a lipid membrane (e.g., *Hidalgo* et al. 1978). It is difficult to obtain meaningful information on the lipid requirement of transport proteins which can be only assayed for activity by measuring uptake rates after reconstitution. The experimental procedure for such studies should be the purification of the transporter, the complete removal of endogeneous lipids, and the incorporation of the protein into proteoliposomes with a defined lipid composition. If the purification of the active transport protein can be achieved and the endogenous lipids can be removed without irreversibly inactivating the transport protein (*Warren* et al. 1974a,b; *Lind* et al. 1981), the transport protein has to be incorporated into proteoliposomes which differ in their lipid composition and/or transmembranous distribution (*Litman* 1973; *Michaelson* et al. 1973; *Berden* et al. 1975; *Kumar* and *Gupta* 1983) but are similar in size, have similar passive permeabilities, and have the same amount of active, properly oriented transporter molecules incorporated. Since delipidation of transport proteins normally leads to inactivation, recovery of transport activity after addition of different lipids may reflect the different effects of lipids on the renaturation of the inactivated transport proteins rather than the effects of lipids on the activity of the transport proteins in the membrane. The role of the lipids during incorporation of the transport proteins, which may vary for different reconstitution procedures, is not understood and hence the effects of added lipids on the proper insertion of transport proteins cannot be distinguished from the lipid effect on the functions of the transport proteins (e.g., *Kagawa* et al. 1973; *Peterson* et al. 1978; *Krämer* and *Klingenberg* 1980; *Nelson* and *Fleischer* 1981; *Krämer* 1982; *Ducis* and *Koepsell* 1983). Findings which show different lipid requirements of transport activities after reconstitution by different reconstitution procedures demonstrate the practical importance of this problem (*Racker* 1977, 1979).

8 Reconstitution of Primary Active Transporters

8.1 (Na^++K^+)-ATPase

The (Na^++K^+)-ATPase, which has been identified as Na^+/K^+ pump (*Dahl* and *Hokin* 1974; *Robinson* and *Flashner* 1979), has been purified from several sources (*Kyte* 1971; *Uesugi* et al. 1971; *Hokin* et al. 1973; *Jørgensen* 1974). In the purified preparations more than 80% of the protein consisted of two polypeptides with molecular weights of about 100 000 and 50 000. An ouabain binding site, a phosphorylation site, and an ATP binding site could be localized on the 100 000-dalton polypeptide, whereas the functional role of the 50 000-dalton polypeptide remained unclear (*Rhee* and *Hokin* 1975; *Koepsell* 1979; *Schuurmans Stekhoven* and *Bonting* 1981; *Freytag* 1983). Reconstitution of purified (Na^++K^+)-ATPase was performed to evaluate whether the components of the purified (Na^++K^+)-ATPase are sufficient and necessary to perform ATP-driven Na^+/K^+ exchange. Furthermore, reconstitution experiments were considered to provide information on the functional role of subunits or of fragments of the (Na^++K^+)-ATPase, which may be formed by protein digestion with proteases. Moreover, it was hoped that in proteoliposomes with low permeability transport kinetics may be measured more accurately than in membrane vesicles or in intact cells.

In 1974 first reports appeared on successful reconstitution of the $(Na^{+}+K^{+})$-ATPase from dog renal medulla and from shark rectal gland (*Goldin* and *Tong* 1974; *Hilden* et al. 1974). Employing the cholate dialysis procedure the authors succeeded in reconstituting ATP-driven Na^{+} uptake. In these early experiments the coupling ratio between Na^{+} and ATP was far below the value of 3, which has been measured in erythrocytes and in squid axon (*Baker* 1965; *Garrahan* and *Glynn* 1967; *Mullins* and *Brinley* 1969). In later reconstitution experiments, in which the dialysis time was reduced, it could be demonstrated that ATP-driven potassium transport was coupled to countertransport of sodium (*Hilden* and *Hokin* 1975; *Sweadner* and *Goldin* 1975). In these experiments the coupling ratio between Na^{+} transported and ATP hydrolyzed was 1.5. For countertransport of Na^{+} and K^{+} a transport ratio of 3:2 was obtained. The same stoichiometry for Na^{+}/K^{+} transport was measured with reconstituted $(Na^{+}+K^{+})$-ATPase from lamb kidney and from beef heart (*Anner* et al. 1977). When the dialysis time was reduced further and phosphatidylcholine was employed for reconstitution, purified $(Na^{+}+K^{+})$-ATPase from canine kidney could be reconstituted so that about 90% of the $(Na^{+}+K^{+})$-ATPase activity in the preparation was associated with liposomal membranes (*Goldin* 1977). In this preparation the stoichiometry of Na^{+} and K^{+} transported to ATP hydrolyzed was 3:2:1. This stoichiometry is the same as has been measured in the intact tissue. The $(Na^{+}+K^{+})$-ATPase molecules in these proteoliposomes were randomly orientated. The ATP-dependent Na^{+} uptake rate was about 0.35 μmol/mg protein and min (23°C). It could be shown that most of the reconstituted $(Na^{+}+K^{+})$-ATPase molecules were engaged in transport. However, they were transporting with a low turnover value of $2 \times s^{-1}$ (23°C) per mole phosphorylated enzyme. With $(Na^{+}+K^{+})$-ATPase reconstituted by cholate dialysis it could be shown by *Forgac* and *Chin* (1981, 1982) that also in the absence of K^{+} the $(Na^{+}+K^{+})$-ATPase is able to promote ATP-driven Na^{+} transport.

In addition to the cholate dialysis procedure other methods were tried for reconstitution. ATP-driven Na^{+} transport could be reconstituted from $(Na^{+}+K^{+})$-ATPase of the electric eel by a sonication procedure (*Racker* and *Fisher* 1975), and from $(Na^{+}+K^{+})$-ATPase of Ehrlich ascites tumor cells by an octylglucoside dilution procedure (*Spector* et al. 1980). Furthermore, *Hokin* and *Dixon* (1979) reconstituted the $(Na^{+}+K^{+})$-ATPase from shark rectal gland by freeze-thaw sonication. After the sonication, the lipid and the Na^{+} concentrations during reconstitution had been optimized, the ATP-dependent Na^{+} and K^{+} transport rates related to reconstituted $(Na^{+}+K^{+})$-ATPase protein were considerably higher than those obtained with the cholate dialysis procedure. The reconstitution was modified by *Karlish* and *Pick* (1981), who performed the freezing, thawing, and sonication steps in the presence of cholate. Factors which were important in achieving an effective reconstitution were: (a) the ratio of cholate to protein during protein solubilization, (b) the ratio of phospholipid to $(Na^+ + K^+)$-ATPase during reconstitution, (c) the sonication time, and (d) the pH during reconstitution. With these proteoliposomes an ATP-dependent Na^+ uptake rate of about 1.6 μmol/mg protein and min (room temperature) was obtained. For the ATP-dependent Na^+/Rb^+ exchange a turnover value of 43 sodium ions per mole phosphoenzyme per second (20°C) was estimated (*Karlish* and *Stein* 1982a).

Performing reconstitution by detergent removal with polystyrene beads, *Cornelius* and *Skou* (1984) reconstituted purified $(Na^+ + K^+)$-ATPase from spiny dogfish rectal glands which had been solubilized with the nonionic detergent dodecyloctaethyleneglycol monoether. For the ATP-driven, K^+-stimulated Na^+ uptake rate measured at 22°C, a value of about 6 μmol/mg protein per minute was found. If the initial uptake rate was related to the inside out oriented $(Na^+ + K^+)$-ATPase protein, a value of 30 μmol/mg per minute was estimated. Recently *Yoda* et al. (1984) solubilized membranous $(Na^+ + K^+)$-ATPase from the electric eel with 3-[(3-cholamidopropyl)-dimethylammonio]-1-propane sulfonate and reconstituted the $(Na^+ + K^+)$-ATPase by detergent removal on a Sephadex G-25 column. Egg phospholipids and cholesterol, which proved to be essential in obtaining $(Na^+ + K^+)$-ATPase liposomes, were added during reconstitution. The ATP-dependent Na^+ transport measured at 37°C in the presence of 70 m*M* Na^+ and 30 m*M* K^+ was about 30 μmol Na^+ per milligram of protein per minute (Fig. 2). For the turnover rate of Na^+/K^+ exchange measured at 37°C a value of about 190 Na^+ ions per mole phosphoenzyme per second was estimated.

The data reported above demonstrate that the conditions employed for reconstitution are critical, but that it is possible to reconstitute the $(Na^+ + K^+)$-ATPase molecules so that the properties of the Na^+/K^+ pump which have been measured in erythrocytes can be obtained. Care must be taken since, by reconstitution, partially active transport molecules may be easily obtained which show nonphysiological transport properties.

One may ask whether the reconstitution experiments on the $(Na^+ + K^+)$-ATPase have increased our knowledge on the $(Na^+ + K^+)$-ATPase. By reconstitution it has been shown that the 100 000- and the 50 000-dalton polypeptides are probably sufficient to obtain ATP-dependent Na^+/K^+ pump activity. It has not, however, been finally proved that the 50 000 polypeptide is also required for pump activity since the 100 000- and the 50 000-dalton polypeptides have not been reconstituted separately. Digestion of the $(Na^+ + K^+)$-ATPase with proteases and successive reconstitution

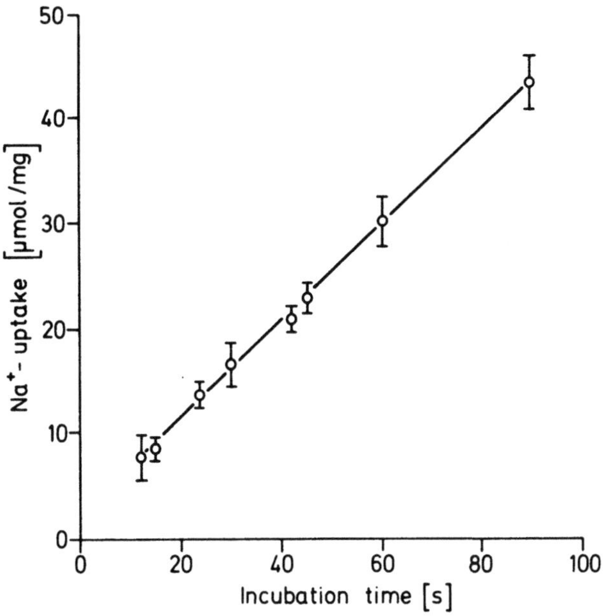

Fig. 2. ATP-dependent, K^+-stimulated Na^+ influx into proteoliposomes which contained purified (Na^++K^+)-ATPase. The proteoliposomes contained 70 m*M* Na^+ and 30 m*M* K^+. They were added to incubation media with or without 1 m*M* ATP, which contained in addition 70 m*M* Na^+, 30 m*M* K^+, 10 m*M* Mg^{2+}, and 20 μM valinomycin (37°C). In the graph the differences in Na^+ uptake after incubation with or without ATP are shown. (Data redrawn from *Yoda* et al. 1984)

has been reported by *Jørgensen* and *Anner* (1979). The experiments are of limited value as the digestion products were not reconstituted separately (*Anner* et al. 1984b). In reconstitution experiments it was shown that, due to the presence of (Na^++K^+)-ATPase, the permeability to Na^+ and K^+ was increased in the absence of ATP (*Anner* 1981). It was found that the permeability to Na^+ and K^+ was increased further if the (Na^++K^+)-ATPase was treated with trypsin (*Anner* 1983, 1985). Unfortunately, the specificity of the permeability increase for Na^+ and K^+ after reconstitution of the (Na^++K^+)-ATPase was not finally proved since effects of ouabain or vanadate on the passive Na^+ and K^+ fluxes were not reported.

The proteoliposomes formed by freeze-thaw reconstitution (*Karlish* and *Pick* 1981) were highly impermeable to cations. Thus it could be demonstrated that the ATP-driven Na^+/K^+ countertransport is rheogenic (*Dixon* and *Hokin* 1980). Furthermore, pump-mediated Rb fluxes in the absence of ATP could be studied much better than in intact tissue (*Karlish* and *Pick* 1981). In these proteoliposomes (a) ATP plus P_i-dependent Rb^+/Rb^+ exchange, (b) passive vanadate sensitive Rb^+/Rb^+ exchange, (c) passive vanadate sensitive net Rb uptake, (d) passive vanadate insensitive net Rb^+ uptake, and (e) passive vanadate sensitive Rb^+/Na^+ exchange could be investigated (see Table 1). Moreover, these proteoliposomes were suited for study of effects of ATP or of phosphate and the combined effects of ATP and phosphate on passive rubidium fluxes (*Karlish* and *Stein* 1982b; *Karlish* et al. 1982). The data were explained by a model in

Table 1. Estimated turnover numbers of the reconstituted ($Na^{+}+K^{+}$)-ATPase working at different transport modes (data from *Karlish* and *Stein* 1982a)[a]

ATP-dependent $^{22}Na^{+}/Rb^{+}$ exchange	43
(ATP+P_i)-stimulated $^{86}Rb^{+}/Rb^{+}$ exchange	7
Vanadate-sensitive $^{86}Rb^{+}/Rb^{+}$ exchange	0.25
Vanadate-sensitive net $^{86}Rb^{+}$ uptake	0.15
Vanadate-sensitive $^{86}Rb^{+}/Na^{+}$ exchange	0.63

[a] Uptake of $^{22}Na^{+}$ and of $^{86}Rb^{+}$ was measured at room temperature; the phosphorylation with (γ-^{32}P)-ATP was measured at 0°C. The turnover numbers are given as mole of Rb^{+} or Na^{+} per second per mole phosphoenzyme. They are corrected with regard to the lack of full saturation of Na^{+} or Rb^{+} or with regard to the inhibition of transport by Mg^{2+} which was present during the measurements

which passive Rb^{+} fluxes occur via conformational transitions of the ($Na^{+}+K^{+}$)-ATPase.

In the case of the ($Na^{+}+K^{+}$)-ATPase important new information has been obtained. Much more work, however, is necessary before the molecular mechanism of the Na^{+}/K^{+} pump is understood.

8.2 Ca^{2+}-ATPase from Erythrocytes

Free Ca^{2+} in the extracellular space is four orders of magnitude higher than in the cytosol. The low intracellular Ca^{2+} concentration is maintained by the action of two different mechanisms: a Ca^{2+}/Na^{+} exchange system and a Ca^{2+} transporting ATPase (for review see *Carafoli* and *Zurini* 1982). The Ca^{2+}-ATPase which appears to be present in plasma membranes of various sources is supposed to be characterized by the following properties: high affinity for Ca^{2+}, inhibition by low concentrations of vanadate, stimulation by calmodulin, and formation of an acyl phosphate in the presence of Ca^{2+} and ATP. Purification experiments have been performed with the Ca^{2+} ATPase from erythrocytes (*Wolf* et al. 1977; *Niggli* et al. 1979; *Haaker* and *Racker* 1979; *Gietzen* et al. 1980b) and from heart sarcolemma (*Caroni* and *Carafoli* 1981). Partial purification of the Ca^{2+}-ATPase from erythrocytes with a considerable increase of the specific activity could be obtained by mixed micelle gel chromatography (*Wolf* et al. 1977; *Haaker* and *Racker* 1979). The ultimate purification was achieved by affinity chromatography with calmodulin (*Niggli* et al. 1979; *Gietzen* et al. 1980b; *Caroni* and *Carafoli* 1981). It was found that the Ca^{2+}-ATPase consisted of a single polypeptide with a molecular weight between 125 000 and 150 000 (*Gietzen* et al. 1980b; *Caroni* and *Carafoli* 1981; *Niggli* et al. 1981b).

In 1979 *Haaker* and *Racker* reconstituted a partially purified Ca^{2+}-ATPase from pig erythrocytes by the freeze-thaw sonication procedure. In the proteoliposomes, which contained oxalate, significant ATP-dependent Ca^{2+} uptake, which could be stimulated by calmodulin, was demonstrated (uptake rate in the presence of calmodulin 0.3 μmol × mg protein^{-1} × min^{-1}). The ratio of Ca^{2+} transported to ATP hydrolyzed was 0.8. Some ATP dependent Ca^{2+} uptake could also be reconstituted from Triton-solubilized pigeon erythrocytes when the detergent was removed with the help of high density lipoprotein (*Yeung* et al. 1979), or from human erythrocytes when the cholate dialysis procedure was employed (*Gietzen* et al. 1980a). After purification of the Ca^{2+}-ATPase by affinity chromatography in the presence of Triton X-100, the transport protein was reconstituted by detergent removal on a Bio-Beads SM-50 column (*Niggli* et al. 1981a,b, 1982) or by cholate dialysis which was performed after Triton X-100 had been exchanged against cholate (*Niggli* et al. 1981a, 1982). Ca^{2+}-ATPase activity and Ca^{2+} uptake in the proteoliposomes were maximally stimulated if the proteoliposomes were formed in the presence of acidic phospholipids, or if they were formed without acidic phospholipids and calmodulin was added. In proteoliposomes formed in the presence of acidic phospholipids, addition of calmodulin had no further stimulating effect on Ca^{2+} uptake (*Niggli* et al. 1981b). For the activated transporter, an initial ATP dependent Ca^{2+} uptake rate of 1.5 μmol Ca^{2+} × mg protein^{-1} × min^{-1} was measured. Stimulation of ATP-driven Ca^{2+} uptake was also observed if the proteoliposomes contained unsaturated fatty acids or if they were treated with trypsin.

Tight coupling between Ca^{2+} uptake and ATPase activity was only observed if the Ca^{2+}-ATPase was reconstituted by cholate dialysis. In these proteoliposomes the Ca^{2+}-ATPase activity could be stimulated nearly tenfold by the Ca^{2+} ionophore A 23187. The ratio of Ca^{2+} accumulated in the vesicles to ATP hydrolyzed was determined to be about 1. This value appears to be questionable since it was measured in proteoliposomes in which the coupling between ATP splitting and Ca^{2+} uptake was not optimal. In reconstituted proteoliposomes two protons were released when one Ca^{2+} atom was taken up. Since Ca^{2+} uptake in the proteoliposomes was electroneutral, it can be assumed that two H^+ ions are pumped out or two hydroxyl ions are pumped in when one Ca^{2+} ion is taken up by the proteoliposomes.

By means of the reconstitution experiments information on the transporter was obtained which could not be gained from measurements on intact membranes. It was shown that the Ca^{2+} transporter is a polypeptide with a molecular weight between 125 000 and 150 000. Effects of phospholipids on the function of the reconstituted transporter could be investigated. Since reconstituted proteoliposomes with relatively low permeability for H^+, OH^-, and Ca^{2+} were formed, data could be obtained which strongly suggest that Ca^{2+} is coupled to H^+ ion countertransport or OH^- ion cotransport.

The efficiency of the different reconstitution procedures cannot be compared, because: (a) the measurements were performed under different experimental conditions; (b) in most of the experiments different protein preparations were used; and (c) the experimental conditions for reconstitution by the different procedures may not have been sufficiently optimized.

9 Reconstitution of Facilitated Diffusion Systems

9.1 D-Glucose Transporter from Erythrocytes and Other Tissues

In red blood cells a facilitated diffusion system for D-glucose has been detected which shows substrate saturation, stereospecificity, competition between sugars, and countertransport of D-glucose (*Miller* 1968). Attempts to identify the D-glucose transporter with D-glucosylisothiocyanate (*Taverna* and *Langdon* 1973), 1-fluoro-2,4,-dinitrobenzene (*Jung* and *Carlson* 1975; *Shanahan* and *Jacquez* 1978), and *N*-ethylmaleimide (*LeFevre* et al. 1975) yielded inconclusive data. By employing impermeant maleimides, *Batt* et al. (1976) demonstrated labeling of the polypeptide band 4.5 (nomenclature *Steck* 1974) which could be protected by D-glucose and the competitive D-glucose transport inhibitor cytochalasin B (*Jung* and *Rampal* 1977). In 1976, the D-glucose transport protein was partially purified in the membrane by extracting extraneous proteins from the erythrocyte membrane (*Kahlenberg* 1976). From these data it was concluded that the D-glucose transporter is identical with band 3 and/or band 4.5 and/or band 7. An important step on the way to identifying and purifying the D-glucose transporter was taken when the erythrocyte membrane was solubilized with Triton X-100 and proteoliposomes were formed by sonication in which D-glucose transport could be measured (*Kasahara* and *Hinkle* 1976; *Zala* and *Kahlenberg* 1976). In 1977 *Kasahara* and *Hinkle* (1977a,b) improved the method of reconstituting the D-glucose transporter by introducing a freeze-thaw step during the sonication procedure. In the same studies, purification of the D-glucose transport protein by ion exchange chromatography on a DEAE column was reported. After ion exchange chromatography, a protein fraction was obtained which consisted of more than 90% of band 4.5. After reconstitution of this protein fraction, the zero *trans* uptake rate of 0.2 m*M* D-glucose was about 10 times higher than after reconstitution of the crude Triton X-100 extract. Since this uptake rate was, however, only about 40% of the zero *trans* uptake rate measured in intact erythrocytes (*Wheeler* and *Hinkle* 1981), uncertainty remained concerning the identification of the D-glucose transporter. In 1978 *Goldin* and *Rhoden* solubilized erythrocyte membranes with cholate and reconstituted the membrane proteins after addition of excess phospholipid and cholesterol. The proteoliposomes, which had a diameter of about 500 Å and which contained not more than one D-glucose transport molecule per proteoliposome, were incubated with D-glucose and were separated by density chromatography due to their ability to perform D-glucose exchange. By this purification procedure proteoliposomes were obtained in which about 35% of the protein was the 4.5 polypeptide.

The final evidence that the purified 4.5 protein is the D-glucose transporter was obtained in three ways: Firstly, it was demonstrated that the specific cytochalasin B binding could be increased about 10 times over that in the erythrocyte membrane if the 4.5 polypeptide was purified by ion exchange chromatography. In this preparation 0.4–0.7 mol of cytochalasin B bound per mole 4.5 polypeptide (*Zoccoli* et al. 1978; *Sogin* and *Hinkle* 1978; *Baldwin* et al. 1979, 1982). Secondly, the erythrocyte membrane was solubilized by octaethyleneglycol *n*-dodecylether, and the band 4.5 polypeptide was purified. Then the detergent was exchanged against cholate, excess lipid was added, and the band 4.5 polypeptide was reconstituted by detergent removal on a Sephadex G-50 column. By this procedure, small unilamellar proteoliposomes were formed which each contained about one transporter as revealed by cytochalasin B binding. By measuring the fraction of the proteoliposomes which showed D-glucose uptake, it was found that almost every transporter molecule was capable of performing transport. Thus it was shown that the low transport rate measured in the proteoliposomes was due to the fact that most of the transporter molecules in the proteoliposomes functioned at a much lower rate than in the red blood cell. Consequently, the D-glucose transport must be mediated by the band 4.5 polypeptide rather than by one of the contaminating proteins in the preparation (*Baldwin* et al. 1981). Thirdly, a preparation of proteoliposomes which contained the purified band 4.5 polypeptide could be prepared in which a maximal exchange rate of 86 μmol D-glucose per mg protein and min was measured (*Wheeler* and *Hinkle* 1981). This figure is about 3 times higher than the maximal rate measured in intact erythrocytes. When compared with erythrocytes, it was found that after reconstitution of the D-glucose transporter, relatively higher uptake rates were measured for D-glucose exchange than for zero *trans* influx or for zero *cis* efflux of D-glucose.

For the solubilization of the D-glucose transporter, different detergents have been employed sucessfully (Triton X-100, octylglucoside, octaethyleneglycol *n*-dodecylether, cholate, deoxycholate). For purification, ion exchange chromatography was the method of choice. Affinity chromatography with lectin (*Fröman* et al. 1981) and chromatography on hydroxyapatite (*Fröman* 1982) did not lead to a comparable purification. Attempts to purify the carrier after it had been reconstituted into proteoliposomes led to some enrichment of D-glucose uptake, but not to the ultimate purification of the D-glucose transporter (*Goldin* and *Rhoden* 1978; *Fröman* et al. 1980).

The D-glucose transport protein could be reconstituted successfully by several procedures (*Jones* and *Nickson* 1981): (a) the sonication procedure (*Kasahara* and *Hinkle* 1976; *Zala* and *Kahlenberg* 1976; *Kahlenberg* and *Zala* 1977), (b) the freeze-thaw sonication procedure (*Kasahara* and *Hinkle* 1977a,b; *Wheeler* and *Hinkle* 1981; *Sase* et al. 1982), (c) the detergent removal procedure (*Goldin* and *Rhoden* 1978; *Baldwin* et al. 1979, 1981, 1982), and (d) a combination of the detergent removal and the freeze-thaw procedure (*Edwards* 1977; *Fröman* et al. 1980; *Acevedo* et al. 1981; *Fröman* 1982). The mean diameters of the proteoliposomes formed by the different methods range from about 300 Å (*Baldwin* et al. 1981) to about 1000 Å (*Kasahara* and *Hinkle* 1977b). Optimal uptake rates of D-glucose were obtained in small proteoliposomes formed by detergent removal, as well as in large proteoliposomes formed by freeze-thaw sonication. Successful reconstitution was performed without addition of extraneous lipids (*Edwards* 1977; *Zoccoli* et al. 1978; *Baldwin* et al. 1981, 1982), after addition of soybean phospholipids (*Zala* and *Kahlenberg* 1976; *Kasahara* and *Hinkle* 1977a; *Baldwin* et al. 1980, 1981; *Wheeler* and *Hinkle* 1981), after addition of red cell lipid extract (*Zala* and *Kahlenberg* 1976; *Kahlenberg* and *Zala* 1977), after addition of egg yolk phospholipids (*Fröman* et al. 1980, 1981; *Fröman* 1982), and after addition of phosphatidylcholine plus cholesterol (*Goldin* and *Rhoden* 1978). During reconstitution, all types of phospholipids plus cholesterol were present in all experiments and nothing can therefore be said about the lipid dependency of the D-glucose transporter.

It has been reported above that the D-glucose transporter from erythrocytes was identified by purification and reconstitution. Furthermore, it has been shown that the reconstitution can be performed by different methods. To elucidate the lipid dependency and the molecular mechanism of this transporter, further reconstitution experiments will be necessary.

In 1980, data from affinity labeling experiments were presented in which maltosyl isothiocyanate was used and which suggested that band 3 may contain the glucose binding site of the D-glucose transporter and that band 4.5, which has been purified for the reconstitution experiments, may be a proteolytic digestion product of band 3 (*Mullins* and *Langdon* 1980a, b). It is unclear why D-glucose protectable labeling of band 3 was obtained with maltosyl isothiocyanate. The hypothesis that the band 4.5 polypeptide is a splitting product of band 3 can, however, be discarded, (a) since antibodies against band 4.5 do not crossreact with band 3 in freshly prepared erythrocyte membranes (*Sogin* and *Hinkle* 1980; *Baldwin* and *Lienhard* 1980; *Allard* and *Lienhard* 1985), (b) since, in fresh erythrocytes, the cytochalasin B binding site on the D-glucose transporter could be identified on band 4.5 (*Carter-Su* et al. 1982; *Shanahan* 1982), and (c) since the D-glucose binding site was identified on band 4.5 with a covalently binding D-glucose analog (*Weber* and *Eichholz* 1985).

It has been shown that the entry of D-glucose into adipocytes (*Czech* 1976; *Avruch* et al. 1972), into muscle (*Holloszy* and *Narahara* 1965), into the trophoblast of the placenta (*Johnson* and *Smith* 1980), and into fibroblasts (*Inui* et al. 1979) occurs via a facilitated diffusion system for D-glucose which is regulated by insulin. The concentration of the D-glucose transporter in adipocyte membranes is more than 100 times lower than in the erythrocyte membrane (*Carter-Su* et al. 1980).

In 1977, adipocyte cell membranes were extracted with dimethylmaleic anhydride and solubilized with cholate, and D-glucose transport was reconstituted by a combined detergent removal–freeze-thaw–sonication procedure in which a 1:1 mixture of phosphatidylcholine and phosphatidylethanolamine was added (*Shanahan* and *Czech* 1977a,b). In the reconstituted proteoliposomes, which contained mainly a polypeptide with a molecular weight of 94 000, the D-glucose uptake rate per unit protein was about 10 times higher than that measured in the dimethylmaleic anhydride-extracted membrane vesicles. Since the D-glucose transporter in the original membrane represents only a small fraction of the proteins (*Pessin* et al. 1982), the possibility that one of the contaminating proteins may be the D-glucose transporter cannot be ruled out. Reconstitution experiments with different lipids showed that the addition of phosphatidylcholine was necessary to obtain transport activity (*Carter-Su* et al. 1980). The D-glucose uptake was measured at different temperatures in proteoliposomes which were formed from phosphatidylcholine plus phosphatidylthanolamine with different fatty acids. It was found that D-glucose uptake increased in parallel with the membrane fluidity of the proteoliposomes (*Melchior* and *Czech* 1979). If cholesterol was added during reconstitution, the phase transition of the proteoliposomes was suppressed and the glucose uptake at 23°C was decreased, whereas the D-glucose uptake at 0°C was increased.

Attempts to purify the cholate-solubilized D-glucose transport protein further by gel chromatography, adsorption chromatography on hydroxyapathite, and affinity chromatography on lectin columns did not lead to a significant increase of the specific activity of D-glucose uptake. The experiments indicated that major polypeptides of the adipocyte membrane with molecular weights of 78 000, 94 000, 116 000, and 140 000 are not part of the transport protein (*Carter-Su* et al. 1980). These purification experiments demonstrate the limitations of conventional purification methods in the case of transport proteins which are present in low concentrations.

Evaluation of the number of D-glucose transporters in fat cells and in muscle cells by the magnitude of cytochalasin B binding revealed that cytochalasin B binding sites are not only present in th plasma membranes, but also in a microsomal fraction (*Wardzala* et al. 1978; *Cushman* and *Wardzala* 1980; *Wardzala* and *Jeanrenaud* 1981). Data were presented which suggested that the microsomal membranes which contained the cytochalasin B binding protein are derived from components of the Golgi apparatus (*Suzuki* and *Kono* 1980; *Simpson* et al. 1983). It was found that after insulin treatment the number of cytochalasin B binding sites in the plasma membranes increased, whereas the number of cytochalasin B binding sites in the microsomal fraction decreased. It was therefore argued that insulin may act by recruiting cytochalasin B binding protein from the mcirosomal fraction (*Cushman* and *Wardzala* 1980; *Wardzala* and *Jeanrenaud* 1981). It was shown that recycling between cytochalasin binding proteins in the presence and absence of insulin is an energy-requiring process which is independent of protein synthesis (*Kono* et al. 1981; *Ezaki* and *Kono* 1982). However, the recruitment of the glucose transport protein may not be the only effect of the insulin action, since it was found that the turnover numbers of plasma membranes from basal and stimulated cells appear to be similar, whereas those measured in intact cells are increased about fourfold after insulin treatment (*Simpson* et al. 1983).

To find out whether the recruitment of cytochalasin B binding sites does actually represent the recruitment of intact D-glucose transport proteins, reconstitution experiments had to be performed. It was found that D-glucose transport could be reconstituted from the microsomal fraction (*Suzuki* and *Kono* 1980; *Watanabe* et al. 1984). However, reconstitution experiments produced conflicting data: one group demonstrated that the action of insulin increased D-glucose uptake in the plasma membrane and reduced it in the microsomal fraction (*Suzuki* and *Kono* 1980), whereas another group found that the total transport measured after reconstitution of a mixture of solubilized plasma membranes and microsomes was increased upon insulin treatment (*Carter-Su* and *Czech* 1980). Since it was also found that the insulin effect in plasma membranes could be mimicked by membrane-fluidizing agents (*Pilch* et al. 1980), it was proposed that insulin may activate transport proteins in the plasma membrane rather than recruit transporters from the microsomal fraction. The different results are probably due to the fact that the efficiency of reconstitution and the turnover number of the reconstituted transporter are dependent on the amount and composition of endogenous lipids, and of other membrane proteins which are associated with the D-glucose transport protein during reconstitution (see below in case of the Na^+ D-glucose cotransporter).

Since antibodies against the purified D-glucose transporter from erythrocytes cross-react with the D-glucose transporter from fat cells, and since cytochalasin B is a competitive inhibitor of D-glucose uptake in erythrocytes and in fat cells, these agents were employed to label the D-glucose transporter in adipocytes, muscle cells, fibroblasts and membrane vesicles from the placenta. Employing D-glucose inhibitable cytochalasin B labeling, the following polypeptides were identified: (a) a 52 000-dalton protein in the microvillus membrane from the placenta (*Johnson* and *Smith* 1982); (b) two polypeptides with molecular weights of 45 000 and 50 000 in membranes from adipocytes (*Shanahan* et al. 1982): (c) a 55 000-dalton polypeptide in virus-transformed fibroblasts (*Shanahan* et al. 1982); and (d) two polypeptides with molecular weights of 46 000 and 52 000 in chicken embryo fibroblasts (*Pessin* et al. 1982). In plasma membranes and in microsomes from rat adipocytes, a 45 000-dalton polypeptide was identified with antibodies directed against the purified D-glucose transporter from erythrocytes (*Lienhard* et al. 1982; *Wheeler* et al. 1982). By means of the antibodies, several polypeptides could be precipitated from normal and Rous sarcoma virus-transformed chicken embryo cells. The amount of two of the precipitated polypeptides with molecular weights of 41 000 and 82 000 was increased during malignant transformation in parallel with an increased D-glucose uptake, suggesting that the 41 000-dalton polypeptide represents the D-glucose transporter and the 82 000-dalton polypeptide a precursor thereof (*Salter* et al. 1982). The differences in the molecular weights of the D-glucose transport protein identified may be due to different degrees of glycosylation of the transport protein (*Lienhard* et al. 1984).

9.2 Anion Exchange Protein from Erythrocytes

The anion exchange protein in the erythrocyte membrane plays a central role in respiration (*Wieth* and *Brahm* 1980). CO_2 diffuses rapidly from the venous blood into the erythrocyte, where bicarbonate anions are formed by catalysis of the carboanhydratase. The anion transport protein transfers bicarbonate into the blood plasma by exchanging it against chloride. In the lung, bicarbonate from the blood plasma (in exchange for chloride), is transported into the erythrocyte, where CO_2 is formed. CO_2 diffuses rapidly into the blood plasma, and through the alveolar epithelium into the expired air.

It could be shown by affinity labeling with specific inhibitors that band 3 protein from the erythrocyte membrane (nomenclature according to *Fairbanks* et al. 1971) is engaged in anion transport. In these experiments radioactively labeled covalently binding anions such as isothiocyanate stilbene disulfonic acids (*Cabantchik* and *Rothstein* 1972, 1974; *Rothstein* et al. 1975; *Passow* et al. 1975; *Zaki* et al. 1975), 1-isothiocyano-4-benzene sulfonate (*Ho* and *Guidotti* 1975), or pyridoxal phosphate (*Cabantchik* et al. 1975) were employed. From digestion experiments with proteolytic enzymes applied to the outside or inside of the red cell membrane, evidence was obtained that band 3 is an asymmetric protein which spans the cell membrane (for review see *Passow* 1986). The band 3 protein is composed of: (a) a hydrophobic transmembranous segment with a molecular weight of 55 000 which contains the outer directed C-terminal and is glycosylated on the outside, and (b) a hydrophilic 42 000-dalton segment which contains the inside located N-terminal and binds membrane proteins like ankyrin and cytosolic proteins like hemoglobin, glyceraldehyde 3-phosphate dehydrogenase, and phosphofructokinase (for references see *Passow* et al. 1986). The hydrophobic membrane spanning 55 000-dalton segment traverses the membrane at least five times (*Tanner* et al. 1980; *Ramjeesingh* et al. 1984) and is involved in anion transport (*Jennings* and *Passow* 1979). It contains several targets of covalently binding chemical modifiers of anion transport. These are at least two lysine residues and three arginyl residues (*Passow* 1986). Since a linear relationship exists between the binding of stilbene disulfonates and the inhibition of anion transport, and since inhibition of anion transport is complete when all band 3 proteins in the erythrocyte membrane are occupied, it is assumed that all band 3 molecules participate equally in the transport process. On the other hand, data on crosslinking of band 3, analytical centrifugation of solubilized band 3, and electron microscopy of erythrocyte membranes suggest that the protein may exist in monomeric, dimeric, tetrameric, and polymeric forms. In the membrane, the dimeric form may predominate and may be able to perform anion transport (*Boodhoo* and *Reithmeier* 1984; *Passow* 1986). A fraction of the band 3 molecules may undergo rotational diffusion around an axis perpendicular to the membrane. The lateral diffusion of band 3 is small, but may be significantly enhanced if the cytoskeleton is detached (*Peters* et al. 1974; *Golan* and *Veatch* 1980, 1982).

The kinetics of anion transport have been studied in detail (for review see *Passow* 1986). It has been shown that the anion transporter is able to transport monovalent and divalent anions. The transporter may perform electrically silent anion exchange or rheogenic net ion uptake. It contains a transfer anion site which may be recruited on the inside or the outside of the erythrocyte membrane. Furthermore, one anion modifier site on the inside and another on the outside of the erythrocyte membrane have been described. The interactions of various side group modifying agents with the transfer and modifier sites have been reviewed by *Passow* (1986).

In order to show that the band 3 protein is sufficient to perform anion transport, this protein has to be purified and anion transport has to be demonstrated. This can be attempted by negative purification of band 3 in erythrocyte ghosts or by purification of solubilized band 3 and reconstitution of the purified protein. During negative purification, proteins were extracted from erythrocyte ghosts by treatment with 0.045% Triton X-100 and with alkaline pH in the presence of EDTA (*Wolosin* et al. 1977). The obtained membrane vesicles, in which 95% of the protein was band 3, were sealed by freezing, thawing, and sonication. In these vesicles, which were uniform in size, anion net uptake and anion exchange were measured. Anion exchange could be inhibited by stilbene disulfonic acids and showed the same pH dependence as the anion transport in intact erythrocytes. The evaluation of transport capacity related to membrane area revealed that in the vesicles about two-thirds of the transport capacity found in intact erythrocytes was preserved. Some negative purification was also obtained when the ghosts were treated with proteases and the split products were removed (*Lepke* and *Passow* 1976).

For complete purification, the anion transporter has to be solubilized, purified, and reconstituted. Band 3 was solubilized with Triton X-100 or dodecyltrimethyl-ammonium bromide. By employing appropriate concentrations of these detergents, band 3 could be selectively removed from the membrane (*Fairbanks* et al. 1971; *Yu* et al. 1973; *Yu* and *Steck* 1975; *Cabantchik* et al. 1977; *Ross* and *McConnell* 1977, 1978; *Wolosin* 1980; *Köhne* et al. 1981; *Lukacovic* et al. 1981; *Darmon* et al. 1983; *van Hoogevest* et al. 1983). If more loosely bound membrane proteins were extracted before solubilization, a highly enriched preparation of band 3 was obtained (*Köhne* et al. 1981). Further purification was achieved by affinity chromatography on a concanavalin A column (*Findlay* 1974; *Ross* and *McConnell* 1977, 1978) or by treatment with *p*-chloromercuribenzoate plus ion exchange chromatography (*Yu* and *Steck* 1975; *Lukacovic* et al. 1981).

In 1975 reconstitution of band 3 was reported. In these experiments band 3 was extracted from membranes with Triton X-100 (*Rothstein* et al. 1975). Triton X-100 was then removed by treatment with a toluene–lecithin mixture, lecithin was added, and the protein was reconstituted by sonication. In the proteoliposomes sulfate efflux was more rapid than in liposomes formed without addition of protein and was slowed down by DIDS. Employing a dialysis procedure *Ross* and *McConnell* (1977, 1978) reported reconstitution of band 3, which had been solubilized with dodecyltrimethyl-ammonium bromide and was partially purified by protein A affinity chromatography. In the proteoliposomes sulfate uptake was more rapid than that of sodium or sucrose. The sulfate uptake was slowed down by addition of 2-(4-aminophenyl)-6-methylbenzene thiazo-3,7-disulfonic acid. However, in contrast to the intact system, no temperature dependence of transport activity was found.

The reconstitution experiments reported so far do not unequivocally prove functional reconstitution of the anion transporter, since increased permeability to sulfate or inhibition of anion transport by high inhibitor concentrations may be nonspecific. This was demonstrated recently by *van Hoogevest* et al. (1983) when they showed that reconstitution of partially purified band 3 not only increased the sulfate permeability but also the permeability for L-glucose. These authors reported further that not only sulfate uptake but also L-glucose uptake was inhibited by 100 μM DIDS. By reconstitution of glycophorin, sulfate permeability was increased; this increase was reduced in the presence of thiocyanate or DIDS. Furthermore *Kaplan* (1972) showed that anion permeability in liposomes was increased by the presence of lysozyme. When a detergent removal–freeze-thaw procedure was employed to reconstitute Triton X-100 extracted band 3, temperature dependence of anion transport was found. For sulfate equilibrium exchange an activation energy of 20 kcal/mol was estimated, which is lower than that in intact cells (*Wolosin* 1980). To achieve reconstitution, lipids were added to the solubilized protein, Triton X-100 was removed with Bio-Beads, and the mixture was frozen and thawed. The sulfate equilibrium exchange in the proteoliposomes could be 50% inhibited by DIDS and was reduced by univalent anions. Furthermore, it was found that the pH dependence of transport was similar to that in intact erythrocytes. When the cholesterol content in the proteoliposomes was reduced, the sulfate transport was increased significantly.

If band 3 was prepared and reconstituted according to *Wolosin* (1980) with the exception that the freezing and thawing steps were omitted (*Köhne* et al. 1981), a higher activation energy of sulfate exchange which was now similar to that in intact erythrocytes, was measured. However, in contrast to intact erythrocytes, the concentration dependence of sulfate exchange was linear up to a concentration of 80 mM sulfate, indicating that the K_D of sulfate equilibrium exchange was increased after reconstitution. Furthermore, in contrast to erythrocytes, where net sulfate uptake is slower than equilibrium exchange, in these proteoliposomes, sulfate trans zero efflux was about the same as sulfate equilibrium exchange. The data show that in these proteoliposomes the anion transport protein does still not work properly.

In 1981 *Lukacovic* et al. reported reconstitution of band 3 which had been purified to more than 95% homogeneity. For purification, band 3 was extracted with Triton X-100 and further purified by ion exchange chromatography and by chromatography on *p*-(chloromercuri)benzamidoethyl agarose gel. For reconstitution, Triton X-100 was removed, multilamellar liposomes were added, and the mixture was sonicated. In these proteoliposomes *trans*-stimulation of sulfate transport was demonstrated. Sulfate efflux was partially inhibited if 10 μM DIDS was added from the outside, and completely inhibited if 10 μM was added before reconstitution. A turnover number was estimated for the band 3 protein which was 20% of that in intact erythrocytes.

From studies on the lipid dependence of anion transport in intact erythrocyte membranes from different species, which contained different concentrations of phospholipids and cholesterol or were enriched in specific lipids, it was found that anion exchange was positively correlated with the phosphatidylcholine concentration and negatively correlated with the sphingomyelin and cholesterol concentration (*Gruber* and *Deuticke* 1973; *Lu* and *Chow* 1982; *Deuticke* et al. 1980; *Grunze* et al. 1980). In studies with monolayers, it was found that high affinity cholesterol binding to band 3 protein which had been purified in the presence of acetic acid was increased by sphingomyelin and decreased by phosphatidylcholine (*Schubert* and *Boss* 1982). Thus it may be speculated that anion transport may be reduced by cholesterol binding to a high affinity binding site. In contrast to the lipid effects observed in intact membranes, studies with reconstituted proteoliposomes demonstrated only the effect of sphingomyelin (*Köhne* et al. 1981, 1983). One reason for this failure may the fact that, before reconstitution, endogenous lipids were not completely removed from the anion-transport protein.

It has been shown above that some, but not all properties, of the anion transporter were preserved after reconstitution, and that functional integrity of the transport protein is highly dependent on the experimental conditions during the reconstitution procedure. Together with experiments on the purification of band 3 within the membrane, the reconstitution experiments have shown that band 3 is sufficient to perform anion transport. To obtain significant new information on the function of the anion transporter, the reconstitution procedure of this transporter will have to be improved further (*Scheuring* et al. 1984).

10 Reconstitution of Na^+-Coupled Transport Systems

In brush-border membranes from intestine and renal proximal tubules, many different Na^+ cotransport systems have been discovered. Properties of these transporters have been detected in studies on intact tissue or on brush-border membrane vesicles (for reviews see *Silbernagl* et al. 1975; *Ullrich* 1979; *Murer* and *Burckhardt* 1983). The Na^+ cotransport proteins represent only a small fraction of the total membrane proteins and are easily denatured so that the purification of these transport proteins is a difficult problem. In the past, identification and purification attempts have focussed on the Na^+ D-glucose cotransporter. Consequently, experiments on this transporter will be reported in detail, whereas the other Na^+ cotransport systems will be mentioned only briefly.

10.1 Cotransporter for Na^+ and D-Glucose

Many attempts have been made to identify the glucose binding site of the Na^+ D-glucose cotransporter. Thus, protecting effects of D-glucose and of the competitive inhibitor phlorizin on modifications of SH-group in brush-border membrane proteins have been studied (*Thomas* 1973; *Lemaire* and *Maestracci* 1978; *Poiree* et al. 1979). These experiments, in which polypeptides with molecular weights of 30 000, 51 000, and 60 000 were identified tentatively, are inconclusive. On the other hand, recent data provide increasing evidence that in intestinal brush-border membranes, a polypeptide with a molecular weight between 72 000 and 75 000 may contain a phlorizin-inhibitable D-glucose binding site of the Na^+ D-glucose cotransporter. In 1981 *Hosang* and co-workers reported experiments using the photoaffinity label, 4-azidophlorizin. They interpreted their results to suggest that a 72 000-dalton protein may be part of the Na^+ D-glucose cotransporter. The same group reported that a monoclonal antibody raised against intestinal brush-border membranes was able to inhibit phlorizin binding and D-glucose uptake in membrane vesicles and was capable of extracting a polypeptide with a molecular weight of 72 000 from solubilized brush-border membranes (*Schmidt* et al. 1983, 1984; *Semenza* et al. 1984). The significance of these data cannot be finally evaluated since important details on the inhibition tests and on the control experiments have not yet been reported. Recently, strong experimental evidence for the involvement of a 75 000-dalton polypeptide in Na^+ D-glucose cotransport in intestinal brush-border membranes has been presented by *Peerce* and *Wright* (1984a,b). These authors demonstrated that inhibition of D-glucose uptake by phenylisothiocyanate (PITC) or by fluorescein isothiocyanate (FITC) could be prevented by Na^+ plus D-glucose. The brush-border membranes were first treated with PITC in the presence of Na^+ and D-glucose. After excess PITC and Na^+ and D-glucose were removed, the membranes were incubated with FITC. 30% of the FITC binding sites were located on a polypeptide with an apparent molecular of 75 000. If D-glucose or phlorizin was present during the incubation with FITC, binding of FITC to this polypeptide was significantly reduced (*Peerce* and *Wright* 1984b). Employing the covalently binding D-glucose analogs

10-*N*-(bromoacetyl)amino-1-decyl-β-D-glucopyranoside and 10-*N*-(N-[4-azido-2-nitrophenyl]-β-alanyl)-amino-1-decyl-β-D-glucopyranoside to identify the D-glucose binding site of the Na^+ D-glucose cotransporter from kidney, which may be different from that of intestine (*Toggenburger* et al. 1982; *Semenza* et al. 1984), a labeling of polypeptides with molecular weights in the range of 65 000 and 85 000 and of about 46 000 was obtained. By addition of D-glucose or phlorizin, which is a competitive inhibitor of Na^+ D-glucose cotransport, the labeling of polypeptides with molecular weights of 74 000 and 82 000 was significantly protected (*Neeb* et al. 1985; *Koepsell* et al. 1986, to be published). Moreover the labeling of the 46 000 polypeptide was increased by Na^+ and D-glucose. The data suggest that in kidney 74 000 and 82 000 dalton polypeptides contain D-glucose binding sites and are components of Na^+-D-glucose cotransporter molecules which may originate from different parts of the nephron (*Turner* and *Moran* 1982). The increase of labeling of a 46 000 dalton polypeptide by Na^+ plus D-glucose can be explained by the assumption that the 46 000 polypeptide is a subunit of the Na^+-D-glucose cotransporter. Binding of D-glucose to one of the D-glucose binding polypeptides may lead to a conformational change of the 46 000 subunit and may alter the accessibility of reactive groups which are in contact with the D-glucose binding site. To confirm that the polypeptides which were identified in intestinal and renal brush-border membranes are actually involved in D-glucose transport, and to find out which of these polypeptides are sufficient to perform Na^+ D-glucose cotransport the Na^+ D-glucose cotransporter has to be purified in an active form and has to be reconstituted into proteoliposomes.

Experiments along these lines were first reported by *Crane* and coworkers (*Crane* et al. 1976a,b). These authors solubilized brush-border membranes from small intestines or from kidney cortex by Triton X-100, removed the detergent with Bio-Beads, added liposomes, and incorporated the proteins by sonication. In the proteoliposomes, the D-glucose uptake rate in the presence of an inwardly directed Na^+ gradient was higher than at K^+ equilibrium. This effect was stereospecific for D-glucose and could be inhibited by 5 m*M* phlorizin. *Crane* and collaborators improved the reconstitution procedure by optimizing the sonication time and by increasing the protein to lipid ratio during incorporation and were able to demonstrate Na^+ D-glucose cotransport in as much as they showed that due to the inwardly directed Na^+ gradient the D-glucose concentration in the liposomes increased transiently over the equilibrium value (*Fairclough* et al. 1979). The Na^+ gradient-dependent uptake rates in the liposomes were, however, very low compared with the uptake rates measured in brush-border membrane vesicles. It was shown that the proteoliposomes in this preparation were mainly monolamellar and had diameters between 10 and 100 nm.

In 1982 successful reconstitution of the Na^+ D-glucose cotransporter was also achieved when brush-border membranes were solubilized with cholate in the presence of Na^+, and the reconstitution was performed by detergent removal on a Sephadex G 50 column. The Na^+ gradient-dependent uptake rates were increased if the membrane proteins were fractionated before reconstitution. This was done by first removing extraneous proteins from the membrane with a low deoxycholate concentration. The membrane was then solubilized with a higher deoxycholate concentration, and finally proteins were removed with hydroxyapatite (*Im* et al. 1982). After these steps the Na^+ gradient-dependent uptake rates were much higher than in the experiments of *Crane* and co-workers, but they were still more than 10 times lower than those measured in intact membrane vesicles.

The reconstitution of the Na^+ cotransporter into proteoliposomes could be significantly improved when we optimized a freeze-thaw reconstitution procedure (*Koepsell* et al. 1983, 1984; *Ducis* and *Koepsell* 1983). In these experiments, purified brush-border membranes from renal proximal tubules were solubilized with deoxycholate in the presence of Na^+ and D-glucose. Then the D-glucose transporter was enriched by precipitating mixed micelles of protein, lipid, and detergent. This material was resolubilized with deoxycholate or octylglucoside and proteoliposomes were formed by detergent removal on a Sephadex G 50 column (*Koepsell* et al. 1983, 1984) or on a DEAE column (*Ducis* and *Koepsell* 1983). In these proteoliposomes, Na^+-dependent high affinity phlorizin binding but only minor Na^+ gradient-dependent D-glucose uptake could be measured.

Proteoliposomes with transport activity were formed by the following procedure: First liposomes which had been formed from either (a) a small amount (reconstitution procedure A), or (b) a large amount of cholesterol plus phosphatidylserine (reconstitution procedure B) were mixed with the proteoliposomes, which had been formed by detergent removal. In both procedures the mixture was heated to 41°C, frozen, and thawed. The thawed material, which consisted of protein–lipid aggregates and some proteoliposomes, was spun down at 150 000 × *g* and resuspended. Finally, lipid–protein aggregates and multilamellar proteoliposomes were removed by a 30-s centrifugation at 8000 × *g*. In procedure A, only some of the proteins which were added during reconstitution were found in the final proteoliposomes. In these proteoliposomes highly active Na^+ gradient-dependent D-glucose uptake was measured. The initial Na^+ gradient-dependent uptake rate was linear for several minutes and the D-glucose concentration in the proteoliposomes increased more than 20 times over the presumed D-glucose equilibrium value if an initial concentration difference of 89 m*M* Na^+ (out > in) was present (*Koepsell* et al. 1983). Under optimal conditions the initial Na^+ gradient-dependent D-glucose uptake rate of 0.1 m*M* D-glucose was as high as in intact membrane vesicles. The proteoliposomes were mainly monolamellar and had a mean diameter of about 0.5 μm. It was demonstrated that addition of cholesterol plus phosphatidylserine was obligatory to obtain transport-active proteoliposomes (*Ducis* and *Koepsell* 1983). The addition of cholesterol was necessary to make the proteoliposomes impermeable to Na^+ and glucose. After reconstitution the D-glucose transporter had the same properties as in intact membrane vesicles. Thus D-glucose uptake was driven by an Na^+ gradient and was stimulated by an inside negative membrane potential. At Na^+ equilibrium, the equilibrium exchange of 0.2 m*M* D-glucose (but not that of L-glucose) was more rapid than at K^+ equilibrium. The initial Na^+-dependent uptake rate of the D-glucose equilibrium exchange was about the same as in intact membrane vesicles. If phlorizin was added to the proteoliposomes only about 60% of the Na^+ gradient-dependent uptake rate was inhibited by 20 μM phlorizin, whereas in brush-border membrane vesicles, which are oriented right side out, the glucose uptake was inhibited 100%. These findings may be interpreted by the hypothesis that the D-glucose transporter is an asymmetric molecule which is bidirectionally oriented in the proteoliposomes, whereas it is unidirectionally oriented in membrane vesicles. Phlorizin may only bind to one side of the D-glucose transporter, whereas D-glucose may be transported in both directions (*Koepsell* and *Madrala* 1986, to be published). By measuring the concentration dependence of phlorizin inhibition of Na^+ D-glucose cotransport in proteoliposomes at different D-glucose concentrations, a competitive type of inhibition was demonstrated and a K_i of 0.3 μM was estimated. In the proteoliposomes, high affinity Na^+-dependent phlorizin binding to the outside of the vesicles could be measured (K_D = 0.5 μM). Since K_i for phlorizin inhibition of Na^+ gradient-dependent D-glucose uptake and K_D of Na^+ gradient-dependent phlorizin binding are about the same, it can be assumed that phlorizin binding and transport measurements are performed at the same protein.

The number of phlorizin binding sites on the outside of the proteoliposomes and V_{max} and K_m of the Na^+ gradient-dependent D-glucose uptake rates were different according to whether the proteoliposomes were formed by reconstitution procedure A or B (see Table 2). It can be seen that V_{max} of glucose uptake is twofold higher if proteoliposomes were formed by procedure A. The number of phlorizin binding sites is, however, about 7 times lower. Thus, in proteoliposomes formed by procedure A, the apparent turnover number of Na^+ gradient-dependent, phlorizin-inhibitable D-glucose uptake per phlorizin binding site is about the same as in intact membrane vesicles, whereas in proteoliposomes formed by procedure B it is 20 times lower. The data suggest that many more D-glucose transport molecules were associated with the proteoliposomes if procedure B was employed for reconstitution. After procedure B, only some of the transport molecules which were associated with the proteo-

Table 2. Correlation between D-glucose uptake and phlorizin binding in brush-border membrane vesicles and in proteoliposomes[a]

	D-Glucose uptake		Phlorizin binding		Turnover number
	V_{max} (nmol × mg^{-1} × s^{-1})	$K_{0.5}$ (mM)	n (nmol/mg)	K_D (μM)	(s^{-1})
Membrane vesicles	1.4	2.5	0.14	0.3	10.0
Proteoliposomes formed by procedure A	0.44	1.0	0.06	0.5	7.3
Proteoliposomes formed by procedure B	0.22	0.8	0.41	1.0	0.5

[a] Membrane vesicles and proteoliposomes were prepared and D-glucose uptake and phlorizin binding measured as described earlier (*Koepsell* et al. 1983, 1984). Initial D-glucose uptake rates at different D-glucose concentrations were measured in the presence of an initial concentration difference of 89 mM Na^+ (out > in) with and without 20 μM phlorizin present. In the table, V_{max} and $K_{0.5}$ values of the phlorizin inhibitable D-glucose uptake rates are indicated. Phlorizin binding at different phlorizin concentrations was measured at K^+ equilibrium or in the presence of an initial concentration difference of 89 mM Na^+ (out > in). The numbers of the Na^+-dependent high affinity phlorizin binding sites (n) and the dissociation constants (K_D) in the different samples are presented

liposomes were able to catalyze D-glucose transport, but after procedure A, most could do so. Data were obtained which show that Na^+ gradient-dependent uptake in proteoliposomes was not only dependent on the employed reconstitution procedure, but probably also on the association of the D-glucose transporter with other proteins and/or lipids. Thus, in different protein fractions, either similar or completely different D-glucose uptake rates were measured according to whether the protein was reconstituted by procedure A or B. With some protein fractions no uptake was obtained if procedure A was used, whereas high D-glucose uptake was found when reconstitution was performed by procedure B (see *Koepsell* et al. 1984). However, with other protein fractions the reverse was true. Our data demonstrate that the measurement of D-glucose uptake after reconstitution is not a reliable method of evaluating the concentration of transporter molecules.

This finding is not surprising for theoretical reasons: Thus, the D-glucose cotransport activity in proteoliposomes should be affected by general properties of the proteoliposomes, by properties of the D-glucose cotransport protein, and by the orientation of the transporter in the membrane. The proteoliposomes should be large and monolamellar, and should have a low passive permeability. Furthermore, the transporter must be present in a sufficiently high concentration, must be incorporated properly, and must be present in an active conformation. Since the experimental conditions during reconstitution affect the properties of the reconstituted proteoliposomes and of the transporter in a complex way, the success of reconstitution experiments cannot be predicted.

The measurement of Na^+-dependent high affinity phlorizin binding (which is only influenced by the concentration and accessibility of active transporter molecules) should be a more reliable method of quantifying the number of active transporter molecules; one may, therefore, wonder why phlorizin binding measurements were not employed in purification experiments. The reason is probably that most transporter molecules are normally inactivated during solubilization, purification, and reconstitution and that the phlorizin binding assay is probably orders of magnitude less sensitive than the assay of D-glucose uptake. Furthermore, we have obtained data indicating that the D-glucose transport protein may lose its phlorizin-binding property in the presence of detergents and that phlorizin binding is highly dependent on the lipid environment and/or on the conformational state of the transport protein. If phorizin binding was measured by the same method in intact vesicles, after solubilization of these vesicles with deoxycholate or after reconstitution of proteoliposomes from the solubilized material it was found that phlorizin binding could not be measured in the presence of deoxycholate but was recovered in the reconstituted proteoliposomes. Moreover, different numbers of phlorizin binding sites were measured if the same protein fraction was precipitated by detergent removal or by polyethylene glycol or if it was frozen with phosphatidylserine and cholesterol. These findings can be explained by the assumption that the Na^+ D-glucose cotransporter denatures during solubilization and may renature if the protein is precipitated or incorporated into lipid membranes. Possibly, for phlorizin binding not only the binding of the glucose moiety to the glucose binding site but also the binding of the phloretin moiety to a hydrophobic protein domain or to the lipid membrane is required. The latter assumption is supported by recent data demonstrating high affinity phlorizin binding to phosphatidylethanolamine monolayers (*Reyes* et al. 1983). To obtain a reliable method of quantifying the concentration of D-glucose transporter molecules during purification experiments, the active protein conformation has to be preserved or recovered. Experiments along these lines are currently being performed in our laboratory.

Inconclusive and contradictory data concerning purification of the Na^+ D-glucose cotransporter have been reported. The reasons for the contradictory results are probably (a) that the reported data are based on the assay of D-glucose uptake in proteoliposomes, (b) that in most of the experiments the assay of polypeptides after purification was insufficient to detect small polypeptide concentrations, and (c) that in some experiments protein digestion was employed.

By employing a phlorizin affinity column for purification of the renal Na^+ D-glucose cotransporter, *Lin* et al. (1981) claimed the enrichment of a polypeptide with a molecular weight between 60 000 and 70 000 in parallel with the enrichment of D-glucose transport activity after reconstitution. These data are not convincing (a) since the authors did not demonstrate that the solubilized protein retained phlorizin binding activity, (b) since the material released from their affinity column by the addition of 0.5 *M* D-glucose contained high concentrations of many other polypeptides, and (c) since the assay of the D-glucose uptake rate was invalid. They measured D-glucose uptake in the presence of a 150 m*M* NaSCN gradient as compared with that at KCl equilibrium and did not correct for the effect of SCN^- to increase the passive permeability of D-glucose (*Ducis* and *Koepsell* 1983). In 1983, *Malathi* and *Preiser* and *Kano-Kameyama* and *Hoshi* claimed the purification of a polypeptide with a molecular weight of 165 000 or 160 000 which dissociates after boiling in the presence of SDS into polypeptides with molecular weights of about 80 000. The data of these authors do not prove the involvement of these proteins in the D-glucose uptake, since in these experiments membrane proteins were digested with proteases before they were solubilized. Therefore, it is not clear whether transport in the proteoliposomes is catalyzed by these protease-resistant proteins or by other proteins which have been split and stick together in the proteoliposomes but appear as low molecular weight digested products in SDS polyacrylamide gels. Furthermore, in the proteoliposomes formed from the purified material, the initial D-glucose uptake rates related to protein were not higher than in intact membrane vesicles. Moreover, additional polypeptides may have been missed after gel electrophoresis, since the gels were stained with the relatively insensitive dye Coomassie Brilliant Blue.

When a protein fraction from renal brush-border membrane was reconstituted (*Koepsell* et al. 1983, 1984; reconstitution procedure A), we found that a protein with an apparent molecular weight of 52 000 was selectively associated with proteoliposomes in which an Na^+ gradient-dependent D-glucose uptake rate nearly as high as in intact membrane vesicles was measured. The Na^+-dependent high affinity phlorizin binding to the outside of these proteoliposomes was about 30% of that measured in membrane vesicles. As estimated from SDS polyacrylamide gels performed according to *Weber* and *Osborn* (1969) and stained with Coomassie Brilliant Blue, more than 80% of the protein in these proteoliposomes consisted of two polypeptides with apparent mean molecular weights of 94 000 and 52 000. Since the initial Na^+ gradient-dependent uptake rates increased in parallel with the incorporation of the 52 000-dalton polypeptide, the hypothesis was raised that this polypeptide may be part of the D-glucose transporter (*Koepsell* et al. 1983). Later experiments, however, revealed this hypothesis to be untenable. After improving the gel

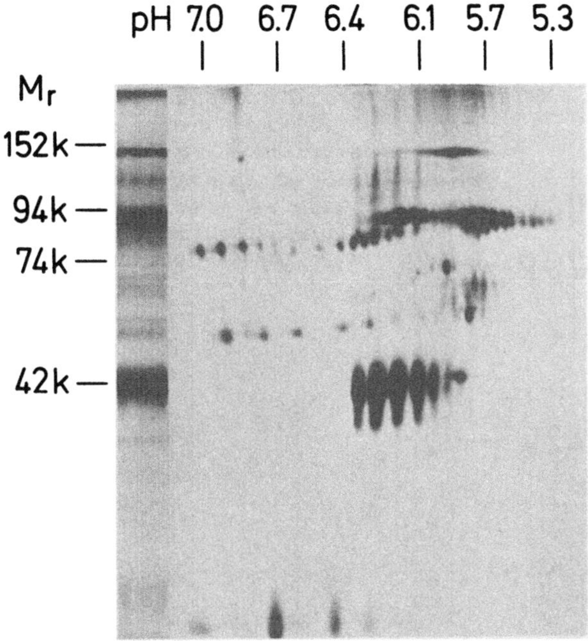

Fig. 3. Polypeptide pattern of proteoliposomes formed by freezing and thawing according to reconstitution procedure A analyzed by one- and two-dimensional gel electrophoresis. From proteoliposomes (formation see LipIIc in *Koepsell* et al. 1984) lipids were extracted by treatment with a mixture of chloroform and methanol. The one- and two-dimensional gel electrophoresis was performed and the gels were stained with silver as described by *Koepsell* et al. (1984)

system and after staining the proteins with silver, many other polypeptides were found in these proteoliposomes. It was found that the protein with the apparent mean molecular weight of 52 000 was migrating more rapidly if the lipids had been extracted before electrophoresis, so that in later experiments an apparent molecular weight between 38 000 and 46 000 was estimated. Furthermore, it could be shown with monoclonal antibodies that this band is composed at least of four polypeptides with different molecular weights and can be separated into five homologous polypeptides with different isoelectric points (*Koepsell* et al. 1984 and unpublished data). Moreover, after improving the solubilization conditions, proteoliposomes were formed which contained the same polypeptides and in which at least five Na^+ cotransport systems could be measured in addition to the Na^+ D-glucose cotransporter (*Koepsell* et al. 1984). In Fig. 3 the polypeptide composition of these proteoliposomes is analyzed by one- and two-dimensional gel electrophoresis.

Figure 4 shows an attempt to purify the Na^+ D-glucose cotransporter in which phlorizin binding measurements were used to quantify the concentration of active transport proteins (unpublished). Brush-border membranes were solubilized with deoxycholate, and mixed micelles of deoxycholate, lipid, and protein were formed by reducing the osmolarity of the buffer. High molecular weight proteins were removed from the mixed micelles by incubating them with 0.2% deoxycholate. To remove further proteins, the mixed micelles were then incubated at pH 12 and spun down.

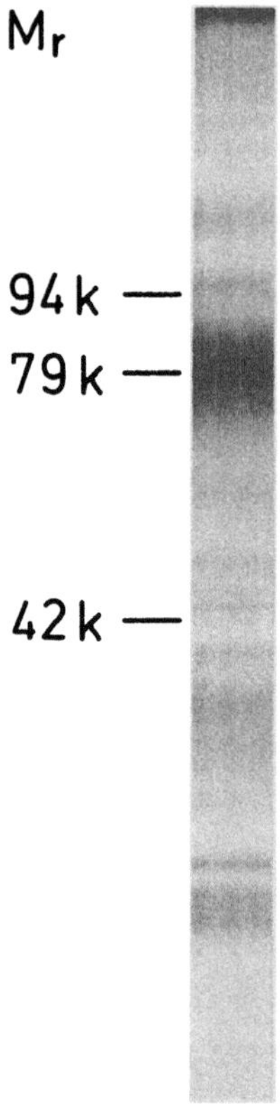

Fig. 4. Polypeptide composition of a protein fraction in which the specific phlorizin binding was about 8 times increased over that measured in intact brush-border membranes. To form this protein fraction mixed micelles were precipitated from solubilized brush-border membranes. Proteins were then removed from the mixed micelles by incubating them with 0.2% deoxycholate and with an alkaline buffer (pH 12). For gel electrophoresis the sample was solubilized with SDS in the presence of urea and mercaptoethanol. The gel electrophoresis was performed and the gels were stained with silver as described by *Koepsell* et al. (1984)

From SDS polyacrylamide gels it can be estimated that the final material consisted to more than 50% of polypeptides with molecular weights between 72 000 and 84 000 (Fig. 4). To renature the Na^+ D-glucose cotransporter the sample was solubilized with octylglucoside and precipitated with polyethylene glycol after addition of phosphatidylserine. By this treatment the number of phlorizin binding sites related to protein was increased threefold to a value of 0.75 nmol × mg protein^{-1}. This value is about 8 times higher than that measured in intact renal brush-border membrane vesicles and about 80 times higher than that measured in brush-border membrane vesicles from intestine (*Tannenbaum* et al. 1977). In this purification procedure, 7% of the phlorizin binding sites from the brush-border membranes which were used for the experiment were recovered.

Some further purification may be possible by employing additional conventional purification methods. Since, however, these methods are usually accompanied by a considerable loss of transport protein, a specific protein separation method such as affinity chromatography with immobilized antibodies may be necessary to achieve the ultimate purification of the transporter.

Table 3. Na^+, K^+ gradient-dependent uptake rates of different Na^+ cotransport systems in proteoliposomes[a]

Transported substances	Uptake rates (pmoles × mg protein^{-1} × s^{-1})
D-Glucose (0.1 m*M*)	51.9 ± 0.1
L-Lactate (0.1 m*M*)	22.3 ± 0.9
DL-Methylsuccinate (0.1 m*M*)	17.0 ± 0.6
L-Alanine (0.1 m*M*)	8.8 ± 0.6
L-Glutamate (25 μ*M*)	31.1 ± 0.9
L-Proline (0.1 m*M*)	3.2 ± 0.1
Glycine (0.1 m*M*)	1.1 ± 0.1

[a] From membranes, which were solubilized with 0.4% deoxycholate in the presence of a high protein concentration, proteoliposomes were formed as recently described (see LipIIc in *Koepsell* et al. 1984). Initial uptake rates were measured at K^+ equilibrium and in the presence of an initial concentration difference of 89 m*M* Na^+ (out > in) and of 89 m*M* K^+ (in > out). In the table the uptake rates stimulated by the inwardly directed Na^+ plus the outwardly directed K^+ gradient over those measured at K^+ equilibrium are presented. The substrate concentrations in the incubaion media are given in parentheses

10.2 Other Na^+ Cotransport Systems

Although reconstitution experiments with solubilized proteins from renal and intestinal membranes have been performed for many years and some attempts to measure L-alanine and phosphate transport in reconstituted proteoliposomes have been reported (*Kinne* and *Faust* 1977), it was only recently that successful reconstitution of Na^+ cotransport proteins other than the D-glucose transporter was demonstrated (*Koepsell* et al. 1984). The successful reconstitution in these experiments was supposedly possible because our reconstitution procedure is superior to those employed by others. Furthermore, we found conditions under which the cotransporters for Na^+ and amino acids or carboxylic acids, which seem to be even more sensitive towards denaturation than the D-glucose transporter, could be solubilized without inactivation. From Table 3 it can be seen that several different Na^+ cotransport activities could be reconstituted if a high concentration of brush-border membranes (10 mg protein/ml) was solubilized with 0.4% deoxycholate. On the other hand, the Na^+ D-glucose cotransport could be reconstituted also if a low concentration of brush-border membrane proteins (1.5 mg protein/ml) was solubilized with 0.2% deoxycholate (*Koepsell* et al. 1984). After solubilization in the presence of a high protein concentration, significant Na^+ gradient-dependent uptake was obtained for a mono- and a dicarboxylic acid, for different neutral amino acids, and for the acidic amino acid L-glutamate. In inhibition experiments it could be shown that monocarboxylic acids, dicarboxylic acids, acidic amino acids, neutral amino acids, and D-glucose are transported by different systems. Furthermore, the inhibition experiments suggest that the neutral amino acids are transported by three different systems. One may be able to transport several neutral amino acids, among others L-alanine, L-leucine, and glycine. The second may accept L-proline and glycine with an apparent K_m between 1 and 10 mM, and the third may be a low affinity transport system for L-proline and glycine with $K_m > 10$ mM. Recently *Wright* and *Peerce* (1984) identified a 100 000-dalton polypeptide as component of the Na^+ proline cotransporter from rabbit intestinal brush-border membranes.

In the proteoliposomes, cotransport of Na^+ with each of the substrates shown in Table 3 is clearly evident since the respective substrate concentration increased over the presumed equilibrium value measured after long incubation times in the presence of an inwardly directed Na^+ gradient (*Koepsell* et al. 1984). In the case of the L-glutamate cotransport system, transport into proteoliposomes was investigated in detail. It could be demonstrated that all properties of this transport system which have been described from measurements in intact membrane vesicles could be also shown in the proteoliposomes. From Fig. 5 it can be seen that the L-glutamate concentration in the proteoliposomes increased over the equilibrium concentration if in the presence of K^+ an inwardly directed Na^+ gradient was applied, or if in the presence of Na^+ an outwardly directed K^+ gradient was applied. These findings suggest that the L-glutamate uptake is driven by both an inwardly directed Na^+ gradient and an outwardly directed gradient for K^+. Proteoliposomes, in contrast to intact brush-border membrane vesicles, are very impermeable for cations and anions. It was therefore possible to demonstrate that uptake of the potassium analogue rubidium can be stimulated by an outwardly directed gradient of L-glutamate (*Koepsell* et al. 1984). In this experiment countertransport of rubidium and L-glutamate was directly demonstrated for the first time. In the proteoliposomes, the L-glutamate uptake rate was linear for several minutes. It could be shown that at pH 7.4 L-glutamate uptake driven by an inwardly directed Na^+ gradient plus an outwardly directed gradient of K^+ was stimulated by an inside negative membrane potential. These experiments show that during cotransport of L-glutamate with Na^+ and countertransport with K^+ a positive charge is transferred. Thus a minimum of three Na^+ ions are cotransported with L-glutamate and one K^+ ion is countertransported. These data demonstrate that some properties of the L-glutamate transporter can be detected better in proteoliposomes than in intact membrane vesicles.

Recently, *Jacobsen* et al. (1984) have isolated a dicarboxylic acid binding protein with an apparent molecular weight of 50 000 using affinity

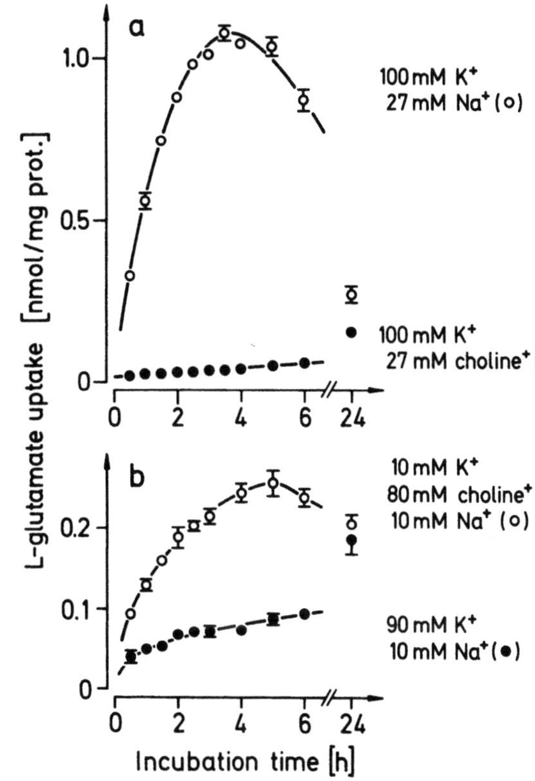

Fig. 5a,b. L-glutamate uptake into proteoliposomes in the presence of (a) an inwardly directed Na^+ gradient ($K_i^+ = K_o^+$) and (b) an outwardly directed gradient of K^+ ($Na_i^+ = Na_o^+$). The proteoliposomes, which were prepared as described by *Koepsell* et al. (1984) (see LipIIc), contained **a** 100 m*M* potassium cyclamate or **b** 90 m*M* potassium cyclamate plus 10 m*M* sodium cyclamate. For transport measurements (see *Koepsell* et al. 1984) the proteoliposomes were incubated with buffers containing 6 *μM* radioactively labeled L-glutamate and the cations indicated in the figure. The anion in the buffers was cyclamate. After different time intervals samples were taken and diluted in ice cold incubation buffers which did not contain L-glutamate, and the radioactivity in the proteoliposomes was measured

chromatography on an L-aspartate–Sepharose column. After treatment with mercaptoethanol, this protein could be split into two subunits with molecular weights of 15 000 and 35 000. Reconstitution experiments are necessary to find out whether this protein is identical with the Na^+-dicarboxylic acid cotransporter.

11 Concluding Remarks

There are several reasons why purification and reconstitution of transport proteins is necessary and cannot be replaced by other methods: The ultimate identification of a transport system can only be performed after purification and reconstitution since it may be composed of more than one protein. It may consist of different polypeptide subunits, may be a protein dimer or oligomer, or may form a functional unit with other proteins. A binding and a transport protein may, for example, work together, as has been described for the glutamine transport system in *E. coli* (*Masters* and *Hong* 1981). Furthermore, the molecular mechanism of transport can only be established after purification and reconstitution since only after purification of active transporters are the following possible: (a) protein digestion experiments, (b) physical studies on the protein, (c) ultrastructural studies, (d) crystallization experiments, and (e) studies on the lipid dependency of the transporters. The interaction of transport proteins with other proteins also has to be studied after purification and reconstitution. Detailed kinetic investigations can be performed with high accuracy and time resolution if rheogenic transport proteins are reconstituted into planar lipid membranes. Unfortunately, reconstitution in planar lipid membranes, which has been demonstrated for ion channels and bacteriorhodopsin (*Miller* 1978; *Schindler* and *Rosenbusch* 1978; *Nelson* et al. 1980; *Schindler* and *Quast* 1980; *Boheim* et al. 1981; *Anholt* 1981; *Latorre* et al. 1982; *Coronado* and *Latorre* 1982; *Darszon* 1983; *Montal* et al. 1984), has not been achieved satisfactorily for transport proteins of mammalian plasma membranes (*Blumenthal* and *Shamoo* 1979; *Hokin* 1981; *Nickson* and *Jones* 1982; *Last* et al. 1983; *Anner* 1985). The purification of transport proteins and their reconstitution into proteoliposomes can be considered an indispensable first step in achieving the reconstitution into planar membranes. Recently, methods have been developed by which transport proteins reconstituted into proteoliposomes can be employed for electrical measurements. Thus, in large proteoliposomes transport proteins may in future be investigated by the patch clamp technique which has been introduced to investigate ion channels in intact membranes (*Neher* and *Sakmann* 1976; *Tank* et al. 1982; *Suarez-Isla* et al. 1983). Furthermore, it has been shown that membrane fragments or proteoliposomes containing transport proteins can be attached to planar lipid membranes and electrical measurements of rheogenic transport processes can be performed (*Bamberg* et al. 1979, 1984; *Fahr* et al. 1981).

To date no general method of solubilizing, purifying, and reconstituting transport proteins from plasma membranes is available. A general procedure to dissociate hydrophobic membrane proteins in an active state is not known. The experimental conditions for solubilization have to be optimized for each individual protein. In so doing, some rules, which have been outlined above, should be followed. Thus to optimize solubilization, different detergents should be tried. With each, a minimal number of different experimental conditions have to be tested. These conditions include: the concentration of detergent, lipid, and protein, the buffer, the pH, and the presence of substrates. Purification of the transport proteins may be relatively easy if the transporters are present in high concentrations. In such cases, relatively ineffective purification procedures may lead to success and the transport proteins may stabilize each other during purification or may introduce enough essential lipids to prevent inactivation. If, however, a transporter is present only in small concentrations, the essential lipids have to be identified and added. For final purification, a specific purification procedure by which the transporter is drastically enriched in one step may be required. This may create problems since (a) affinity columns with covalently bound ligands can only be applied in some favorable cases, and (b) the affinity purification with antibodies has to be developed much further to be applied successfully.

Several methods for reconstitution have been developed. The different procedures yield proteoliposomes with different characteristics. Furthermore, experimental procedures have been established to increase the size of the proteoliposomes and to make them impermeable. In several cases some of the procedures to produce proteoliposomes, such as the cholate dialysis procedure and the freeze-thaw sonication procedure, have been applied with more success than others. From the reported reconstitution examples we learn, however, that a reconstitution procedure will only give good results if it has been optimized for the specific transport protein under investigation. Thus, if one reconstitution procedure leads to better results than another it may be because the experimental conditions have been optimized more thoroughly. Above, many examples have been reported where a reconstituted transport protein only shows some of the properties of the intact system. Reconstituted proteins should, therefore, be used only for extended investigations if they show the same kinetic characteristics as the intact system.

Above I have pointed out that one difficulty in purification experiments is due to the fact that the measurement of transport rates in reconstituted protein fractions is not a reliable method of quantifying the number of active transport proteins in a preparation. For quantification, binding measurements are supposedly superior, provided a specific high affinity ligand is available and the binding assay can be optimized sufficiently.

The experience gained so far gives the impression that for identification of transport proteins, purification and reconstitution may not be the best method and that the identification may be more quickly achieved by affinity labeling experiments. We should not, however, forget that such types of experiment were often unsuccessful and have in many cases led to conflicting results. Since in labeling experiments binding proteins and not complete transporter molecules are identified, purification and reconstitution have to be performed even if a transport protein has been identified by affinity labeling.

Although investigators performing reconstitution experiments may have lost much of their initial euphoria as the limitations and inherent difficulties have become apparent, one should keep in mind that purification and reconstitution is an obligatory step in understanding the molecular basis of transport. I hope that the reader of this review has gained awareness of the methodological difficulties but has also been convinced of the necessity of purification experiments, and may finally share my confidence that, given the present state of knowledge, purification and reconstitution of all transport proteins will be achieved if sufficient patience and effort are employed.

References

Acevedo F, Lundahl P, Fröman G (1981) The stereospecific D-glucose transport protein in cholate extracts of human erythocyte membranes. Molecular sieve chromatography and estimation of molecular weight. Biochim Biophys Acta 648:254–262

Albertsson P-Å (1973) Application of the phase partition method to a hydrophobic membrane protein, phospholipase A1 from *Escherichia coli.* Biochemistry 12: 2525–2530

Allard WJ, Lienhard GE (1985) Monoclonal antibodies to the glucose transporter from human erythrocytes. Identification of the transporter as M_r = 55,000 protein. J Biol Chem 260:8668–8675

Altendorf K, Lukas M, Kohl B, Müller CR, Sanderman H Jr (1977) Isolation and purification of bacterial membrane proteins by the use of organic solvents: The lactose permease and the carbodiimide-reactive protein of the adenosinetriphosphatase complex of *Escherichia coli.* J Supramol Struct 6:229–238

Amanuma H, Motojima K, Yamaguchi A, Anraku Y (1977) Solubilization of a functionally active proline carrier from membranes of *Escherichia coli* with an organic solvent. Biochem Biophys Res Commun 74:366–373

Andersen JP, Le Maire M, Kragh-Hansen U, Champeil P, Møller JV (1983) Perturbation of the structure and function of a membranous Ca^{2+}-ATPase by non-solubilizing concentrations of a non-ionic detergent. Eur J Biochem 134:205–214

Anholt R (1981) Reconstitution of acetylcholine receptors in model membranes. TIBS (Trends in Biological Sciences) 6:288–291

Anner BM (1981) A K-selective cation channel formed by Na,K-ATP-ase in liposomes. Biochem Int 2:365–371

Anner BM (1983) A bar model for the pump and channel function of the reconstituted Na^+, K^+-ATPase. FEBS Lett 158:7–11

Anner BM (1985) Interaction of Na^++K^+)-ATPase with artificial membranes. II. Expression of partial transport reactions. Biochim Biophys Acta 822:335–353

Anner BM, Lane LK, Schwartz A, Pitts BJR (1977) A reconstituted Na^++ K^+ pump in liposomes containing purified (Na^++K^+)-ATPase from kidney medulla. Biochim Biophys Acta 467:340–345

Anner BM, Robertson JD, Ting-Beall HP (1984a) Characterization of (Na^++K^+)-ATPase liposomes. I. Effect on enzyme concentration and modification on liposome size, intramembrane particle formation and Na^+,K^+-transport. Biochim Biophys Acta 773:253–261

Anner BM, Ting-Beall HP, Robertson JD (1984b) Characterization of (Na^++K^+)-ATPase liposomes. II. Effect of α-subunit digestion on intramembrane particle formation and Na^+,K^+-transport. Biochim Biophys Acta 773:262–270

Aronson PS (1978) Energy-dependence of phorizin binding to isolated renal microvillus membranes. Evidence concerning the mechanism of coupling between the electrochemical Na^+ gradient and sugar transport. J Membr Biol 42:81–98

Avruch J, Carter JR, Martin DB (1972) The effect of insulin on the metabolism of adipose tissue. In: Steiner DF, Freinkel N (eds) Handbook of physiology, sect 7, vol 1. American Physiological Society, Washington, pp 545–562

Baker PF (1965) Phosphorus metabolism of intact crab nerve and its relation to the active transport of ions. J Physiol (Lond) 180:383–423
Baldwin JM, Lienhard GE, Baldwin SA (1980) The monosaccharide transport system of the human erythrocyte. Orientation upon reconstitution. Biochim Biophys Acta 599:699–714
Baldwin JM, Gorga JC, Lienhard GE (1981) The monosaccharide transporter of the human erythrocyte. Transport activity upon reconstitution. J Biol Chem 256: 3685–3689
Baldwin SA, Lienhard GE (1980) Immunological identification of the human erythrocyte monosaccharide transporter. Biochem Biophys Res Commun 94:1401–1408
Baldwin SA, Baldwin JM, Gorga FR, Lienhard GE (1979) Purification of the cytochalasin B binding component of the human erythrocyte monosaccharide transport system. Biochim Biophys Acta 552:183–188
Baldwin SA, Baldwin JM, Lienhard GE (1982) Monosaccharide transporter of the human erythrocyte. Characterization of an improved preparation. Biochemistry 21:3836–3842
Bamberg E, Apell H-J, Dencher NA, Sperling W, Stieve H, Läuger P (1979) Photocurrents generated by bacteriorhodopsin on planar bilayer membranes. Biophys Struct Mech 5:277–292
Bamberg E, Hegemann P, Oesterhelt D (1984) The chromoprotein of halorhodopsin is the light-driven electrogenic chloride pump in *Halobacterium halobium*. Biochemistry 23:6216–6221
Bangham JA, Lea EJA (1978) The interaction of detergents with bilayer lipid membranes. Biochim Biophys Acta 511:388–396
Baron C, Thompson TE (1975) Solubilization of bacterial membrane proteins using alkyl glucosides and dioctanoyl phosphatidylcholine. Biochim Biophys Acta 382: 276–285
Batt ER, Abbott RE, Schachter D (1976) Impermeant maleimides. Identification of an exofacial component of the human erythrocyte hexose transport mechanism. J Biol Chem 251:7184–7190
Becker R, Helenius A, Simons K (1975) Solubilization of the Semliki Forest virus membrane with sodium dodecyl sulfate. Biochemistry 14:1835–1841
Berden JA, Barker RW, Radda GK (1975) NMR studies on phospholipid bilayers. Some factors affecting lipid distribution. Biochim Biophys Acta 375:186–208
Blumenthal R, Shamoo AE (1979) Incorporation of transport molecules into black lipid membranes. In: O'Brien RD (ed) The receptors, vol 1. Plenum, New York, pp 215–245
Boheim G, Hanke W, Barrantes FJ, Eibl H, Sakman B, Fels G, Maelicke A (1981) Agonist-activated ionic channels in acetylcholine receptor reconstituted into planar lipid bilayers. Proc Natl Acad Sci USA 78:3586–3590
Boodhoo A, Reithmeier RAF (1984) Characterization of matrix-bound band 3, the anion transport protein from human erythrocyte membranes. J Biol Chem 259: 785–790
Bordier C (1981) Phase separation of integral membrane proteins in Triton X-114 solution. J Biol Chem 256:1604–1607
Brotherus JR, Jost PC, Griffith OH, Keana JFW, Hokin LE (1980) Charge selectivity at the lipid-protein interface of membranous Na,K-ATPase. Proc Natl Acad Sci USA 77:272–276
Bruni Y, van Dijck PWM, de Gier J (1975) The role of phospholipid acyl chains in the activation of mitochondrial ATPase complex. Biochim Biophys Acta 406:315–328
Brunner J, Skrabal P, Hauser H (1976) Single bilayer vesicles prepared without sonication. Physico-chemical properties. Biochim Biophys Acta 455:322–331
Brunner J, Hauser H, Semenza G (1978) Single bilayer lipid-protein vesicles formed from phosphatidylcholine and small intestinal sucrase isomaltase. J Biol Chem 253:7538–7546
Burkhard RK, Stolzenberg GE (1972) Interaction between sodium dodecyl sulfate and ferricytochrome c. Biochemistry 11:1672–1677

Cabantchik ZI, Rothstein A (1972) The nature of the membrane sites controlling anion permeability of human red blood cells as determined by studies with disulfonic stilbene derivatives. J Membr Biol 10:311–330
Cabantchik ZI, Rothstein A (1974) Membrane proteins related to anion permeability of human red blood cells. I. Localization of dilsulfonic stilbene binding sites in proteins involved in permeation. J Membr Biol 15:207–226
Cabantchik ZI, Balshin M, Breuer W, Rothstein A (1975) Pyridoxal phosphate. An anionic probe for protein amino groups exposed on the outer and inner surfaces of intact human red blood cells. J Biol Chem 250:5130–5136
Cabantchik ZI, Wolosin JM, Ginsburg H, Zemel O (1977) Structural and functional properties of the anion transport system isolated from human erythrocyte membranes. FEBS Symp 42:328–345
Carafoli E, Zurini M (1982) The Ca^{2+}-pumping ATPase of plasma membranes. Purification, reconstitution and properties. Biochim Biophys Acta 683:279–301
Carey MC, Small DM (1972) Micelle formation by bile salts. Physical-chemical and thermodynamic considerations. Arch Intern Med 130:506–527
Caroni P, Carafoli E (1981) The Ca^{2+}-pumping ATPase of heart sarcolemma. Characterization, calmodulin dependence, and partial purification. J Biol Chem 256: 3263–3270
Carson SD, Konigsberg WH (1981) Phenyl-sepharose chromatography of membrane proteins solubilized in Triton X-100. Anal Biochem 116:398–401
Carter-Su C, Czech MP (1980) Reconstitution of D-glucose transport activity from cytoplasmic membranes. Evidence against recruitment of cytoplasmic membrane transporters into the plasma membrane as the sole action of insulin. J Biol Chem 255:10382–10386
Carter-Su C, Pillion DJ, Czech MP (1980) Reconstituted D-glucose transport from the adipocyte plasma membrane. Chromatographic resolution of transport activity from membrane glycoproteins using immobilized concanavalin A. Biochemistry 19: 2374–2385
Carter-Su C, Pessin JE, Mora R, Gitomer W, Czech MP (1982) Photoaffinity labeling of the human erythrocyte D-glucose transporter. J Biol Chem 257:5419–5425
Charnock JS, Bashford CL (1975) A fluorescent probe study of the lipid mobility of membranes containing sodium- and potassium-dependent adenosine triphosphatase. Mol Pharmacol 11:766–774
Chiesi M, Peterson SW, Acuto O (1978) Reconstitution of a Ca^{2+}-transporting ATPase system from Triton X-100-solubilized sarcoplasmic reticulum. Arch Biochem Biophys 189:132–136
Christiansen K, Carlsen J (1983) Reconstitution of a protein into lipid vesicles using natural detergents. Biochim Biophys Acta 735:225–233
Clarke S (1981) Direct renaturation of the dodecyl sulfate complexes of proteins with Triton X-100. Biochim Biophys Acta 670:195–202
Colowick SP, Womack FC (1969) Binding of diffusible molecules by macromolecules: Rapid measurement by rate of dialysis. J Biol Chem 244:774–777
Corda D, Pasternak C, Shinitzky M (1982) Increase in lipid microviscosity of unilamellar vesicles upon the creation of transmembrane potential. J Membr Biol 65: 235–242
Cornelius F, Skou JC (1984) Reconstitution of $(Na^{+}+K^{+})$-ATPase into phospholipid vesicles with full recovery of its specific activity. Biochim Biophys Acta 772: 357–373
Coronado R, Latorre R (1982) Detection of K^{+} and Cl^{-} channels from calf cardiac sarcolemma in planar lipid bilayer membranes. Nature 298:849–852
Crane LJ, Lampen JO (1974) Bacillus licheniformis 749/C plasma membrane penicillinase, a hydrophobic polar protein. Arch Biochem Biophys 160:655–666
Crane RK, Malathi P, Preiser H (1976a) Reconstitution of specific Na^{+}-dependent D-glucose transport in liposomes by Triton X-100-extracted proteins from purified brush border membranes of rabbit kidney cortex. FEBS Lett 67:214–216

Crane RK, Malathi P, Preiser H (1976b) Reconstitution of specific Na^+-dependent D-glucose transport in liposomes by Triton X-100 extracted proteins from purified brush border membranes of hamster small intestine. Biochem Biophys Res Commun 71:1010–1016

Cresswell P (1979) Deoxycholic acid-coupled poly(L-lysyl) agarose. An amphipathic matrix with binding affinity for integral membrane proteins. J Biol Chem 254: 414–419

Cushman SW, Wardzala LJ (1980) Potential mechanism of insulin action on glucose transport in the isolated rat adipose cell. Apparent translocation of intracellular transport systems to the plasma membrane. J Biol Chem 255:4758–4762

Czech MP (1976) Regulation of the D-glucose transport system in isolated fat cells. Mol Cell Biochem 11:51–63

Da Cruz MEM, Kinne R, Lin JT (1983) Temperature dependence of D-glucose transport in reconstituted liposomes. Biochim Biophys Acta 732:691–698

Dahl JL, Hokin LE (1974) The sodium-potassium adenosinetriphosphatase. Annu Rev Biochem 43:327–356

Darmon A, Zangvill M, Cabantchik ZI (1983) New approaches for the reconstitution and functional assay of membrane transport proteins. Application to the anion transporter of human erythrocytes. Biochim Biophys Acta 727:77–88

Darszon A (1983) Strategies in the reassembly of membrane proteins into lipid bilayer systems and their functional assay. J Bioenerg Biomembr 15:321–334

Darszon A, Vandenberg CA, Ellisman MH, Montal M (1979) Incorporation of membrane proteins into large single bilayer vesicles. Application to rhodopsin. J Cell Biol 81:446–452

Deamer D, Bangham AD (1976) Large volume liposomes by an ether vaporization method. Biochim Biophys Acta 443:629–634

Dean WL, Gray RD (1982) Relationship between state of aggregation and catalytic activity for cytochrome $P\text{-}450_{LM2}$ and NADPH-cytochrome P-450 reductase. J Biol Chem 257:14679–14685

Dean WL, Tanford C (1977) Reactivation of lipid-depleted Ca^{2+}-ATPase by a nonionic detergent. J Biol Chem 252:3551–3553

De Gier J, Mandersloot JG, van Deenen LLM (1968) Lipid composition and permeability of liposomes. Biochim Biophys Acta 150:666–675

De Gier J, Mandersloot JG, van Deenen LLM (1969) The role of cholesterol in lipid membranes. Biochim Biophys Acta 173:143–145

De Kruyff B, Demel RA, van Deenen LLM (1972) The effect of cholesterol and epicholesterol incorporation on the permeability and on the phase transition of intact *Acholeplasma laidlawii* cell membranes and derived liposomes. Biochim Biophys Acta 255:331–347

De Kruyff B, de Greef WJ, van Eyk RVW, Demel RA, van Deenen LLM (1973) The effect of different fatty acid and sterol composition on the erythritol flux through the cell membrane of *Acholeplasma laidlawii*. Biochim Biophys Acta 298:479–499

De Kruyff B, van Dijck PWM, Demel RA, Schuijff A, Brants F, van Deenen LLM (1974) Non-random distribution of cholesterol in phosphatidylcholine bilayers. Biochim Biophys Acta 356:1–7

Demel RA, de Kruyff B (1976) The function of sterols in membranes. Biochim Biophys Acta 457:109–132

Demel RA, Bruckdorfer KR, van Deenen LLM (1972) The effect of sterol structure on the permeability of lipomes to glucose, glycerol and Rb^+. Biochim Biophys Acta 255:321–330

Dennis EA (1974) Formation and characterization of mixed micelles of the nonionic surfactant Triton X-100 with egg, dipalmitoyl, and dimyristoyl phosphatidylcholines. Arch Biochem Biophys 165:764–773

De Pinto V, Tommasino M, Palmieri F, Kadenbach B (1982) Purification of the active mitochondrial phosphate carrier by affinity chromatography with an organomercurial agarose column. FEBS Lett 148:103–106

Deuticke B, Gunze M, Haest CWM (1980) Influence of membrane lipids on ion and non electrolyte transport through the erythrocyte membrane. In: Lassen UV, Ussing HH, Wieth JO (eds) Membrane transport in erythrocytes. Munksgaard, Copenhagen, pp 143–156

Dixon JF, Hokin LE (1980) The reconstituted (Na,K)-ATPase is electrogenic. J Biol Chem 255:10681–10686

Ducis I, Koepsell H (1983) A simple liposomal system to reconstitute and assay highly efficient Na^+/D-glucose cotransport from kidney brush-border membranes. Biochim Biophys Acta 730:119–129

Duppel W, Dahl G (1976) Effect of phase transition on the distribution of membrane-associated particles in microsomes. Biochim Biophys Acta 426:408–417

Dwiggins CW Jr, Bolen RJ, Dunning HN (1960) Ultracentrifugal determination of the micellar character of non-ionic detergent solutions. J Phys Chem 64:1175–1178

Edwards PAW (1977) Reconstitution of glucose-transporting vesicles from erythrocyte membranes disaggregated in detergent. Biochem J 164:125–129

Engelman DM, Terry TM, Morowitz HJ (1967) Characterization of the plasma membrane of *Mycoplasma laidlawii*. I. Sodium dodecyl sulfate solubilization. Biochim Biophys Acta 135:381–390

Evans E, Kwok R (1982) Mechanial calorimetry of large dimyristoylphosphatidylcholine vesicles in the phase transition region. Biochemistry 21:4874–4879

Eytan GD (1982) Use of liposomes for reconstitution of biological functions. Biochim Biophys Acta 694:185–202

Eytan GD, Racker E (1977) Selective incorporation of membrane proteins into proteoliposomes of different compositions. J Biol Chem 252:3208–3213

Eytan GD, Matheson MJ, Racker E (1975) Incorporation of biologically active proteins into liposomes. FEBS Lett 57:121–125

Eytan GD, Matheson MJ, Racker E (1976) Incorporation of mitochondrial membrane proteins into liposomes containing acidic phospholipids. J Biol Chem 251: 6831–6837

Ezaki O, Kono T (1982) Effects of temperature on basal and insulin-stimulated glucose transport activities in fat cells. Further support for the translocation hypothesis of insulin action. J Biol Chem 257:14306–14310

Fahr A, Läuger P, Bamberg E (1981) Photocurrent kinetics of purple-membrane sheets bound to planar bilayer membranes. J Membr Biol 60:51–62

Fairbanks G, Steck TL, Wallach DFH (1971) Electrophoretic analysis of the major polypeptides of the human erythrocyte membrane. Biochemistry 10:2606–2617

Fairclough P, Malathi P, Preiser H, Crane RK (1979) Reconstitution into liposomes of glucose active transport from the rabbit renal proximal tubule. Characteristics of the system. Biochim Biophys Acta 553:295–306

Fenster LJ, Copenhaver JH Jr (1967) Phosphatidyl serine requirement of (Na^+-K^+)-activated adenosine triphosphatase from rat kidney and brain. Biochim Biophys Acta 137:406–408

Findlay JBC (1974) The receptor proteins for concanavalin A and lens culinaris phytohemagglutinin in the membrane of the human erythrocyte. J Biol Chem 249:4398–4403

Finkelstein A, Cass A (1967) Effect of cholesterol on the water permeability of thin lipid membranes. Nature 216:717–718

Forgac M, Chin G (1981) K^+-independent active transport of Na^+ by the (Na^+ and K^+)-stimulated adenosine triphosphatase. J Biol Chem 256:3645–3646

Forgac M, Chin G (1982) Na^+ transport by the (Na^+)-stimulated adenosine triphosphatase. J Biol Chem 257:5652–5655

Freytag JW (1983) The (Na^+,K^+)ATPase exhibits enzymic activity in the absence of the glycoprotein subunit. FEBS Lett 159:280–284

Fröman G (1982) Hydroxyapatite chromatography of the D-glucose transport protein of human erythrocyte membranes. FEBS Lett 143:220–224

Fröman G, Acevedo F, Lundahl P, Hjertén S (1980) The glucose transport activity of human erythrocyte membranes. Reconstitution in phospholipid liposomes and fractionation by molecular sieve and ion exchange chromatography. Biochim Biophys Acta 600:489–501

Fröman G, Lundahl P, Acevedo F (1981) Partial purification of the D-glucose transport protein from human erythrocyte membranes by affinity chromatography on wheat germ lectin-sepharose. FEBS Lett 129:100–104

Futai M, Kimura H (1977) Inducible membrane-bound L-lactate dehydrogenase from *Escherichia coli*. Purification and properties. J Biol Chem 252:5820–5827

Gal A, Braun S, Feder D, Levitzki A (1983) Reconstitution of a functional β-adrenergic receptor using cholate and a novel method for its functional assay. Eur J Biochem 134:391–396

Garrahan PJ, Glynn IM (1967) The stoichiometry of the sodium pump. J Physiol (Lond) 192:217–235

Gavish M, Amit T, Barkey RJ, Youdim MBH (1983) Characterization of CHAPS-solubilized prolactin receptors induced by estradiol in liver of male rats. Unmasking of cryptic sites by CHAPS. FEBS Lett 154:81–86

Gazzotti P, Peterson SW (1977) Lipid requirement of membrane-bound enzymes. J Bioenerg Biomembr 9:373–386

Geck P, Pietrzyk C, Burckhardt B-C, Pfeiffer B, Heinz E (1980) Electrically silent cotransport of Na^+,K^+ and Cl^- in Ehrlich cells. Biochim Biophys Acta 600:432–447

Gietzen K, Seiler S, Fleischer S, Wolf HU (1980a) Reconstitution of the Ca^{2+}-transport system of human erythrocytes. Biochem J 188:47–54

Gietzen K, Tejčka M, Wolf HU (1980b) Calmodulin affinity chromatography yields a functional purified erythrocyte (Ca^{2+}+Mg^{2+})-dependent adenosine triphosphatase. Biochem J 189:81–88

Gitler C (1972) Plasticity of biological membranes. Annu Rev Biophys Bioeng 1:51–92

Golan DE, Veatch W (1980) Lateral mobility of band 3 in the human erythrocyte membrane studied by fluorescence photobleaching recovery: evidence for control by cytoskeletal interactions. Proc Natl Acad Sci 77:2537–2541

Golan DE, Veatch WR (1982) Lateral mobility of band 3 in the human erythrocyte membrane: Control by ankyrin-mediated interactions. Biophys J 37:177a

Goldin SM (1977) Active transport of sodium and potassium ions by the sodium and potassium ion-activated adenosine triphosphatase from renal medulla. J Biol Chem 252:5630–5642

Goldin SM, Rhoden V (1978) Reconstitution and "transport specificity fractionation" of the human erythrocyte glucose transport system. J Biol Chem 253:2575–2583

Goldin SM, Tong SW (1974) Reconstitution of active transport catalyzed by the purified sodium and potassium ion-stimulated adenosine triphosphatase from canine renal medulla. J Biol Chem 249:5907–5915

Gomez-Fernandez JC, Goni FM, Bach D, Restall CJ, Chapman D (1980) Protein-lipid interaction. Biophysical studies of (Ca^{2+}+Mg^{2+})-ATPase reconstituted systems. Biochim Biophys Acta 598:502–516

Gonenne A, Ernst R (1978) Solubilization of membrane proteins by sulfobetaines, novel zwitterionic surfactants. Anal Biochem 87:28–38

Goodman SL, Isern de Caldentey M, Wheeler KP (1978) A simple and rapid method for the reversible removal of lipids from a membrane-bound enzyme. Biochem J 169:305–311

Greenberg M, Tsong TY (1984) Detergent solubilization and affinity purification of a local anesthetic binding protein from mammalian axonal membranes. J Biol Chem 259:13241–13245

Greger R, Schlatter E (1981) Presence of luminal K^+, a prerequisite for active NaCl transport in the cortical thick ascending limb of Henle's loop of rabbit kidney. Pflügers Arch 392:92–94

Grover AK, Slotboom AJ, de Haas GH, Hammes GG (1975) Lipid specificity of β-hydroxybutyrate dehydrogenase activation. J Biol Chem 250:31–38

Gruber W, Deuticke B (1973) Comparative aspects of phosphate transfer across mammalian erythrocyte membranes. J Membr Biol 13:19–36
Grunze M, Deuticke B (1974) Changes of membrane permeability due to extensive cholesterol depletion in mammalian erythrocytes. Biochim Biophys Acta 356:125–130
Grunze M, Forst B, Deuticke B (1980) Dual effect of membrane cholesterol on simple and mediated transport processes in human erythrocytes. Biochim Biophys Acta 600:860–869
Guidotti G (1972) Membrane proteins. Annu Rev Biochem 41:731–752
Haaker H, Racker E (1979) Purification and reconstitution of the Ca^{2+}-ATPase from plasma membranes of pig erythrocytes. J Biol Chem 254:6598–6602
Hatefi Y, Hanstein WG (1969) Solubilization of particulate proteins and nonelectrolytes by chaotropic agents. Proc Natl Acad Sci 62:1129–1136
Hayase K, Hayano S (1978a) Effect of alcohols on the critical micelle concentration decrease in the aqueous sodium dodecyl sulfate solution. J Colloid Interface Sci 63:446–451
Hayase K, Hayano S (1978b) Critical micelle concentration reduction of aqueous sodium dodecyl sulfate solutions upon the addition of alcohols. Bull Chem Soc Jpn 51:933–934
Hekman M, Feder D, Keenan AK, Gal A, Klein HW, Pfeuffer T, Levitzki A, Helmreich EJM (1984) Reconstitution of β-adrenergic receptor with components of adenylate cyclase. EMBO J 3:3339–3345
Helenius A, Simons K (1972) The binding of detergents to lipophilic and hydrophilic proteins. J Biol Chem 247:3656–3661
Helenius A, Simons K (1975) Solubilization of membranes by detergents. Biochim Biophys Acta 415:29–79
Helenius A, Söderlund H (1973) Stepwise dissociation of the Semliki Forest virus membrane with Triton X-100. Biochim Biophys Acta 307:287–300
Helenius A, Fries E, Kartenbeck J (1977) Reconstitution of Semliki Forest virus membrane. J Cell Biol 75:866–880
Helenius A, McCaslin DR, Fries E, Tanford C (1979) Properties of detergents. Methods Enzymol 56:734–749
Helenius A, Sarvas M, Simons K (1981) Asymmetric and symmetric membrane reconstitution by detergent elimination. Studies with Semliki-Forest-virus spike glycoprotein and penicillinase from the membrane of *Bacillus licheniformis*. Eur J Biochem 116:27–35
Hidalgo C, Thomas DD, Ikemoto N (1978) Effect of the lipid environment on protein motion and enzymatic activity of the sarcoplasmic reticulum calcium ATPase. J Biol Chem 253:6879–6887
Hilden S, Hokin LE (1975) Active potassium transport coupled to active sodium transport in vesicles reconstituted from purified sodium and potassium ion-activated adenosine triphosphatase from the rectal gland of *Squalus acanthias*. J Biol Chem 250:6296–6303
Hilden S, Rhee HM, Hokin LE (1974) Sodium transport by phospholipid vesicles containing purified sodium and potassium ion-activated adenosine trihosphatase. J Biol Chem 249:7432–7440
Hjelmeland LM (1980) A nondenaturing zwitterionic detergent for membrane biochemistry: Design and synthesis. Proc Natl Acad Sci USA 77:6368–6370
Hjelmeland LM, Chrambach A (1984) Solubilization of functional membrane proteins. Methods Enzymol 104:305–318
Hjertén S (1983) A study on the renaturation of membrane proteins after solubilization in SDS or following polyacrylamide gel electrophoresis in SDS, with special reference to a phosphatase from *Acholeplasma laidlawii*. Biochim Biophys Acta 736:130–136
Ho MK, Guidotti G (1975) A membrane protein from human erythrocytes involved in anion exchange. J Biol Chem 250:675–683

Hofstee BHJ (1973) Protein binding by agarose carrying hydrophobic groups in conjunction with charges. Biochem Biophys Res Commun 50:751–757

Hokin LE (1981) Reconstitution of "carriers" in artificial membranes. J Membr Biol 60:77–93

Hokin LE, Dixon JF (1979) Parameters of reconstituted Na^+ and K^+ transport in liposomes in which purified Na,K-ATPase is incorporated by "freeze-thaw-sonication". In: Skou JC, Nørby JG (eds) Na,K-ATPase structure and kinetics. Academic, London, pp 47–67

Hokin LE, Dahl JL, Deupree JD, Dixon JF, Hackney JF, Perdue JF (1973) Studies on the characterization of the sodium-potassium transport adenosine triphosphatase. X. Purification of the enzyme from the rectal gland of Squalus acanthias. J Biol Chem 248:2593–2605

Holloszy JO, Narahara HT (1965) Studies of tissue permeability. X. Changes in permeability to 3-methylglucose associated with contraction of isolated frog muscle. J Biol Chem 240:3493–3500

Homcy C, Wrenn S, Haber E (1978) Affinity purification of cardiac adenylate cyclase: Dependence on prior hydrophobic resolution. Proc Natl Acad Sci USA 75:59–63

Hong K, Hubbell WL (1973) Lipid requirements for rhodopsin regenerability. Biochemistry 12:4517–4523

Hosang M, Gibbs EM, Diedrich DF, Semenza G (1981) Photoaffinity labeling and identification of (a component of) the small-intestinal Na^+,D-glucose transporter using 4-azidophlorizin. FEBS Lett 130:244–248

Howlett GJ, Wardrop AJ (1978) Dissociation and reconstitution of human erythrocyte membrane proteins using 3,4,5,6-tetrahydrophthalic anhydride. Arch Biochem Biophys 188:429–437

Huang C-H (1969) Studies on phosphatidylcholine vesicles. Formation and physical characteristics. Biochemistry 8:344–352

Huang K-S, Bayley H, Liao M-J, London E, Khorana HG (1981) Refolding of an integral membrane protein. Denaturation, renaturation, and reconstitution of intact bacteriorhodopsin and two proteolytic fragments. J Biol Chem 256:3802–3809

Ikemoto N (1982) Structure and function of the calcium pump protein of sarcoplasmic reticulum. Annu Rev Physiol 44:297–317

Im WB, Ling KY, Faust RG (1982) Partial purification of the Na^+-dependent D-glucose transport system from renal brush border membranes. J Membr Biol 65:131–137

Inoue K (1974) Permeability properties of liposomes prepared from dipalmitoyllecithin, dimyristoyllecithin, egg lecithin, rat liver lecithin and beef brain sphingomyelin. Biochim Biophys Acta 339:390–402

Inui K-I, Moller DE, Tillotson LG, Isselbacher KJ (1979) Stereospecific hexose transport by membrane vesicles from mouse fibroblasts: Membrane vesicles retain increased hexose transport associated with viral transformation. Proc Natl Acad Sci USA 76:3972–3976

Jackson ML, Litman BJ (1985) Rhodopsin-egg phosphatidylcholine reconstitution by an octyl glucoside dilution procedure. Biochim Biophys Acta 812:369–376

Jackson ML, Schmidt CF, Lichtenberg D, Litman BJ, Albert AD (1982) Solubilization of phosphatidylcholine bilayers by octyl glucoside. Biochemistry 21:4576–4582

Jacobsen C, Røigaard-Petersen H, Jørgensen KE, Sheikh MI (1984) Isolation and partial purification of dicarboxylic acid binding protein from luminal-membrane vesicles of rabbit kidney cortex. Biochim Biophys Acta 773:173–179

Jennings ML, Passow H (1979) Anion transport across the erythrocyte membrane, in situ proteolysis of band 3 protein, and cross-linking of proteolytic fragments by 4,4′-diisothiocyano-dihydrostilbene-2,2′-disulfonate. Biochim Biophys Acta 554:498–519

Jennissen HP, Heilmeyer LMG Jr (1975) General aspects of hydrophobic chromatography. Adsorption and elution characteristics of some skeletal muscle enzymes. Biochemistry 14:754–759

Jilka RL, Martonosi AN, Tillack TW (1975) Effect of the purified [Mg^{2+}+Ca^{2+}]-activated ATPase of sarcoplasmic reticulum upon the passive Ca^{2+} permeability and ultrastructure of phospholipid vesicles. J Biol Chem 250:7511–7524
Johnson LW, Smith CH (1980) Monosaccharide transport across microvillous membrane of human placenta. Am J Physiol 238:C160–C168
Johnson LW, Smith CH (1982) Identification of the glucose transport protein of the microvillous membrane of human placenta by photoaffinity labelling. Biochem Biophys Res Commun 109:408–413
Jones MN, Nickson JK (1981) Monosaccharide transport proteins of the human erythrocyte membrane. Biochim Biophys Acta 650:1–20
Jørgensen PL (1974) Purification and characterization of (Na^++K^+)-ATPase. III. Purification from the outer medulla of mammalian kidney after selective removal of membrane components by sodium dodecylsulphate. Biochim Biophys Acta 356: 36–52
Jørgensen PL, Anner BM (1979) Purification and characterization of (Na^++K^+)-ATPase. VIII. Altered Na^+:K^+ transport ratio in vesicles reconstituted with purified (Na^++K^+)-ATPase that has been selectively modified with trypsin in presence of NaCl. Biochim Biophys Acta 555:485–492
Jung CY, Carlson LM (1975) Glucose transport carrier in human erythrocyte membranes. Dinitrophenylation of a membrane component modified by D-glucose. J Biol Chem 250:3217–3220
Jung CY, Rampal AL (1977) Cytochalasin B binding sites and glucose transport carrier in human erythrocyte ghosts. J Biol Chem 252:5456–5463
Jung CY, Carlson LM, Balzer CJ (1973) Effect of proteolytic digestion on glucose transport carrier of human erythrocyte ghosts. Biochim Biophys Acta 298:108–114
Kagawa Y, Racker E (1971) Partial resolution of the enzymes catalyzing oxidative phosphorylation. XXV. Reconstitution of vesicles catalyzing $^{32}P_i$-adenosine triphosphate exchange. J Biol Chem 246:5477–5487
Kagawa Y, Kandrach A, Racker E (1973) Partial resolution of the enzymes catalyzing oxidative phosphorylation. XXVI. Specificity of phospholipids required for energy transfer reactions. J Biol Chem 248:676–684
Kahane I, Razin S (1971) Characterization of the mycoplasma membrane proteins. I. Reaggregation of solubilized membrane proteins of *Acholeplasma laidlawii*. Biochim Biophys Acta 249:159–168
Kahlenberg A (1976) Partial purification of a membrane protein from human erythrocytes involved in glucose transport. J Biol Chem 251:1582–1590
Kahlenberg A, Zala CA (1977) Reconstitution of D-glucose transport in vesicles composed of lipids and intrinsic protein (Zone 4.5) of the human erythrocyte membrane. J Supramol Struct 7:287–300
Kano-Kameyama A, Hoshi T (1983) Purification and reconstitution of Na^+/D-glucose cotransport carriers from guinea pig small intestine. Jpn J Physiol 33:955–970
Kapitza HG, Rüppel DA, Galla H-J, Sackmann E (1984) Lateral diffusion of lipids and glycophorin in solid phosphatidylcholine bilayers. The role of structural defects. Biophys J 45:577–587
Kaplan JH (1972) Anion diffusion across artificial lipid membranes: The effects of lysozyme on anion diffusion from phospholipid liposomes. Biochim Biophys Acta 290:339–347
Karlish SJD, Pick U (1981) Sidedness of the effects of sodium and potassium ions on the conformational state of the sodium-potassium pump. J Physiol (Lond) 312: 505–529
Karlish SJD, Stein WD (1982a) Passive rubidium fluxes mediated by Na-K-ATPase reconstituted into phospholipid vesicles when ATP- and phosphate-free. J Physiol (Lond) 328:295–316
Karlish SJD, Stein WD (1982b) Effects of ATP or phosphate on passive rubidium fluxes mediated by Na-K-ATPase reconstituted into phospholipid vesicles. J Physiol 317–331

Karlish SJD, Lieb WR, Stein WD (1982) Combined effects of ATP and phosphate on rubidium exchange mediated by Na-K-ATPase reconstituted into phospholipid vesicles. J Physiol (Lond) 328:333–350

Kasahara M, Hinkle PC (1976) Reconstitution of D-glucose transport catalyzed by a protein fraction from human erythrocytes in sonicated liposomes. Proc Natl Acad Sci USA 73:396–400

Kasahara M, Hinkle PC (1977a) Reconstitution and purification of the D-glucose transporter from human erythrocytes. J Biol Chem 252:7384–7390

Kasahara M, Hinkle PC (1977b) Reconstitution and purification of the D-glucose transport protein from human erythrocytes. FEBS Symp 42:346–350

Kelleher DJ, Rashidbaigi A, Ruoho AE, Johnson GL (1983) Rapid vesicle reconstitution of alprenolol-sepharose-purified β_1-adrenergic receptors. Interaction of the purified receptor with N_s. J Biol Chem 258:12881–12885

Kimelberg HK, Papahadjopoulos D (1971) Phospholipid-protein interactions: Membrane permeability correlated with monolayer "penetration". Biochim Biophys Acta 233:805–809

Kimelberg HK, Papahadjopoulos D (1972) Phospholipid requirements for $(Na^+ + K^+)$-ATPase activity: Head-group specificity and fatty acid fluidity. Biochim Biophys Acta 282:277–292

Kimelberg HK, Papahadjopoulos D (1974) Effects of phospholipid acyl chain fluidity, phase transitions, and cholesterol on $(Na^+ + K^+)$-stimulated adenosine triphosphatase. J Biol Chem 249:1071–1080

King RG, Marchbanks RM (1982) The effect of pH on the size of liposomes formed by cholate dialysis. Biochim Biophys Acta 691:183–187

Kinne R, Faust RG (1977) Incorporation of D-glucose-, L-alanine- and phosphate-transport systems from rat renal brush-border membranes into liposomes. Biochem J 168:311–314

Kirkpatrick FH, Sandberg HE (1973) Effect of anionic surfactants, nonionic surfactants and neutral salts on the conformation of spin-labeled erythrocyte membrane proteins. Biochim Biophys Acta 298:209–218

Kirkpatrick FH, Gordesky SE, Marinetti GV (1974) Differential solubilization of proteins, phospholipids, and cholesterol of erythrocyte membranes by detergents. Biochim Biophys Acta 345:154–161

Klappauf E, Schubert D (1979) Interactions of band 3-protein from human erythrocyte membranes with cholesterol and cholesterol analogues. Hoppe Seylers Z Physiol Chem 360:1225–1235

Klip A, Grinstein S, Semenza G (1979) Partial purification of the sugar carrier of intestinal brush border membranes. Enrichment of the phlorizin-binding component by selective extractions. J Membr Biol 51:47–73

Knowles AF, Racker E (1975) Properties of a reconstituted calcium pump. J Biol Chem 250:3538–3544

Knowles AF, Eytan E, Racker E (1976) Phospholipid-protein interactions in the Ca^{2+}-adenosine triphosphatase of sarcoplasmic reticulum. J Biol Chem 251:5161–5165

Köhne W, Haest CWM, Deuticke B (1981) Mediated transport of anions in band 3-phospholipid vesicles. Biochim Biophys Acta 664:108–120

Köhne W, Deuticke B, Haest CWM (1983) Phospholipid dependence of the anion transport system of the human erythrocyte membrane. Studies on reconstituted band 3/lipid vesicles. Biochim Biophys Acta 730:139–150

Koepsell H (1979) Conformational changes of membrane-bound $(Na^+ - K^+)$-ATPase as revealed by trypsin digestion. J Membr Biol 48:69–94

Koepsell H, Madrala A (1986) Interaction of phlorizin with the Na^+-D-glucose cotransporter from intestine and kidney. In: Anner B, Roberts GKC, Burgen A (eds) Topics in molecular pharmacology. Elsevier Science Publishers B.V., Amsterdam (in press)

Koepsell H, Menuhr H, Wissmüller TF, Ducis I, Haase W (1980) Reconstitution of D-glucose transport and high-affinity phlorizin binding after solubilization of kidney brush border proteins. Ann NY Acad Sci 358:267–281

Koepsell H, Menuhr H, Ducis I, Wissmüller TF (1983) Partial purification and reconstitution of the Na^+-D-glucose cotransport protein from pig renal proximal tubules. J Biol Chem 258:1888–1894

Koepsell H, Korn K, Ferguson D, Menuhr H, Ollig D, Haase W (1984) Reconstitution and partial purification of several Na^+ cotransport systems from renal brush-border membranes. Properties of the L-glutamate transporter in proteoliposomes. J Biol Chem 259:6548–6558

Koepsell H, Neeb M, Madrala A (1986) The renal D-glucose/sodium symporter: attempts at identification, isolation and reconstitution. In: Alvarado F, van Os CH (eds) INSERM 26 (in press)

Kono T, Suzuki K, Dansey LE, Robinson FW, Blevins TL (1981) Energy-dependent and protein synthesis-independent recycling of the insulin-sensitive glucose transport mechanism in fat cells. J Biol Chem 256:6400–6407

Krämer R (1982) Cholesterol as activator of ADP-ATP exchange in reconstituted liposomes and in mitochondria. Biochim Biophys Acta 693:296–304

Krämer R, Klingenberg M (1977) Reconstitution of inhibitor binding properties of the isolated adenosine 5′-diphosphate,adenosine 5′-triphosphate carrier-linked binding protein. Biochemistry 16:4954–4961

Krämer R, Klingenberg M (1980) Enhancement of reconstituted ADP,ATP exchange activity by phosphatidylethanolamine and by anionic phospholipids. FEBS Lett 119:257–260

Kumar A, Gupta CM (1983) Effect of altered polar headgroup of phosphatidylethanolamines on transbilayer aminophospholipid distribution in sonicated vesicles. Biochim Biophys Acta 730:1–9

Kuriyama K (1962) Temperature dependence of micellar weight of non-ionic surfactant in the presence of various additives. Kolloid Z Z Polym 181:144–149

Kusaka I, Kanai K (1978) Purification and characterization of alanine carrier isolated from H-proteins of *Bacillus subtilis*. Eur J Biochem 83:307–311

Kushner LM, Hubbard WD (1954) Viscometric and turbidimetric measurements on dilute aqueous solutions of a non-ionic detergent. J Phys Chem 58:1163–1167

Kyte J (1971) Purification of the sodium- and potassium-dependent adenosine triphosphatase from canine renal medulla. J Biol Chem 246:4157–4165

Lacks SA, Springhorn SS (1980) Renaturation of enzymes after polyacrylamide gel electrophoresis in the presence of sodium dodecyl sulfate. J Biol Chem 255:7467–7473

Ladbrooke BD, Chapman D (1969) Thermal analysis of lipids, proteins and biological membranes. A review and summary of some recent studies. Chem Phys Lipids 3:304–367

Ladbrooke BD, Williams RM, Chapman D (1968) Studies on lecithin-cholesterol-water interactions by differential scanning calorimetry and X-ray diffraction. Biochim Biophys Acta 150:333–340

Last TA, Gantzer ML, Tyler CE (1983) Ion-gated channel induced in planar bilayers by incorporation of (Na^+,K^+)-ATPase. J Biol Chem 258:2399–2404

Latorre R, Vergara C, Hidalgo C (1982) Reconstitution in planar lipid bilayers of a Ca^{2+}-dependent K^+ channel from transverse tubule membranes isolated from rabbit skeletal muscle. Proc Natl Acad Sci USA 79:805–809

LeFevre PG, D'Angelo G, Masiak SJ (1975) Use of cytochalasin B to allow differential *N*-ethylmaleimide-labeling of a membrane sugar-transport component. Fed Proc 34:238

Lemaire J, Maestracci D (1978) Labeling of a glucose binding protein in the rabbit intestinal brush border membrane. Can J Physiol Pharmacol 56:760–770

Lentz BR, Clubb KW, Alford DR, Höchli M, Meissner G (1985) Phase behavior of membrane reconstituted from dipentadecanoylphosphatidylcholine and the Mg^{2+}-dependent, Ca^{2+}-stimulated adenosinetriphosphatase of sarcoplasmic reticulum: Evidence for a disrupted lipid domain surrounding protein. Biochemistry 24:433–442

Lepke S, Passow H (1976) Effects of incorporated trypsin on anion exchange and membrane proteins in human red blood cell ghosts. Biochim Biophys Acta 455:353–370

Levey GS (1971) Restoration of glucagon responsiveness of solubilized myocardial adenyl cyclase by phosphatidylserine. Biochem Biophys Res Commun 43:108–113
Liang S-M, Winter CG (1977) Digitonin-induced changes in subunit arrangement in relation to some in vitro activities of the (Na^+, K^+)-ATPase. J Biol Chem 252: 8278–8284
Lichtenberg D, Zilberman Y, Greenzaid P, Zamir S (1979) Structural and kinetic studies on the solubilization of lecithin by sodium deoxycholate. Biochemistry 18: 3517–3525
Lichtenberg D, Robson RJ, Dennis EA (1983) Solubilization of phospholipids by detergents. Structural and kinetic aspects. Biochim Biophys Acta 737:285–304
Lienhard GE, Kim HH, Ransome KJ, Gorga JC (1982) Immunological identification of an insulin-responsive glucose transporter. Biochim Biophys Res Commun 105: 1150–1156
Lienhard GE, Crabb JH, Ransome KJ (1984) Endoglycosidase F cleaves the oligosaccharides from the glucose transporter of the human erythrocyte. Biochim Biophys Acta 769:404–410
Liljas L, Lundahl P, Hjertén S (1974) Selective solubilization with Tween 20 of proteins from water-extracted human erythrocyte membranes. Analysis by gel electrophoresis in dodecylsulfate and in Tween 20. Biochim Biophys Acta 352:327–337
Lin JT, Riedel S, Kinne R (1979) The use of octyl β-D-glucoside as detergent for hog kidney brush border membrane. Biochim Biophys Acta 557:179–187
Lin JT, da Cruz MEM, Riedel S, Kinne R (1981) Partial purification of hog kidney sodium-D-glucose cotransport system by affinity chromatography on a phlorizin polymer. Biochim Biophys Acta 640:43–54
Lin JT, Schwarc K, Stroh A (1984) Chromatofocussing and centrifugal reconstitution as tools for the separation and characterization of the Na^+-cotransport systems of the brush-border membrane. Biochim Biophys Acta 774:254–260
Lind C, Höjeberg B, Khorana HG (1981) Reconstitution of delipidated bacteriorhodopsin with endogenous polar lipids. J Biol Chem 256:8298–8305
Liscia DS, Alhadi T, Vonderhaar BK (1982) Solubilization of active prolactin receptors by a nondenaturing zwitterionic detergent. J Biol Chem 257:9401–9405
Litman BJ (1973) Lipid model membranes. Characterization of mixed phospholipid vesicles. Biochemistry 12:2545–2554
Lookman T, Pink DA, Grundke EW, Zuckermann MJ, de Verteuil F (1982) Phase separation in lipid bilayers containing integral proteins. Computer simulation studies. Biochemistry 21:5593–5601
Lotan R, Nicolson GL (1979) Purification of cell membrane glycoproteins by lectin affinity chromatography. Biochim Biophys Acta 559:329–376
Lu Y-B, Chow EI (1982) Bicarbonate/chloride transport kinetics of 37°C and its relationship to membrane lipids in mammalian erythrocytes. Biochim Biophys Acta 689:485–489
Lukacovic MF, Feinstein MB, Sha'afi RI, Perrie S (1981) Purification of stabilized band 3 protein of the human erythrocyte membrane and its reconstitution into liposomes. Biochemistry 20:3145–3151
MacDonald RI, MacDonald RC (1983) Lipid mixing during freeze-thawing of liposomal membranes as monitored by fluorescence energy transfer. Biochim Biophys Acta 735:243–251
McLennan DH (1970) Purification and properties of an adenosine triphosphatase from sarcoplasmic reticulum. J Biol Chem 245:4508–4518
MacLennan DH, Khanna VK, Stewart PS (1976) Restoration of calcium transport in the trypsin-treated (Ca^{2+}+Mg^{2+})-dependent adenosine triphosphatase of sarcoplasmic reticulum exposed to sodium dodecyl sulfate. J Biol Chem 251:7271–7274
Makino S, Reynolds JA, Tanford C (1973) The binding of deoxycholate and Triton X-100 to proteins. J Biol Chem 248:4926–4932
Malathi P, Preiser H (1983) Isolation of the sodium-dependent D-glucose transport protein from brush-border membranes. Biochim Biophys Acta 735:314–324

Mandrell RE, Zollinger WD (1984) Use of a zwitterionic detergent for the restoration of the antibody-binding capacity of electroblotted meningococcal outer membrane proteins. J Immunol Methods 67:1–11
Mankowich AM (1954) Micellar molecular weights of selected surface active agents. J Phys Chem 58:1027–1030
Martonosi A (1968) Sarcoplasmic reticulum. IV. Solubilization of microsomal adenosine triphosphatase. J Biol Chem 243:71–81
Masters PS, Hong J-S (1981) Reconstitution of binding protein dependent active transport of glutamine in spheroplasts of *Escherichia coli*. Biochemistry 20:4900–4904
Mazer NA, Benedek GB, Carey MC (1976) An investigation of the micellar phase of sodium dodecyl sulfate in aqueous sodium chloride solutions using quasielastic light scattering spectroscopy. J Phys Chem 80:1075–1085
McEnery MW, Buhle EL Jr, Aebi U, Pedersen PL (1984) Proton ATPase of rat liver mitochondria. Preparation and visualization of a functional complex using the novel zwitterionic detergent 3-[(3-cholamidopropyl)dimethylammonio]-1-propanesulfonate. J Biol Chem 259:4642–4651
McFarland BH, Inesi G (1971) Solubilization of sarcoplasmic reticulum with Triton X-100. Arch Biochem Biophys 145:456–464
McMurchie EJ, Raison JK (1979) Membrane lipid fluidity and its effect on the activation energy of membrane-associated enzymes. Biochim Biophys Acta 554:364–374
Melchior DL, Czech MP (1979) Sensitivity of the adipocyte D-glucose transport system to membrane fluidity in reconstituted vesicles. J Biol Chem 254:8744–8747
Michaelson DM, Horwitz AF, Klein MP (1973) Transbilayer asymmetry and surface homogeneity of mixed phospholipids in cosonicated vesicles. Biochemistry 12: 2637–2645
Miller C (1978) Voltage-gated cation conductance channel from fragmented sarcoplasmic reticulum: Steady-state electrical properties. J Membr Biol 40:1–23
Miller C, Racker E (1976) Fusion of phospholipid vesicles reconstituted with cytochrome c oxidase and mitochondrial hydrophobic protein. J Membr Biol 26:319–333
Miller DM (1968) The kinetics of selective biological transport. III. Erythrocyte-monosaccharide transport data. Biophys J 8:1329–1338
Milsmann MHW, Schwendener RA, Weder H-G (1978) The preparation of large single bilayer liposomes by a fast and controlled dialysis. Biochim Biophys Acta 512: 147–155
Mimms LT, Zampighi G, Nozaki Y, Tanford C, Reynolds JA (1981) Phospholipid vesicle formation and transmembrane protein incorporation using octyl glucoside. Biochemistry 20:833–840
Montal M (1976) Experimental membranes and mechanisms of bioenergy transductions. Annu Rev Biophys Bioeng 5:119–175
Montal M, Labarca P, Fredkin DR, Suarez-Isla BA (1984) Channel properties of the purified acetylcholine receptor from *Torpedo californica* reconstituted in planar lipid bilayer membranes. Biophys J 45:165–174
Morgan TE, Hanahan DJ (1966) Solubilization and characterization of a lipoprotein from erythrocyte stroma. Biochemistry 5:1050–1059
Mullins LJ, Brinley FJ Jr (1969) Potassium fluxes in dialyzed squid axons. J Gen Physiol 53:704–740
Mullins RE, Langdon RG (1980a) Maltosyl isothiocyanate: An affinity label for the glucose transporter of the human erythrocyte membrane. 1. Inhibition of glucose transport. Biochemistry 19:1199–1205
Mullins RE, Langdon RG (1980b) Maltosyl isothiocyanate: An affinity label for the glucose transporter of the human erythrocyte membrane. 2. Identification of the transporter. Biochemistry 19:1205–1212
Murer H, Burckhardt G (1983) Membrane transport of anions across epithelia of mammalian small intestine and kidney proximal tubule. Rev Physiol Biochem Pharmacol 96:1–51

Neeb M, Fasold H, Koepsell H (1985) Identification of the D-glucose binding polypeptide of the renal Na^+-D-glucose cotransporter with a covalently binding D-glucose analog. FEBS Lett 182:139–144

Ne'eman Z, Kahane I, Kovartovsky J, Razin S (1972) Characterization of the myoplasma membrane proteins. III. Gel filtration and immunological characterization of *Acholeplasma laidlawii* membrane proteins. Biochim Biophys Acta 266:255–268

Neher E, Sakmann B (1976) Single-channel currents recorded from membrane of denervated frog muscle fibres. Nature 260:799–802

Nelson BD, Fleischer S (1981) Phospholipid requirements for the reconstitution of complex-III vesicles exhibiting controlled electron transport. Biochem J 194:783–787

Nelson N, Anholt R, Lindstrom J, Montal M (1980) Reconstitution of purified acetylcholine receptors with functional ion channels in planar lipid bilayers. Proc Natl Acad Sci USA 77:3057–3061

Nickson JK, Jones MN (1982) The reconstitution of the human erythrocyte sugar transporter in planar bilayer membranes. Biochim Biophys Acta 690:31–40

Niggli V, Penniston JT, Carafoli E (1979) Purification of the (Ca^{2+}-Mg^{2+})-ATPase from human erythrocyte membranes using a calmodulin affinity column. J Biol Chem 254:9955–9958

Niggli V, Adunyah ES, Carafoli E (1981a) Acidic phospholipids, unsaturated fatty acids, and limited proteolysis mimic the effect of calmodulin on the purified erythrocyte Ca^{2+}-ATPase. J Biol Chem 256:8588-8592

Niggli V, Adunyah ES, Penniston JT, Carafoli E (1981b) Purified (Ca^{2+}-Mg^{2+})-ATPase of the erythrocyte membrane. Reconstitution and effect of calmodulin and phospholipids. J Biol Chem 256:395–401

Niggli V, Sigel E, Carafoli E (1982) The purified Ca^{2+} pump of human erythrocyte membranes catalyzes an electroneutral Ca^{2+}-H^+ exchange in reconstituted liposomal systems. J Biol Chem 257:2350–2356

Oku N, MacDonald RC (1983) Differential effects of alkali metal chlorides on formation of giant liposomes by freezing and thawing and dialysis. Biochemistry 22:855–863

Osborne HB, Sardet C, Helenius A (1974) Bovine rhodopsin: Characterization of the complex formed with Triton X-100. Eur J Biochem 44:383–390

O'Shea PS, Feuerstein-Thelen S, Azzi A (1984) Membrane-potential-dependent changes of the lipid microviscosity of mitochondria and phospholipid vesicles. Biochem J 220:795–801

Osipow L, Snell FD, York WC, Finchler A (1956) Methods of preparation . . . Fatty acid esters of sucrose. Industrial and Engineering Chemistry 48:1459–1462

Ottolenghi P (1979) The relipidation of delipidated Na,K-ATPase. An analysis of complex formation with dioleoylphosphatidylcholine and with dioleoylphosphatidylethanolamine. Eur J Biochem 99:113–131

Papahadjopoulos D (1971) Na^+-K^+ discrimination by "pure" phospholipid membranes. Biochim Biophys Acta 241:254–259

Papahadjopoulos D, Nir S, Ohki S (1971) Permeability properties of phospholipid membranes: Effect of cholesterol and temperature. Biochim Biophys Acta 266:561–583

Papahadjopoulos D, Jacobson K, Nir S, Isac T (1973) Phase transitions in phospholipid vesicles. Fluorescence polarization and permeability measurements concerning the effect of temperature and cholesterol. Biochim Biophys Acta 311:330–348

Papahadjopoulos D, Vail WJ, Jacobson K, Poste G (1975) Cochleate lipid cylinders: Formation by fusion of unilamellar lipid vesicles. Biochim Biophys Acta 394:483–491

Papazian D, Rahamimoff H, Goldin SM (1979) Reconstitution and purification by "transport specificity fractionation" of an ATP-dependent calcium transport component from synaptosome-derived vesicles. Proc Natl Acad Sci USA 76:3708–3712

Pappert G, Schubert D (1983) The state of association of band 3 protein of the human erythrocyte membrane in solutions of nonionic detergents. Biochim Biophys Acta 730:32–40

Passow H (1986) Molecular aspects of the band 3 protein-mediated anion transport across the red blood cell membrane. Rev Physiol Biochem Pharmacol 103:61–203

Passow H, Fasold H, Zaki L, Schuhmann B, Lepke S (1975) Membrane proteins and anion exchange in human erythrocytes. FEBS Proc Meet 35:197–214

Pearlstein E (1977) Isolation and partial characterization of plasma membrane glycoproteins from normal and transformed mammalian cells employing plant lectin affinity chromatography. Exp Cell Res 109:95–103

Pearlstein E, Seaver J (1976) Non-lytic, non-ionic detergent extraction of plasma membrane constituents from normal and transformed fibroblasts. Biochim Biophys Acta 426:589–597

Peerce BE, Wright EM (1984a) Conformational changes in the intestinal brush border sodium-glucose cotransporter labeled with fluorescein isothiocyanate. Proc Natl Acad Sci USA 81:2223–2226

Peerce BE, Wright EM (1984b) Sodium-induced conformational changes in the glucose transporter of intestinal brush borders. J Biol Chem 259:14105–14112

Pessin JE, Tillotson LG, Yamada K, Gitomer W, Carter-Su C, Mora R, Isselbacher KJ, Czech MP (1982) Identification of the stereospecific hexose transporter from starved and fed chicken embryo fibroblasts. Proc Natl Acad Sci USA 79:2286–2290

Peters R, Peters J, Tews KH, Bähr W (1974) A microfluorimetric study of translational diffusion in erythrocyte membranes. Biochim Biophys Acta 367:282–294

Peterson SW, Ronner P, Carafoli E (1978) Partial purification and reconstitution of the (Ca^{2+}+Mg^{2+})-ATPase of erythrocyte membranes. Arch Biochem Biophys 186: 202–210

Pick U (1981) Liposomes with a large trapping capacity prepared by freezing and thawing of sonicated phospholipid mixtures. Arch Biochem Biophys 212:186–194

Pick U, Bassilian S (1982) Activation of magnesium ion specific adenosinetriphospatase in chloroplast coupling factor 1 by octyl glucoside. Biochemistry 21:6144–6152

Pick U, Racker E (1979) Purification and reconstitution of the *N,N'*-dicyclohexylcarbodiimide-sensitive ATPase complex from spinach chloroplasts. J Biol Chem 254:2793–2799

Pilch PF, Thompson PA, Czech MP (1980) Coordinate modulation of D-glucose transport activity and bilayer fluidity in plasma membranes derived from control and insulin-treated adipocytes. Proc Natl Acad Sci USA 77:915–918

Pitt-Rivers R, Impiombato FSA (1968) The binding of sodium dodecyl sulphate to various proteins. Biochem J 109:825–830

Poiree JC, Mengual R, Sudaka P (1979) Identification of a protein component of horse kidney brush border D-glucose transport system. Biochem Biophys Res Commun 90:1387–1392

Racker E (1972) Reconstitution of cytochrome oxidase vesicles and conferral of sensitivity to energy transfer inhibitors. J Membr Biol 10:221–235

Racker E (1973) A new procedure for the reconstitution of biologically active phospholipid vesicles. Biochem Biophys Res Commun 55:224–230

Racker E (1977) Perspectives and limitations of resolutions-reconstitution experiments. J Supramol Struct 6:215–228

Racker E (1979) Reconstitution of membrane processes. Methods Enzymol 55:699–711

Racker E, Eytan E (1973) Reconstitution of an efficient calcium pump without detergents. Biochem Biophys Res Commun 55:174–178

Racker E, Fisher LW (1975) Reconstitution of an ATP-dependent sodium pump with an ATPase from electric eel and pure phospholipids. Biochem Biophys Res Commun 67:1144–1150

Racker E, Kandrach A (1973) Partial resolution of the enzymes catalyzing oxidative phosphorylation. XXXIX. Reconstitution of the third segment of oxidative phosphorylation. J Biol Chem 248:5841–5847
Racker E, Chien T-F, Kandrach A (1975) A cholate-dilution procedure for the reconstitution of the Ca^{++} pump, $^{32}P_i$-ATP exchange, and oxidative phosphorylation. FEBS Lett 57:14–18
Racker E, Violand B, O'Neal S, Alfonzo M, Telford J (1979) Reconstitution, a way of biochemical research; some new approaches to membrane-bound enzymes. Arch Biochem Biophys 198:470–477
Ramaswamy K, Bhattacharyya BR, Crane RK (1976) Studies on the transport of aliphatic glucosides by hamster small intestine in vitro. Biochim Biophys Acta 433: 32–38
Ramjeesingh M, Gaarn A, Rothstein A (1984) Pepsin cleavage of band 3 produces its membrane-crossing domains. Biochim Biophys Acta 769:381–389
Razin S (1972) Reconstitution of biological membranes. Biochim Biophys Acta 265: 241–296
Razin S, Ne'eman Z, Ohad I (1969) Selective reaggregation of solubilized mycoplasma-membrane proteins and the kinetics of membrane reformation. Biochim Biophys Acta 193:277–293
Reyes J, Greco F, Motais R, Latorre R (1983) Phloretin and phloretin analogs: Mode of action in planar lipid bilayers and monolayers. J Membr Biol 72:93–103
Reynolds JA, Tanford C (1970) Binding of dodecyl sulfate to proteins at high binding ratios. Possible implications for the state of proteins in biological membranes. Proc Natl Acad Sci USA 66:1002–1003
Reynolds JA, Nozaki Y, Tanford C (1983) Gel-exclusion chromatography on S1000 sephacryl: Application to phospholipid vesicles. Anal Biochem 130:471–474
Rhee HM, Hokin LE (1975) Inhibition of the purified sodium-potassium activated adenosinetriphosphatase from the rectal gland of *Squalus acanthias* by antibody against the glycoprotein subunit. Biochem Biophys Res Commun 63:1139–1145
Rich PR, Clarke SD (1982) Reconstitution of cytochrome bc_1 complex into lipid vesicles and the restoration of uncoupler sensitivity. FEBS Lett 148:54–58
Richards CD (1978) Anesthetics and membranes. Int Rev Biochem 19:157-220
Riepl RG, Vidaver GA (1978) Effects of Triton X-100 treatments on the composition and activities of membrane vesicles from pigeon erythrocytes. Biochim Biophys Acta 507:94–106
Rigaud JL, Bluzat A, Buschlen S (1983) Incorporation of bacteriorhodopsin into large unilamellar liposomes by reverse phase evaporation. Biochem Biophys Res Commun 111:373–382
Robinson FW, Blevins TL, Suzuki K, Kono T (1982) An improved method of reconstitution of adipocyte glucose transport activity. Anal Biochem 122:10–19
Robinson JB Jr, Strottmann JM, Wick DG, Stellwagen E (1980) Affinity chromatography in nonionic detergent solutions. Proc Natl Acad Sci USA 77:5847–5851
Robinson JD, Flashner MS (1979) The $(Na^{+}+K^{+})$-activated ATPase. Enzymatic and transport properties. Biochim Biophys Acta 549:145–176
Robinson NC, Capaldi RA (1977) Interaction of detergents with cytochrome c oxidase. Biochemistry 16:375–381
Rohn R (1983) Einfluß von Proteasen auf das Transportverhalten von Bürstensaummembranvesikeln des Dünndarmes und des proximalen Tubulus der Niere von Ratten. Thesis, Johann Wolfgang Goethe-Universität, Frankfurt/Main
Rosenberg SA, McIntosh JR (1968) Erythrocyte membranes: Effects of sonication. Biochim Biophys Acta 163:285–289
Rosengren J, Påhlman S, Glad M, Hjertén S (1975) Hydrophobic interaction chromatography on noncharged sepharose derivatives. Binding of a model protein, related to ionic strength, hydrophobicity of the substituent, and degree of substitution (determind by NMR). Biochim Biophys Acta 412:51–61

Ross AH, McConnell HM (1977) Reconstitution of band 3, the erythrocyte anion exchange protein. Biochem Biophys Res Commun 74:1318–1325

Ross AH, McConnell HM (1978) Reconstitution of the erythrocyte anion channel. J Biol Chem 253:4777–4782

Rothman JE, Lenard J (1977) Membrane asymmetry. The nature of membrane asymmetry provides clues to the puzzle of how membranes are assembled. Science 195:743–753

Rothstein A, Cabantchik ZI, Balshin M, Juliano R (1975) Enhancement of anion permeability in lecithin vesicles by hydrophobic proteins extracted from red blood cell membranes. Biochem Biophys Res Commun 64:144–150

Rottem S, Cirillo VP, de Kruyff B, Shinitzky M, Razin S (1973) Cholesterol in mycoplasma membranes. Correlation of enzymic and transport activities with physical state of lipids in membranes of *Mycoplasma mycoides* var. *Capri* adapted to grow with low cholesterol concentrations. Biochim Biophys Acta 323:509–519

Salter DW, Baldwin SA, Lienhard GE, Weber MJ (1982) Proteins antigenically related to the human erythrocyte glucose transporter in normal and Rous sarcoma virus-transformed chicken embryo fibroblasts. Proc Natl Acad Sci USA 79:1540–1544

Sandermann H Jr (1978) Regulation of membrane enzymes by lipids. Biochim Biophys Acta 515:209–237

Sandermann H, Strominger JL (1974) C_{55}-isoprenoid alcohol phosphokinase: An enzyme soluble in organic solvents. Methods Enzymol 32:439–446

Sandermann H Jr, Bavoil P, Nikaido H (1978) Phage lambda receptor protein from *Escherichia coli.* Solubilization and purification in an aprotic solvent. FEBS Lett 95:107–110

Sardet C, Tardieu A, Luzzati V (1976) Shape and size of bovine rhodopsin: A small-angle X-ray scattering study of a rhodopsin-detergent complex. J Mol Biol 105: 383–407

Sase S, Anraku Y, Nagano M, Osumi M, Kasahara M (1982) Random distribution of the glucose transporter of human erythrocytes in reconstituted liposomes. J Biol Chem 257:11100–11105

Scheuring U, Kollewe K, Schubert D (1984) A new method for the reconstitution of the anion transport system of the human erythrocyte membrane. Hoppe Seylers Z Physiol Chem 365:1056–1057

Schieren H, Rudolph S, Finkelstein M, Coleman P, Weissmann G (1978) Comparison of large unilamellar vesicles prepared by a petroleum ether vaporization method with multilamellar vesicles. ESR, diffusion and entrapment analyses. Biochim Biophys Acta 542:137–153

Schindler H, Quast U (1980) Functional acetylcholine receptor from *Torpedo marmorata* in planar membranes. Proc Natl Acad Sci USA 77:3052–3056

Schindler H, Rosenbusch JP (1978) Matrix protein from *Escherichia coli* outer membranes forms voltage-controlled channels in lipid bilayers. Proc Natl Acad Sci USA 75:3751–3755

Schmidt UM, Eddy B, Fraser CM, Venter JC, Semenza G (1983) Isolation of (a subunit of) the Na^+/D-glucose cotransporter(s) of rabbit intestinal brush border membranes using monoclonal antibodies. FEBS Lett 161:279–283

Schmidt UM, Eddy B, Fraser CM, Semenza G, Venter JC (1984) Immunoaffinity purification of the Na^+/D-glucose cotransport protein using monoclonal antibodies. Fed Proc 43:1085

Schöpp W, Grunow M, Tauchert H, Aurich H (1976) Specific hydrophobic chromatography of alcohol dehydrogenases on 10-carboxydecyl-sepharose. FEBS Lett 68: 198–202

Schraw WP, Regen DM (1983) Reconstitution of the D-glucose transporter of bovine thymocyte plasma membrane: Partial purification of transport activity by chromatography on agarose lentil lectin and agarose ethanethiol. Arch Biochem Biophys 220:214–224

Schubert D, Boss K (1982) Band 3 protein-cholesterol interactions in erythrocyte membranes. Possible role in anion transport and dependency on membrane phospholipid. FEBS Lett 150:4–8
Schubert D, Domning B (1978) A new method for the preparation of band 3, the main integral protein of the human erythrocyte membrane. Hoppe Seylers Z Physiol Chem 359:507–515
Schuurmans Stekhoven F, Bonting SL (1981) Transport adenosine triphosphatases: Properties and functions. Physiol Rev 61:1–76
Scott HL Jr, Coe TL (1983) A theoretical study of lipid-protein interactions in bilayers. Biophys J 42:219–224
Seelig A, Seelig J (1974) The dynamic structure of fatty acyl chains in a phospholipid bilayer measured by deuterium magnetic resonance. Biochemistry 13:4839–4845
Semenza G, Kessler M, Hosang M, Weber J, Schmidt U (1984) Biochemistry of the Na^+, D-glucose cotransporter of the small-intestinal brush-border membrane. The state of the art in 1984. Biochim Biophys Acta 779:343–379
Shanahan MF (1982) Cytochalasin B. A natural photoaffinity ligand for labeling the human erythrocyte glucose transporter. J Biol Chem 257:7290–7293
Shanahan MF, Czech MP (1977a) Partial purification of the D-glucose transport system in rat adipocyte plasma membranes. J Biol Chem 252:6554–6561
Shanahan MF, Czech MP (1977b) Purification and reconstitution of the adipocyte plasma membrane D-glucose transport system. J Biol Chem 252:8341–8343
Shanahan MF, Jacquez JA (1978) Differential labeling of components in human erythrocyte membranes associated with the transport of glucose. Membr Biochem 1:239–267
Shanahan MF, Olson SA, Weber MJ, Lienhard GE, Gorga JC (1982) Photolabeling of glucose-sensitive cytochalasin B binding proteins in erythrocyte, fibroblast and adipocyte membranes. Biochem Biophys Res Commun 107:38–43
Shimshick EJ, McConnell HM (1973) Lateral phase separations in binary mixtures of cholesterol and phospholipids. Biochem Biophys Res Commun 53:446–451
Shinitzky M, Inbar M (1976) Microviscosity parameters and protein mobility in biological membranes. Biochim Biophys Acta 433:133–149
Shinoda K (1954) The effect of alcohols on the critical micelle concentrations of fatty acid soaps and the critical micelle concentration of soap mixtures. J Phys Chem 58: 1136–1141
Shinoda K, Yamaguchi T, Hori R (1961) The surface tension and the critical micelle concentration in aqueous solution of β-D-alkyl glucosides and their mixtures. Bull Chem Soc Jpn 34:237–241
Sigrist H, Sigrist-Nelson K, Gitler C (1977) Single-phase butanol extraction: A new tool for proteolipid isolation. Biochem Biophys Res Commun 74:178–184
Silbernagl S, Foulkes EC, Deetjen P (1975) Renal transport of amino acids. Rev Physiol Biochem Pharmacol 74;105–167
Simmonds RJ, Yon RJ (1976) Protein chromatography on adsorbents with hydrophobic and ionic groups. Chromatography of dodecyl sulphate-solubilized proteins of the human erythrocyte membrane on *N*-(3-carboxypropionyl)aminodecyl-sepharose. Biochem J 157:153–159
Simon SA, McIntosh TJ, Latorre R (1982) Influence of cholesterol on water penetration into bilayers. Science 216:65–67
Simonds WF, Koski G, Streaty RA, Hjelmeland LM, Klee WA (1980) Solubilization of active opiate receptors. Proc Natl Acad Sci USA 77:4623–4627
Simons K, Helenius A, Garoff H (1973) Solubilization of the membrane proteins from Semliki Forest virus with Triton X100. J Mol Biol 80;119–133
Simons K, Helenius A, Leonard K, Sarvas M, Gething MJ (1978) Formation of protein micelles from amphiphilic membrane proteins. Proc Natl Acad Sci USA 75:5306–5310

Simpson IA, Yver DR, Hissin PJ, Wardzala LJ, Karnieli E, Salans LB, Cushman SW (1983) Insulin-stimulated translocation of glucose transporters in the isolated rat adipose cells: Characterization of subcellular fractions. Biochim Biophys Acta 763: 393–407

Singer SJ (1974) The molecular organization of membranes. Annu Rev Biochem 43: 805–833

Singer SJ (1977) Thermodynamics, the structure of integral membrane proteins, and transport. J Supramol Struct 6:313–323

Singer SJ, Nicolson GL (1972) The fluid mosaic model of the structure of cell membranes. Cell membranes are viewed as two-dimensional solutions of oriented globular proteins and lipids. Science 175:720–731

Small DM, Penkett SA, Chapman D (1969) Studies on simple and mixed bile salt micelles by nuclear magnetic resonance spectroscopy. Biochim Biophys Acta 176: 178–189

Sogin DC, Hinkle PC (1978) Characterization of the glucose transporter from human erythrocytes. J Supramol Struct 8:447–453

Sogin DC, Hinkle PC (1980) Immunological identification of the human erythrocyte glucose transporter. Proc Natl Acad Sci USA 77:5725–5729

Spector M, O'Neal S, Racker E (1980) Reconstitution of the $Na^{+}K^{+}$ pump of Ehrlich ascites tumor and enhancement of efficiency by quercetin. J Biol Chem 255:5504–5507

Spiess M, Hauser H, Rosenbusch JP, Semenza G (1981) Hydrodynamic properties of phospholipid vesicles and of sucrase isomaltase-phospholipid vesicles. J Biol Chem 256:8977–8982

Stasny JT, Crane FL (1964) The effect of sonic oscillation on the structure and function of beef heart mitochondria. J Cell Biol 22:49–62

Steck TL (1974) The organisation of proteins in the human red blood cell membrane. J Cell Biol 62:1–19

Stier A, Sackmann E (1973) Spin labels as enzyme substrates. Heterogeneous lipid distribution in liver microsomal membranes. Biochim Biophys Acta 311:400–408

Stipani I, Palmieri F (1983) Purification of the active mitochondrial tricarboxylate carrier by hydroxylapatite chromatography. FEBS Lett 161:269–274

Stoeckenius W, Engelman DM (1969) Current models for the structure of biological membranes. J Cell Biol 42:613–646

Stubbs GW, Litman BJ (1978) Effect of alterations in the amphipathic microenvironment on the conformational stability of bovine ospin. 1. Mechanism of solubilization of disk membranes by the nonionic detergent, octyl glucoside. Biochemistry 17:215–219

Stubbs GW, Smith HG Jr, Litman BJ (1976) Alkyl glucosides as effective solubilizing agents for bovine rhodopsin. A comparison with several commonly used detergents. Biochim Biophys Acta 425:46–56

Suarez-Isla BA, Wan K, Lindstrom J, Montal M (1983) Single-channel recordings from purified acetylcholine receptors reconstituted in bilayers formed at the tip of patch pipets. Biochemistry 22:2319–2323

Suzuki K, Kono T (1980) Evidence that insulin causes translocation of glucose transport activity to the plasma membrane from an intracellular storage site. Proc Natl Acad Sci USA 77:2542–2545

Sweadner KJ, Goldin SM (1975) Reconstitution of active ion transport by the sodium and potassium ion-stimulated adenosine triphosphatase from canine brain. J Biol Chem 250:4022–4024

Szoka F Jr, Papahadjopoulos D (1978) Procedure for preparation of liposomes with large internal aqueous space and high capture by reverse-phase evaporation. Proc Natl Acad Sci USA 75:4194–4198

Tanford C (1972) Hydrophobic free energy, micelle formation and the association of proteins with amphiphiles. J Mol Biol 67:59–74

Tanford C, Reynolds JA (1976) Characterization of membrane proteins in detergent solutions. Biochim Biophys Acta 457:133–170

Tank DW, Miller C, Webb WW (1982) Isolated-patch recording from liposomes containing functionally reconstituted chloride channels from *Torpedo electroplax*. Proc Natl Acad Sci USA 79:7749–7753

Tannenbaum C, Toggenburger G, Kessler M, Rothstein A, Semenza G (1977) High-affinity phlorizin binding to brush border membranes from small intestine: Identity with (a part of) the glucose transport system, dependence on the Na^+-gradient, partial purification. J Supramol Struct 6:519–533

Tanner MJA, Williams DG, Jenkins RE (1980) Structure of the erythrocyte anion transport proteins. Ann NY Acad Sci 341:455–464

Taverna RD, Langdon RG (1973) D-Glucosyl isothiocyanate, an affinity label for the glucose transport proteins of the human erythrocyte membrane. Biochem Biophys Res Commun 54:593–599

Thomas L (1973) Isolation of *N*-ethylmaleimide-labelled phlorizin-sensitive D-glucose binding protein of brush border membrane from rat kidney cortex. Biochim Biophys Acta 291:454–464

Toggenburger G, Kessler M, Rothstein A, Semenza G, Tannenbaum C (1978) Similarity in effects of Na^+ gradients and membrane potentials on D-glucose transport by, and phlorizin binding to, vesicles derived from brush borders of rabbit intestinal mucosal cells. J Membr Biol 40:269–290

Toggenburger G, Kessler M, Semenza G (1982) Phlorizin as a probe of the small-intestinal Na^+, D-glucose cotransporter. A model. Biochim Biophys Acta 688: 557–571

Touraille S, Briand Y, Durand R, Bonnafous J-C, Mani J-C (1981) Purification of a phosphate carrier in pig heart mitochondria by affinity chromatography on mersalyl-ultrogel. FEBS Lett 128:142–144

Träuble H, Eibl H (1974) Electrostatic effects on lipid phase transitions: Membrane structure and ionic environment. Proc Natl Acad Sci USA 71:214–219

Turner RJ, Moran A (1982) Heterogeneity of sodium-dependent D-glucose transport sites along the proximal tubule: evidence from vesicle studies. Am J Physiol 242: F406–F414

Tzartos SJ, Changeux J-P (1984) Lipid-dependent recovery of α-bungarotoxin and monoclonal antibody binding to the purified α-subunit from *Torpedo marmorata* acetylcholine receptor. Enhancement by noncompetitive channel blockers. J Biol Chem 259:11512–11519

Uesugi S, Dulak NC, Dixon JF, Hexum TD, Dahl JL, Perdue JF, Hokin LE (1971) Studies on the characterization of the sodium-potassium transport adenosine triphosphatase. VI. Large scale partial purification and properties of a lubrol-solubilized bovine brain enzyme. J Biol Chem 246:531–543

Ullrich KJ (1979) Sugar, amino acid and Na^+ cotransport in the proximal tubule. Annu Rev Physiol 41:181–195

Unden G, Kröger A (1982) Reconstitution in liposomes of the electron-transport chain catalyzing fumarate reduction by formate. Biochim Biophys Acta 682:258–263

Utsumi K, Okimasu E, Morimoto YM, Nishihara Y, Miyahara M (1982) Selective interaction of cytoskeletal proteins with liposomes. FEBS Lett 141:176–180

Van der Steen ATM, de Kruijff B, de Gier J (1982) Glycophorin incorporation increases the bilayer permeability of large unilamellar vesicles in a lipid-dependent manner. Biochim Biophys Acta 691:13–23

Van Hoogevest P, van Duijn G, Batenburg AM, de Kruijff B, de Gier J (1983) The anion permeability of vesicles reconstituted with intrinsic proteins from the human erythrocyte membrane. Biochim Biophys Acta 734:1–17

Van Renswoude J, Kempf C (1984) Purification of integral membrane proteins. Methods Enzymol 104:329–339

Vasseur M, Ferard G, Pousse A (1978) Rat intestinal brush border enzymes release by deoxycholate in vivo. Pflügers Arch 373:133–138

Verkleij AJ, Zwaal RFA, Roelofsen B, Comfurius P, Kastelijn D, van Deenen LLM (1973) The asymmetric distribution of phospholipids in the human red cell membrane. A combined study using phospholipases and freeze-etch electron microscopy. Biochim Biophys Acta 323:178–193

Verkleij AJ, Ververgaert PHJTh, de Kruyff B, van Deenen LLM (1974) The distribution of cholesterol in bilayers of phosphatidylcholines as visualized by freeze fracturing. Biochim Biophys Acta 373:495–501

Wacker H, Müller F, Semenza G (1976) Incorporation of hydrophobic aminopeptidase from hog kidney into egg lecithin liposomes: Number and orientation of aminopeptidase molecules in the lecithin vesicles. FEBS Lett 68:145–152

Wagner SL, Dean WL, Gray RD (1984) Effect of a zwitterionic detergent on the state of aggregation and catalytic activity of cytochrome P-450$_{LM_2}$ and NADPH-cytochrome P-450 reductase. J Biol Chem 259:2390–2395

Wardzala LJ, Jeanrenaud B (1981) Potential mechanism of insulin action on glucose transport in the isolated rat diaphragm. Apparent translocation of intracellular transport units to the plasma membrane. J Biol Chem 256:7090–7093

Wardzala LJ, Cushman SW, Salans LB (1978) Mechanism of insulin action on glucose transport in the isolated rat adipose cell. Enhancement of the number of functional transport systems. J Biol Chem 253:8002–8005

Warren GB, Toon PA, Birdsall NJM, Lee AG, Metcalfe JC (1974a) Complete control of the lipid environment of membrane-bound proteins: Application to a calcium transport system. FEBS Lett 41:122–124

Warren GB, Toon PA, Birdsall NJM, Lee AG, Metcalfe JC (1974b)Reconstitution of a calcium pump using defined membrane components. Proc Natl Acad Sci USA 71: 622–626

Watanabe T, Smith MM, Robinson FW, Kono T (1984) Insulin action on glucose transport in cardiac muscle. J Biol Chem 259:13117–13122

Weber K, Kuter DJ (1971) Reversible denaturation of enzymes by sodium dodecyl sulfate. J Biol Chem 246:4504–4509

Weber K, Osborn M (1969) The reliability of molecular weight determinations by dodecyl sulfate-polyacrylamide gel electrophoresis. J Biol Chem 244:4406–4412

Weber TM, Eichholz A (1985) Characterization of a photosensitive glucose derivative. A photoaffinity reagent for the erythrocyte hexose transporter. Biochim Biophys Acta 812:503–511

Wheeler TJ, Hinkle PC (1981) Kinetic properties of the reconstituted glucose transporter from human erythrocytes. J Biol Chem 256:8907–8914

Wheeler TJ, Simpson IA, Sogin DC, Hinkle PC, Cushman SW (1982) Detection of the rat adipose cell glucose transporter with antibody against the human red cell glucose transporter. Biochem Biophys Res Commun 105:89–95

Wieth JO, Brahm J (1980) Kinetics of bicarbonate exchange in human red cells – physiological implications. In: Lassen UV, Ussing HH, Wieth JO (eds) Membrane transport in erythrocytes. Munksgaard, Copenhagen, pp 467–487

Wilchek M, Miron T, Kohn J (1984) Affinity chromatography. Methods Enzymol 104: 3–55

Wolf HU, Dieckvoss G, Lichtner R (1977) Purification and properties of high Ca^{2+}-ATPase of human erythrocyte membranes. Acta Biol Med Ger 36:847–858

Wolosin JM (1980) A procedure for membrane-protein reconstitution and the functional reconstitution of the anion transport system of the human-erythrocyte membrane. Biochem J 189:35–44

Wolosin JM, Ginsburg H, Cabantchik ZI (1977) Functional characterization of anion transport system isolated from human erythrocyte membranes. J Biol Chem 252: 2419–2427

Womack MD, Kendall DA, MacDonald RC (1983) Detergent effects on enzyme activity and solubilization of lipid bilayer membranes. Biochim Biophys Acta 733:210–215
Wright EM, Peerce BE (1984) Identification and conformational changes of the intestinal proline carrier. J Biol Chem 259:14993–14996
Yeagle PL (1983) Cholesterol modulation of $(Na^{+}+K^{+})$-ATPase ATP hydrolyzing activity in the human erythrocyte. Biochim Biophys Acta 727:39–44
Yeagle PL, Bensen J, Boni L, Hui SW (1982) Molecular packing of cholesterol in phospholipid vesicles as probed by dehydroergosterol. Biochim Biophys Acta 692:139–146
Yeung WK, Weisman G, Vidaver GA (1979) Active Ca^{2+} transport by vesicles reconstituted from Triton X-100-solubilized pigeon erythrocyte membrane. Biochim Biophys Acta 555:249–258
Yoda A, Clark AW, Yoda S (1984) Reconstitution of $(Na^{+}+K^{+})$-ATPase proteoliposomes having the same turnover rate as the membranous enzyme. Biochim Biophys Acta 778:332–340
Yu J, Steck TL (1975) Isolation and characterization of band 3, the predominant polypeptide of the human erythrocyte membrane. J Biol Chem 250:9170–9175
Yu J, Fischman DA, Steck TL (1973) Selective solubilization of proteins and phospholipids from red blood cell membranes by nonionic detergents. J Supramol Struct 1:233–248
Yuli I, Wilbrandt W, Shinitzky M (1981) Glucose transport through cell membrane of modified lipid fluidity. Biochemistry 20:4250–4256
Zaki L, Fasold H, Schuhmann B, Passow H (1975) Chemical modification of membrane proteins in relation to inhibition of anion exchange in human red blood cells. J Cell Physiol 86:471–494
Zala CA, Kahlenberg A (1976) Reconstitution of D-glucose transport in vesicles composed of lipids and a partially purified protein from the human erythrocyte membrane. Biochem Biophys Res Commun 72:866–874
Zoccoli MA, Baldwin SA, Lienhard GE (1978) The monosaccharide transport system of the human erythrocyte. Solubilization and characterization on the basis of cytochalasin B binding. J Biol Chem 253:6923–6930
Zumbuehl O, Weder HG (1981) Liposomes of controllable size in the range of 40 to 180 nm by defined dialysis of lipid/detergent mixed micelles. Biochim Biophys Acta 640:252–262
Zwaal RFA, Roelofsen B, Colley CM (1973) Localization of red cell membrane constituents. Biochim Biophys Acta 300:159–182

Rev. Physiol. Biochem. Pharmacol., Vol. 104

Renal Erythropoietin: Properties and Production

WOLFGANG JELKMANN

Dedicated to Professor James W. Fisher
on the occasion of his recent 60th birthday

Contents

Institut für Physiologie, Medizinische Universität zu Lübeck, D–2400-Lübeck

1 Introduction

Erythropoiesis counterbalances the permanent loss of aged red blood cells. Erythropoiesis increases after phlebotomy and subsides after transfusion. Hence red cell mass, blood hemoglobin concentration, and oxygen transport capacity are relatively constant. In addition, erythropoiesis may be stimulated by lowered arterial oxygen tension, e.g., by residence at high altitude. Clearly, erythropoietic activity is regulated on the basis of feedback mechanisms which involve tissue oxygen tension. It is reasonable to assume that both hormonal and cellular factors play an important role in these mechanisms.

The existence of a humoral erythropoiesis-regulating factor was first proposed by *Carnot* and *Deflandre* (1906), who found that the injection of small volumes of serum from anemic rabbits caused polycythemia in normal animals. The humoral erythropoietic factor was initially called "hemopoietine." The more specific term "erythropoietin" was introduced by *Bonsdorff* and *Jalavisto* (1948). In fact, it is now known that several distinct glycoproteins are concerned with the production of the different types of blood cells: erythropoietin with red cells, colony-stimulating factor(s) with granulocytes (*Brennan* et al. 1980; *Metcalf* 1981), and thrombopoietin with platelets (*McDonald* 1981; *McDonald* and *Shadduck* 1982).

Erythropoietin is mainly produced in the kidney. *Jacobson* et al. (1957a) first showed that bilateral nephrectomy in adult mammals prevented an increase in plasma erythropoietin following bleeding. Subsequently, erythropoietic activity was detected in the blood perfusate of isolated kidneys (*Kuratowska* et al. 1961; *Fisher* and *Birdwell* 1961) and in cultures of kidney tissue (*Ozawa* 1967; *McDonald* et al. 1969). Immunofluorescence studies with antisera to erythropoietin revealed staining in the glomerular tuft of hypoxic kidneys (*Fisher* et al. 1965a; *Frenkel* et al. 1968). However, until recently it was not clear whether renal cells directly produce erythropoietin or whether they produce an enzymic activity ("erythrogenin") that liberates erythropoietin from an α-globulin in plasma (*Gordon* et al. 1967a). Recent evidence suggests that the kidney can directly produce erythropoietin. *Erslev* (1974) has reported erythropoietin synthesis by isolated kidneys perfused with serum-free medium. Significant amounts of erythropoietin were extracted from hypoxic kidneys flushed free of blood (*Jelkmann* and *Bauer* 1981).

In view of the several previous reviews on erythropoietin (*Krantz* and *Jacobson* 1970; *Fisher* 1972; *Gordon* 1973; *Fried* 1975; *Graber* and *Krantz* 1978; *Sherwood* 1984) it is not intended here to provide a complete survey of the early literature. In addition, *Erslev* (1980) has recalled the story of early erythropoiesis research in a delightful article.

The exploitation of new technologies has resulted in major advances in erythropoietin research, especially during the past 15 years. Important progress has been made in the purification of erythropoietin. In 1968, *Goldwasser* and *Kung* calculated on the basis of their experience with the purification of sheep erythropoietin that 3250 l urine from anemic patients would be required to prepare 10 mg purified human erythropoietin. The authors also noted that "This would represent about three years' daily collection from a single patient, or one month's collection from 36 patients, which does not seem to be an impossible goal." The authors have indeed achieved their goal (*Miyake* et al. 1977). The availability of pure human erythropoietin has significantly improved the possibility of studying cellular effects of the hormone. It has also enabled investigators to develop more reliable radioimmunoassays for erythropoietin.

Tissue cultures have provided a new approach to the study of the production of erythropoietin. The initial work was done on organ cultures. Only recently have homogeneous renal cell lines been utilized (*Kurtz* et al. 1983a; *Sytkowski* et al. 1983). Cell culture studies have several objectives: production of erythropoietin for therapeutic uses, identification of the erythropoietin-producing renal cell, and investigation of the cellular mechanisms of erythropoietin synthesis.

Notworthy investigations also include those which have approached the question of how renal hypoxia may induce the elaboration of erythropoietin. Current concepts propose that renal prostaglandins and cyclic AMP are involved in this reaction (cf. *Fisher* 1980, 1983).

The recent studies on the chemistry and production of erythropoietin are dealt with in more detail in the present monograph. Furthermore, information will be provided on assays and actions of erythropoietin. Although the coverage includes primarily laboratory experiments, some relevant clinical findings are also reported. Additional information on the pathophysiology of erythropoiesis may be found in the monographs by *Fisher* (1977) and *Dunn* (1983).

2 Regulation of Erythropoiesis

Blood cells are continuously generated in the hemopoietic organs. This process involves a sequence of cell proliferation, differentiation, and maturation. Three main developmental stages of hemopoietic cells are generally distinguished: (a) pluripotent hemopoietic stem cells; (b) progenitors committed to a distinct blood cell line; and (c) differentiated, morphologically identifiable precursor cells. At least 12 replications occur along the erythropoietic pathway from the most primitive stem cell to the reticulocyte (*Schofield* and *Lajtha* 1976). Hereby, the progression in differentiation is associated with a loss of proliferative potential. Erythropoietin enhances the production of red cells mainly by amplifying the pool size of erythrocytic progenitors (*Chui* et al. 1971; *Reissmann* and *Udupa* 1972; *Papayannopoulou* and *Finch* 1972). The high erythropoietin titers in stress erythropoiesis can cause macrocytosis due to a skipping of terminal cell divisions (*Stohlman* 1967).

Most of the information about the different stages of erythropoietic cells and their control by erythropoietin has been derived from in vitro studies in which the cells together with their progeny were cultured in semisolid media containing fibrin (*Stephenson* et al. 1971), methylcellulose (*Iscove* et al. 1974), or agar (*Johnson* and *Metcalf* 1977). Adult bone marrow, fetal liver, or murine spleen have been used most commonly as a source of erythropoietic cells for culture. Because the nomenclature of the various hemopoietic cells and their potential growth factors is not completely standardized, and actually sometimes confusing, the reader is referred to a synopsis of the most important terms employed in cell culture studies of hemopoiesis (*Lajtha* 1979).

2.1 The Erythropoietic Progenitor Cells

The most primitive hemopoietic stem cells seem to produce progenitors of all types of blood cells, including the myelopoietic and lymphopoietic lines (*Abramson* et al. 1977). The murine myelopoietic pluripotent stem cell in progeny is called "colony-forming unit spleen" (CFU-S), since *Till* and *McCulloch* (1961) found that mouse bone marrow cells can repopulate the spleen of heavily irradiated mice to form erythrocytic, granulocytic, megakaryocytic, and mixed colonies. CFU-Ss have the capacity for self-renewal and for differentiation. CFU-Ss are usually in a resting state (G_o phase). They can be triggered into in vitro proliferation by various neurotransmitter and hormonal agents, including β_1-adrenergic and cholinergic agonists, testosterone (*Byron* 1975; *Jepson* et al. 1974), and prostaglandin

E_2 (*Fehér* and *Gidáli* 1974). Thymus cells or T-lymphocytes augment the in vivo proliferation and differentiation of CFU-Ss. This effect is possibly mediated by the lymphokine interleukin-3 (*Garland* and *Crompton* 1983). Although almost all knowledge about the pluripotent stem cells is based on findings in the mouse, a similar clonal origin of human hemopoietic cells has been made likely by the demonstration of common chromosomal markers among different cell lineages (cf. *Cline* and *Golde* 1979; *Quesenberry* and *Levitt* 1979; *Spivak* and *Graber* 1980).

How CFU-Ss are committed to the erythrocytic or any other cell line remains to be clarified. Present concepts are: (a) the stochastic model, which proposes that stem cell commitment is a random event (*Till* et al. 1964); (b) the hemopoietic inductive microenvironment concept (*Trentin* 1970); and (c) the instructional model, which proposes that stem cells are committed to one of the sublines by inducers like erythropoietin during distinct periods of their cell cycle (*Goldwasser* 1975). *Goldwasser* (1981) and *Ogawa* et al. (1983) have recently discussed arguments for and against these concepts.

The principal compartment of erythropoietin-responsive cells (ERCs) is exhibited by the sequence of erythrocytic progenitors (Fig. 1). A primitive population of cells close to the CFU-S is called burst-forming unit erythroid (BFU-E), because it produces large multiclustered bursts of hemoglobin-synthesizing cells when cultured with high doses of erythropoietin (*Axelrad* et al. 1974; *Iscove* and *Sieber* 1975). Two different types of BFU-E have been detected in mouse bone marrow: a primitive (day 8) BFU-E that gives rise to bursts of several hundred erythroblasts after 8–10 days in culture, and a mature (day 3) BFU-E that gives rise to early bursts of 50–200 erythroblasts (*Gregory* 1976). Human bone marrow cultures also contain two subclasses of BFU-E that show peak growth after 17–20 and 10–12 days respectively (*Gregory* and *Eaves* 1977). The primitive type of BFU-E was also found in human (*Clarke* and *Housman* 1977; *Nathan* et al. 1978) and murine blood (*Hara* and *Ogawa* 1977).

A third and more differentiated erythrocytic progenitor is the colony-forming unit erythroid (CFU-E). This cell is present in erythropoietic tissues but is very rare in blood (*Eaves* and *Eaves* 1984). Upon microscopic evaluation, isolated CFU-Es appear similar to basophilic erythroblasts (*Nijhof* and *Wierenga* 1983). CFU-Es require low concentrations of erythropoietin for in vitro growth (*Gregory* 1976). CFU-Es produce small erythroid clusters of 8–60 erythroblasts after 2–4 days in cultures of murine (*Stephenson* et al. 1971) and after 7–8 days in cultures of human cells (*Gregory* and *Eaves* 1977). Apart from their growth characteristics, the three erythropoietic progenitors differ in a number of physical and biological parameters (*Heath* et al. 1976; *Gregory* 1976; *Gregory* and *Eaves* 1978).

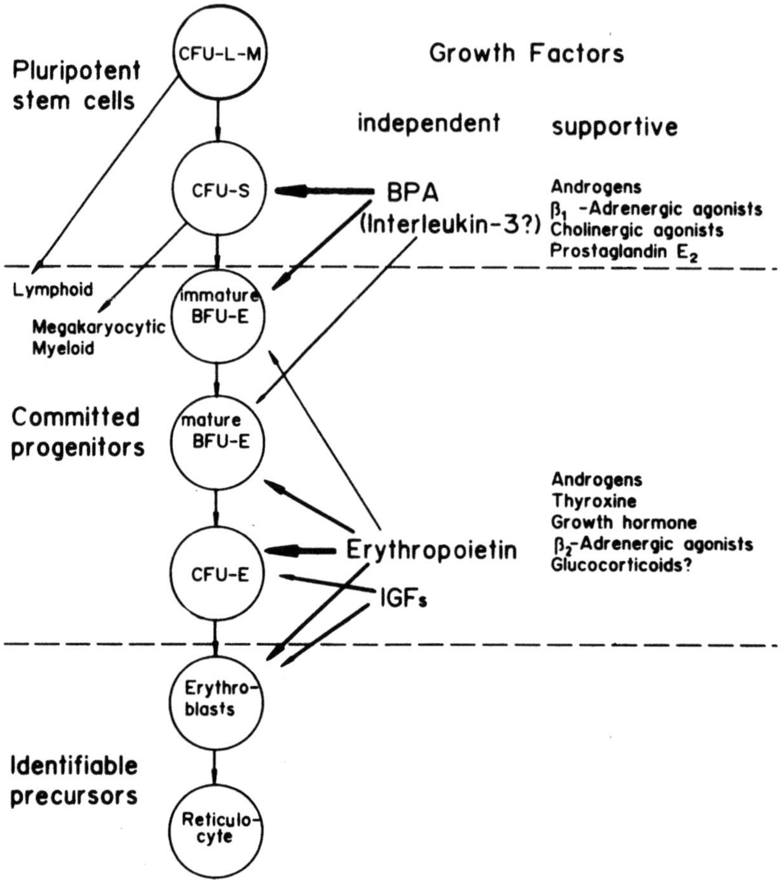

Fig. 1. Scheme of erythropoiesis and plausible sites of action of hormonal and neurotransmitter agents. *BPA*, burst-promoting activity; *IGFs*, insulin-like growth factors; *CFU-L-M*, colony-forming unit lymphoid-myeloid; *CFU-S*, colony-forming unit spleen; *BFU-E*, burst-forming unit erythroid; *CFU-E*, colony-forming unit erythroid

Finally, CFU-Es feed into the pool of morphologically identifiable erythrocytic cells. After the transition from the pro- to the basophilic erythroblast, globin mRNA becomes detectable (*Harrison* et al. 1974). Hemoglobin synthesis is then initiated, and continues up to the reticulocytic stage. Cell divisions cease after the polychromatophilic stage (*Denton* and *Arnstein* 1973).

2.2 Hormonal Influences

Primitive BFU-Es are relatively insensitive to erythropoietin. Their in vivo pool size was little influenced by erythroid perturbations and changes in the plasma erythropoietin titer (*Iscove* 1977; *Adamson* et al. 1978; *Udupa* and *Reissmann* 1979). Primitive BFU-Es did not require erythropoietin for in vitro survival (*Iscove* and *Guilbert* 1978). Yet they could be triggered into proliferation with purified erythropoietin (*Braun* et al. 1979; *Dessypris* and *Krantz* 1984). High concentrations of erythropoietin were necessary for in vitro growth of erythroid bursts (*Gregory* 1976). The erythropoietin requirement was reduced by the addition of conditioned media derived from leukocytes (*Aye* 1977), irradiated bone marrow cells (*Wagemaker* 1978), and mitogen-stimulated spleen cells (*Iscove* 1978; *Metcalf* and *Johnson* 1979). The activity in these media is generally called "burst-promoting activity" (BPA) according to *Iscove* (1978) (or "burst-feeding activity" or "burst-promoting factor"). It has not been clearly established whether T-lymphocytes (*Nathan* et al. 1978) or blood and tissue macrophages (*Zuckerman* 1981) are the main cellular source of BPA. BPA has not been purified as yet and may not represent a single entity (*Eliason* and *Van Zant* 1983). Note, however, that pure interleukin-3 seems to possess the same properties as partially purified BPA (*Rich* et al. 1983). An erythroid-potentiating activity (EPA) has been purified from human T-lymphocyte-conditioned medium which, by contrast with BPA, stimulated the growth not only of BFU-Es but also of late erythroid progenitors of the CFU-E type (*Westbrook* et al. 1984).

The in vivo pool of mature BFU-Es is affected by erythroid perturbations. *Adamson* et al. (1978) observed an increase in the number and proliferation rate of BFU-Es in anemic mice. The reverse, i.e., suppression of BFU-Es in plethora, has been confirmed by some investigators (*Adamson* et al. 1978; *Peschle* et al. 1979) but not by others (*Gregory* and *Eaves* 1978). Mature BFU-Es were much more sensitive to erythropoietin in vitro than immature BFU-Es (*Gregory* 1976).

The CFU-E pool size in vivo is largely correlated with plasma erythropoietin activity. CFU-Es were suppressed in plethora and activated in anemia (*Gregory* et al. 1974; *Iscove* 1977; *Gregory* and *Eaves* 1978; *Adamson* et al. 1978; *Peschle* et al. 1979). The erythropoietin concentration needed for maximal in vitro growth of CFU-Es was almost two orders of magnitude lower than for mature BFU-Es (*Gregory* 1976) and was not affected by BPA (*Gregory* and *Eaves* 1977). The number of erythroid clusters that grew from CFU-Es was closely correlated with erythropoietin concentration in culture (*McLeod* et al. 1974; *Iscove* et al. 1974). This relationship has been utilized for in vitro assay of erythropoietin (see Sect. 3.2). The steep increase in the in vitro dose response curve is in the range of physiological plasma erythropoietin titers (1–100 mU/ml). The erythropoietin/CFU-E colony relationship has been confirmed with pure erythropoietin in serum-free cultures (*Iscove* et al. 1980).

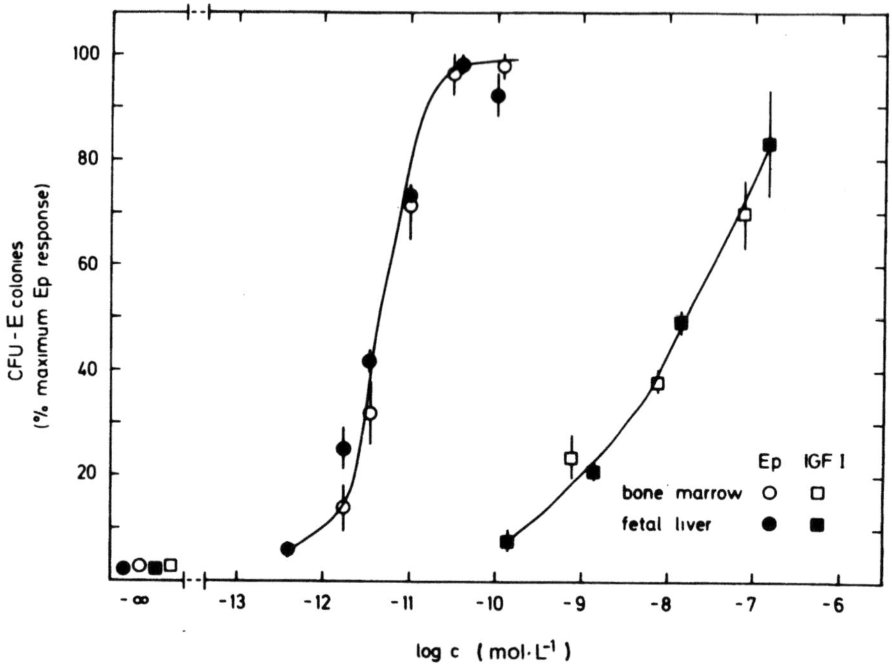

Fig. 2. Effects of erythropoietin (*Ep*) and insulin-like growth factor I (*IGF I*) on the growth of erythroid clusters from murine fetal liver and adult bone marrow CFU-Es. Hormones concentration is given on a molar basis. (*Kurtz* et al. 1982a)

Until recently, it was assumed that erythropoietin would be strictly required for in vitro growth of CFU-Es. However, *Kurtz* et al. (1982a) have shown that pure insulin-like growth factor I (IGF I or somatomedin C) at physiological concentrations induced the growth of erythroid clusters from murine bone marrow and fetal liver CFU-Es in the absence of erythropoietin, that is, in a serum-free culture system (Fig. 2). Insulin at very high concentrations had a similar effect, probably by activating IGF I receptors (*Kurtz* et al. 1983b). Future studies may show how IGF I relates to the poorly defined erythropoietic activities detected in spleen (*Fagg* 1981) or thymocyte-conditioned media (*Maeda* et al. 1983) and in serum (*Udupa* and *Reissmann* 1980; *Krystal* 1983; *Congote* 1984a). These activities were distinct from erythropoietin and associated side by side with small polypeptides (*Congote* 1984a) and proteins with molecular weights ranging from 30 000 (*Fagg* 1981; *Maeda* et al. 1983) to 130 000 daltons (*Krystal* 1983). Noteworthily, the polypeptide IGF I can form complexes with plasma proteins of 150 000 and 50 000–70 000 daltons that may dissociate to monomers of 30 000 daltons (*Zapf* et al. 1975; *Daughaday* et al. 1982). In addition, IGF I (*Kurtz* et al. 1985a) and IGF II (*Congote* 1985) may be of significance for fetal erythropoiesis.

Several hormones stimulate in vivo erythropoiesis. These include androgens, glucocorticoids, thyroxine, and growth hormone (cf. *Fisher* and *Gross* 1977). In addition, catecholamines may activate erythropoiesis through a β_2-adrenergic mechanism (*Fink* and *Fisher* 1977a; *Jelkmann* and *Bauer* 1980). In vitro studies indicate that these hormonal agents not only enhance renal erythropoietin production (see Sect. 6.3) but also directly stimulate erythropoietic tissues. The following hormonal agents augmented the effect of erythropoietin on in vitro CFU-E growth but were ineffective in the absence of erythropoietin: testosterone (*Moriyama* and *Fisher* 1974; *Singer* et al. 1976; *Singer* and *Adamson* 1976), thyroxine (*Popovic* et al. 1977; *Golde* et al. 1977a), growth hormone (*Golde* et al. 1977b), and β_2-adrenergic agonists (*Przala* et al. 1977; *Brown* and *Adamson* 1977a). Thyroxine (*Dainiak* et al. 1978), growth hormone (*Hoffman* et al. 1979), and β_2-adrenergic agonists (*Beckman* and *Fisher* 1979) also stimulated the in vitro growth of BFU-Es. β_2-Adrenergic agonists probably act through surface receptors linked to adenyl cyclase (*Brown* and *Adamson* 1977a). The cyclic AMP responsive cell compartment overlapped with the erythropoietin responsive cell compartment but was not identical with it as a whole (*Brown* and *Adamson* 1977a,b).

Glucocorticoids were studied with controversial results. Both stimulatory effects on CFU-Es (*Golde* et al. 1976) and BFU-Es (*Hoffman* et al. 1979) and inhibitory effects on CFU-Es (*Singer* et al. 1976) and BFU-Es (*Zalman* et al. 1979; *Leung* and *Gidari* 1982) have been reported. In addition, studies on the effects of parathyroid hormone on in vitro erythropoiesis yielded conflicting results (*Dunn* and *Trent* 1981; *Meytes* et al. 1981; *McGonigle* et al. 1984c), which is noteworthy with a view to hyperparathyroidism secondary to renal insufficiency.

Prostaglandins seem to enhance the growth of CFU-Es and BFU-Es indirectly by releasing erythropoietic factors from other cells, because the effect of prostaglandins was abolished when adherent cells were removed (*Rossi* et al. 1980). Similarly, the in vitro stimulation of CFU-E and BFU-E growth by platelet-derived growth factor (PDGF) appears to be mediated by adherent bone marrow cells, because blood BFU-Es were unaffected by PDGF (*Dainiak* et al. 1983).

2.3 Cellular Mechanisms of the Action of Erythropoietin

Erythropoietin is thought to bind to a protein receptor on the surface of its target cells (*Chang* et al. 1974; *Krantz* and *Goldwasser* 1984). Erythropoietin attached to glass beads did not lose its in vitro ability to stimulate heme synthesis (*Dukes* and *Einbinder* 1975). Erythropoietin immobilized on agarose-bound lectin stimulated the growth of CFU-Es like unbound erythropoietin (*Roodman* et al. 1981). In addition, *Goldwasser* (1975) failed to detect radioactivity in the cytoplasm of cells after exposure to ^{125}I-labeled erythropoietin. *Lafferty* et al. (1980) incubated fetal mouse liver cells with erythropoietin and subsequently with antiserum to erythropoietin. The erythropoietin antibody complex was visualized on cell surfaces with a gold-labeled IgG reagent. The surface labeling was most extensive on primitive nonidentifiable cells and on early polychromatophilic erythroblasts, and decreased with the maturation stage of the cells. Similar results were obtained with human bone marrow cells (*Lafferty* et al. 1981).

The intracellular events following receptor binding of erythropoietin are not well understood. The earliest nuclear effect induced by erythropoietin is apparently on transcription, as RNA synthesis is rapidly stimulated. The effect of erythropoietin on transcription is possibly mediated by a cytoplasmic protein (*Chang* and *Goldwasser* 1973) that acts as an intracellular second messenger on RNA polymerases (*Goldwasser* and *Inana* 1978). The in vivo administration of erythropoietin resulted in a sequential activation of nuclear RNA polymerases in the spleen of polycythemic mice (*Piantadosi* et al. 1976). It is assumed that different types of RNAs are subsequently produced (*Goldwasser* 1975), including globin mRNA (*Terada* et al. 1972; *Ramirez* et al. 1975). Following RNA synthesis, DNA synthesis (*Rudolph* and *Perretta* 1967; *Dukes* 1968; *Orlic* et al. 1968), cell division and, in erythroblasts, iron uptake and hemoglobin synthesis were initiated (*Gallien-Lartigue* and *Goldwasser* 1964; *Cole* and *Paul* 1966; *Nakao* et al. 1966). Purified human erythropoietin exerted all of these effects in rat and mouse bone marrow cultures (*Weiss* and *Goldwasser* 1981; *Datta* and *Dukes* 1982).

Several other cellular processes are stimulated in association with the action of erythropoietin: acetylation and methylation of nuclear proteins (*Takaku* et al. 1969; *Spivak* and *Peck* 1979), incorporation of glucosamine (*Dukes* et al. 1964), transmembrane flux of calcium (*Misiti* and *Spivak* 1979; *Sawyer* and *Krantz* 1984), synthesis of membrane proteins (*Tong* and *Goldwasser* 1981) and production of cyclic nucleotides. Elevated cyclic AMP levels were observed in erythropoietin-stimulated cells by some authors (*Chiuini* et al. 1979) but not by others (*Graber* et al. 1974; *Rodgers* et al. 1976). Effects of erythropoietin on cyclic guanosine monophosphate (GMP) levels seemed to be secondary to erythropoietin-induced cell cycle changes (*White* and *George* 1981).

In summary, erythropoietin stimulates erythropoiesis mainly by amplifying the pool of erythroid-committed cells. The erythropoietin sensitivity of erythropoietic progenitors increases with their state of differentiation. Cell culture, immunohistochemical, and in vivo studies suggest that the CFU-E is the primary target cell of the hormone. Erythropoietin probably binds to a membrane receptor on its target cells. Possibly, a cytoplasmic protein acts as the intracellular messenger. The intracellular effects of erythropoietin include RNA synthesis, DNA synthesis, and cell division, thus increasing the progeny of hemoglobin-producing cells.

3 Standardization, Assays, and Normal Plasma Values

In vivo and in vitro bioassays, hemagglutination inhibition assay, and radioimmunoassay have been used to determine erythropoietin activities. Because the results obtained with these assays may differ significantly (*Dukes* et al. 1970; *Sherwood* and *Goldwasser* 1979; *Lange* et al. 1980), erythropoietin data should be clearly identified by the assay system employed. *Heim* (1980) has recommended a unified terminology using subscripts which specify the assay system (e.g., EP_{FMLC} for erythropoietin assayed in the fetal mouse liver cell system). A comparison of erythropoietin titers in normal human serum resulting from the different assay techniques is found in the article by *Lange* et al. (1980). Furthermore, *Dunn* and *Lange* (1980) have summarized the essential criteria for valid quantitation of bioassay results.

Erythropoietin activities are usually expressed in international units (IU). One unit of erythropoietin was originally defined as the activity which produced the same erythropoiesis-stimulating effect as 5 μmol cobalt. The international reference preparation of erythropoietin from human urine (*Annable* et al. 1972) for the calibration of laboratory standards is available in ampoules containing 10 IU from the Division of Biological Standards, National Institute for Medical Research, Holly Hill, Hampstead, London NW3 6RB, United Kingdom. Erythropoietin for laboratory usage can be prepared from urine of anemic patients or from plasma of hypoxic animals. Commercial erythropoietin preparations are available (cf. *Levine* et al. 1983).

3.1 In Vivo Bioassays

In in vivo bioassays test samples are injected into animals whose endogenous erythropoietin production is suppressed by an absolute or relative plethora of red blood cells. The erythropoietic potency of the samples is determined by measurement of the rate of incorporation of radioactive iron into newly formed red blood cells. Erythropoietin titers are calculated from log-dose response or log-dose log-response curves obtained with standardized erythropoietin preparations. The assay has been performed in hypophysectomized or fasting rats (*Fried* et al. 1957), hypertransfused mice (*Jacobson* et al. 1957b), and mice made polycythemic by exposure to hypobaric hypoxia (*Cotes* and *Bangham* 1961), normobaric hypoxia (*McDonald* and *Lange* 1967), or carbon monoxide (*Fogh* 1966).

With minor modifications, the procedure in mice exposed to low pressure has been most widely adopted as follows (*Weintraub* et al. 1963; *Kazal* and *Erslev* 1975a; *Fink* and *Fisher* 1977a; *Dunn* and *Lange* 1980). Mice are intermittently exposed to low pressure corresponding to about 7000 m altitude for at least 2 weeks and thereafter maintained at normal pressure for several days. Standard erythropoietin or test samples are then administered on 1–3 days, and this is followed by the injection of radioiron (^{59}Fe); 48–72 h later the mice are bled and the percentage of ^{59}Fe-incorporation into the blood is determined. The assay seems to be relatively specific for erythropoietin and at present is accepted as the international reference determination. Nevertheless, it is not sufficiently sensitive (> 25 mU/mouse), in that it does not enable one to measure erythropoietin in unconcentrated normal plasma or urine samples. In addition, the assay requires many test mice and is expensive and time-consuming. Erythropoietic agents such as cobalt or androgens and contaminants such as endotoxin (*Zuckerman* et al. 1979) can interfere with the assay. Inappropriate dose response relationships may be obtained with crude human serum samples (*Dunn* and *Napier* 1978). Therefore, in order to perform valid measurements, it is necessary to prove that the dose response curves to the test material and to the erythropoietin standard are parallel. According to *Erslev* et al. (1979), normal human plasma titers of in vivo active erythropoietin range from > 5 mU/ml to 18 mU/ml with a mean of 8 mU/ml. These values have been based on assays of highly concentrated extracts of boiled plasma samples.

3.2 In Vitro Bioassays

Erythropoietin activities are here assessed in cultures of erythropoietic cells. The assay is generally carried out in cultures from rat (*Bartley* and *Powsner* 1974; *Goldwasser* et al. 1975) or mouse (*McLeod* et al. 1974; *Iscove* et al. 1974) bone marrow or, most commonly, 12–15 day fetal mouse liver (*Stephenson* and *Axelrad* 1971; *Wardle* et al. 1973; *Dunn* et al. 1975; *Rich* and *Kubanek* 1979). The following parameters have been used for in vitro assay of erythropoietin: ^{59}Fe-incorporation into heme (*Ward* 1967; *Stephenson* and *Axelrad* 1971; *Wardle* et al. 1973; *Bartley* and *Powsner* 1974; *Dunn* et al. 1975; *Goldwasser* et al. 1975), ^{14}C-glucosamine uptake (*Dukes* et al. 1970), ^{3}H-thymidine (*Fredrickson* et al. 1977; *Brandan* et al. 1981), or ^{125}I-deoxyuridine (*Napier* and *Evans* 1980;

Dunn and *Gibson* 1983) incorporation into DNA, and growth of erythroid clusters from CFU-Es (*McLeod* et al. 1974; *Iscove* et al. 1974).

In vitro assays of erythropoietin are rapid, economic, and sensitive (limit: 1–5 mU/ml culture medium). In addition, they require much less test material than in vivo assays. Note that certain filtration-sterilization membranes bind erythropoietin (*Lowy* and *Keighley* 1968; *Gallicchio* and *Murphy* 1981). In vitro assays can detect deglycosylated erythropoietin, which is inactive in vivo (*Lukowsky* and *Painter* 1972; *Goldwasser* et al. 1974). Endotoxin, which is ubiquitously present in erythropoietin preparations, significantly inhibited erythropoiesis in vitro (*Zuckerman* et al. 1979).

Plasma samples are often toxic to cultures. Toxins could be partly removed from sera by prolonged dialysis (*Hågå* and *Falkanger* 1979) or heat inactivation (*Von Preyss* and *Goudsmit* 1976; *Brandan* et al. 1981). Plasma samples from patients with chronic renal failure contained substances which inhibited erythropoiesis (*Wallner* et al. 1978; *Ohno* et al. 1978; *Radtke* et al. 1981). According to *Dunn* and *Napier* (1975), the nonspecific erythropoietic agents iron, folic acid, and vitamin B_{12} did not significantly affect the in vitro assay of erythropoietin. Other investigators have reported interference with transferrin-bound serum iron (*De Klerk* et al. 1978; *Radtke* et al. 1978a). Serum samples (*Udupa* and *Reissmann* 1980; *Firkin* and *Russell* 1983; *Krystal* 1983; *Congote* 1984a) and tissue extracts (*Congote* 1983) contained proteins and polypeptides that stimulated in vitro erythropoiesis but were distinct from erythropoietin. Finally, the possibility must be excluded that test samples trigger endogenous erythropoietin production by fetal liver cells (*Zucali* and *Mirand* 1978) or by cocultured macrophages (*Rich* et al. 1982).

Most in vitro measurements of plasma titers of erythropoietin in normal humans ranged from 100 to 1000 mU/ml (*Dunn* et al. 1975; *Napier* et al. 1977; *Krystal* et al. 1981). *De Klerk* et al. (1978) and *Firkin* and *Russell* (1983) have developed modified in vitro bioassays which resulted in erythropoietin titers below 50 mU/ml in normal human plasma. The modifications included mathematical correction for the effects of serum iron (*De Klerk* et al. 1978) and experimental correction for the effects of nonspecific erythropoietic contaminants by assay of dose responses of standard erythropoietin dissolved in the test sera.

3.3 Immunological Assays

Immunological tests employed for assay of erythropoietin include agar double diffusion (*Goudsmit* et al. 1967), hemagglutination-inhibition (*Lange* et al. 1969); and radioimmunoassay (*Lertora* et al. 1975). Most of the studies were carried out with neutralizing or hemagglutinating antisera raised in rabbits to relatively crude erythropoietin preparations (cf. *Garcia* 1977; *Lange* and *Ichiki* 1977). A method for labeling erythropoietin with ^{125}I has been described by *Murphy* (1976). However, as valid immunoassays require the use of either pure antigen or monospecific antibody, attempts at assay of erythropoietin remained questionable until pure human hormone became available (*Miyake* et al. 1977). Commercial hemagglutination-inhibition test kits (*Jordan* et al. 1975) proved to measure contaminants rather than erythropoietin, at least in European laboratories (*Kolk-Vegter* et al. 1975; *Omran* and *Neumann* 1979; *De Klerk* et al. 1980; *Essers* and *Heilmann* 1981). In addition, the present immunoassay systems are not readily suited for the measurement of erythropoietin in animal studies. There can be significant species differences in the reactivity of antierythropoietin with the hormone (*Garcia* 1977; *Sherwood* and *Goldwasser* 1979). Hybridomas secreting monoclonal antibodies against human erythropoietin have been raised (*Weiss* et al. 1982; *Lee-Huang* 1982; *Yanagawa* et al. 1984a), which may prove useful in radioimmunological assays. *Sue* and *Sytkowski* (1983) immunized rabbits with a synthetic polypeptide composed of the putative 26 NH_2-terminal amino acids of human erythropoietin. The resulting antiserum bound human erythropoietin but did not neutralize its biological activity. It is known that the biologically active site of erythropoietin is not necessarily identical with the antigenic site, as hormone complexed with F $(ab')_2$ fragments retained its biological activity (*Schooley* 1980). *Sytkowski* and *Sue* (1984) have already employed the synthetic erythropoietin fragment as radiolabeled tracer together with antiserum against this fragment in a radioimmunoassay for human erythropoietin. The significance of these studies is evident, for the synthetic derivative of erythropoietin may become more readily available to investigators than purified erythropoietin.

For assay of erythropoietin in human blood, several investigators have described erythropoietin radioimmunoassays in which radiolabeled pure erythropoietin and antisera from rabbits immunized with less pure erythropoietin were used (*Garcia* et al. 1979; *Sherwood* and *Goldwasser* 1979; *Zaroulis* et al. 1981; *Miller* et al. 1981; *Garcia* et al. 1982; *Cotes* 1982; *Birgegård* et al. 1982; *Rege* et al. 1982). Mean serum titers of immunoreactive erythropoietin in normal adult humans have ranged from 13 to 28 mU/ml. Erythropoietin titers were neither sex nor age related (*Miller* et al. 1981; *Garcia* et al. 1982; *Cotes* 1982; *Rege* et al. 1982). Diurnal fluctuations of serum erythropoietin were observed in one case (*Cotes* and *Brozovic* 1982) but not in a study of a greater number of individuals (*Miller* et al. 1981). Radioimmunoassays may detect erythropoietin fragments that are devoid of in vivo biological activity (*Sherwood* and *Goldwasser* 1979). Accordingly, lowered serum titers of immunoreactive erythropoietin were found in patients with renal insufficiency following a single hemodialysis treatment, while regular erythropoietin is not expected to pass through dialysis membranes (*McGonigle* et al. 1984a). Note, too, that some of the antisera to erythropoietin used for radioimmunoassay reacted with desialated erythropoietin with the same ability as with the intact hormone (*Garcia* et al. 1979). *Cohen* et al. (1985) have demonstrated a good correlation between radioimmunoassay and polycythemic mouse bioassay for erythropoietin in human serum and urine.

In conclusion, in vivo and in vitro bioassays and radioimmunoassays have proved useful for the determination of erythropoietin. Each of these techniques has its own potentials but also certain limitations. The in vivo bioassay is at present regarded as the reference assay. Almost all of the results summarized in this article are based on this assay. Reliable radioimmunoassays have been developed recently. However, immunoassays are not applicable for routine measurements as yet, mainly because pure human erythropoietin is in short supply. In addition, even if pure human erythropoietin is available, accurate immunoassay of the hormone in nonhuman samples is not self-evident because of species differences in the antigen-antibody reaction.

4 Chemistry and Purification

Erythropoietin is a relatively heat (*Rambach* et al. 1957) and pH stable α_2-globulin (*Espada* et al. 1973; Table 1). Isoelectric focusing of human plasma erythropoietin in polyacrylamide gel yielded several glycoprotein components of similar molecular size and antigenetic properties (*Espada* et al. 1974). Thus there may be a microheterogeneity in the carbohydrate portion of the hormone. The isoelectric point values of the major components were in a pH range between 3 and 4 (*Dorado* et al. 1972). The in vivo erythropoietin activity is destroyed when the terminal sialic acid residues are removed by neuraminidase treatment or acid hydrolysis. The sialic acid of erythropoietin is necessary for its in vivo stability but not for its effect on the target cells (*Lukowsky* and *Painter* 1972; *Goldwasser* et al. 1974). Even completely deglycosylated erythropoietin stimulated the in vitro growth of erythropoietic cells (*Dordal* et al. 1985). Measurements of the molecular weight of erythropoietin have ranged from 23 000 daltons by means of gel electrophoresis in the presence of sodium dodecyl sulfate (SDS) (*Dorado* et al. 1974) to 62 000 daltons by Sephadex gel filtration (*Olesen* and *Fogh* 1968; *Lukowsky* and *Painter* 1968). It has been proposed that erythropoietin can exist in monomeric or polymeric forms (*Shelton* et al. 1975; *Neal* et al. 1978) and form complexes with

Table 1. Characterization of human erythropoietin[a]

Acidic glycoprotein	70% protein 30% carbohydrate
Isoelectric point:	pH 3.5–4.0
Molecular weight:	34 000 (SDS gel electrophoresis)
Stability:	pH stable (range pH 3–9) heat resistant (e.g., to boiling for 5 min)
Specific activity:	70 400 IU/mg protein

[a] For further details see *Espada* (1977), *Goldwasser* (1981), *Sherwood* (1984), and *Jacobs* et al. (1985)

other proteins (*Sytkowski* 1980). *Jacobs* et al. (1985) have recently described the complete nucleotide and deduced amino-acid sequence of human fetal liver erythropoietin cDNA. According to this study, the protein backbone of human erythropoietin is composed of 166 amino acids and has a molecular weight of 18 398 daltons.

Erythropoietin has been purified to apparent homogeneity from two sources, sheep plasma and human urine. *Goldwasser* and co-workers initially isolated erythropoietin from anemic sheep plasma. The starting material with a potency of 0.007 IU/mg protein was purified in a number of purification steps up to 8500 IU/mg (*Goldwasser* et al. 1962; *Goldwasser* and *Kung* 1971). The molecular weight of sheep erythropoietin was determined to be in the range of 46 000 daltons by means of SDS gel electrophoresis (*Goldwasser* and *Kung* 1972). The carbohydrate portion of sheep erythropoietin amounted to 24% and contained sialic acid, mannose, galactose, *N*-acetylglucosamine, and glucose (*Goldwasser* 1981).

In 1977, *Miyake* et al. succeeded in isolating 10 mg pure erythropoietin with a potency of 70 400 IU/mg protein in 21% yield from 2550 liters of urine from patients with aplastic anemia. The purification procedure included ion exchange chromatography, ethanol precipitation, gel filtration, and adsorption chromatography. The purified material produced a single band on gel electrophoresis with Triton X-100 at pH 6 and with SDS and mercaptoethanol at pH 7. Pure human erythropoietin was initially reported to have a molecular weight of 39 000 daltons, but reexaminations showed it to be closer to 34 000 daltons (*Wang* et al. 1985). The sugar content amounts to about 30% (*Dordal* et al. 1985). After the final hydroxylapatite fractionation, *Miyake* et al. (1977) obtained two fractions (α and β) of similar potency and molecular weight but slightly different electrophoretic mobility at pH 9. The overall amino acid composition of α- and β-erythropoietin is similar (*Goldwasser* 1981). However, their carbohydrate composition differs. *Dordal* et al. (1985) have identified the following sugars in α-(β-) erythropoietin (mol/mol protein): fucose 4 (4), galactose 11 (11), mannose 9 (8), *N*-acetylglucosamine 12 (9), and sialic acid 16 (12).

A dexterous method to isolate erythropoietin has been developed recently by *Yanagawa* et al. (1984b). These authors applied human urinary erythropoietin with a specific activity of 14 IU/mg protein to an immunoadsorbent column containing fixed monoclonal antibody against erythropoietin. The bound hormone was then released from the column in 75% yield with acetate buffer at pH 2.5. This single step product had a specific activity of 59 000 IU/mg protein. Additional gel filtration on Sephadex G-100 yielded erythropoietin with a potency of 81 600 IU/mg protein. This procedure may enable investigators to prepare larger amounts of pure erythropoietin for laboratory and clinical purposes.

Numerous techniques have been described for the initial concentration of erythropoietin from human urine, including precipitation with ethanol, acetone, benzoic acid, or heat, adsorption with kaolin, ultrafiltration, dialysis against Carbowax, and diethylaminoethyl (DEAE)-cellulose chromatography (cf. *Espada* 1977). *Napier* (1975) compared several of these techniques and found the following most suitable. Urine should be collected with phenol (1 g/l) and frozen as soon as possible. The material may then be concentrated by ethanol precipitation or ultrafiltration. Protease activity in phenol-treated urine was eliminated by heating at 80°C for 5 min without loss of erythropoietin activity (*Mok* and *Spivak* 1982). *Krystal* et al. (1984) have recently employed chromatography with carboxymethylated Cibacron Blue F3GA-agarose (CM Affi-Gel Blue) and, in a single-step procedure, obtained erythropoietin from human urine which had a specific activity of about 200 IU/mg protein and was apparently free of toxins and hemopoietic growth factors other than erythropoietin.

Other techniques which have proved useful at least for partial purification of erythropoietin include preparative electrophoresis and electrofocusing (*Dorado* et al. 1972; *Lukowsky* and *Painter* 1972; *Ikegami* et al. 1973), isotachophoresis (*Thorn* et al. 1981), hydrophobic interaction chromatography on Phenyl-Sepharose CL 4B (*Lee-Huang* 1980), and high-performance liquid chromatography (*Draganac* et al. 1983; *Congote* 1984b). Affinity chromatography with agarose-bound lectins is especially suited to handle small quantities of erythropoietin. This is important in the preparation of culture-grade erythropoietin, which should be free of other growth factors and of toxins. Chromatography on agarose concanavalin A has been used to remove colony-stimulating activity from erythropoietin (*Iscove* et al. 1974). Absorption with *Limulus* lysate removed endotoxin from crude erythropoietin (*Graber* et al. 1979; *Zuckerman* et al. 1979). Erythropoietin was bound to phytohemagglutinin (*Sieber* 1977;

Spivak et al. 1977) and wheat germ agglutinin (*Spivak* et al. 1977). The hormone could be released from these lectins with *N*-acetylgalactosamine or sialic acid and *N,N*-diacetylchitobiose respectively (*Spivak* et al. 1978). Finally, chromatography on *Ricinus communis* agglutinin I was used to separate sialated from desialated erythropoietin, because the latter was bound to the lectin through its free galactosyl residues (*Kurtz* et al. 1983a).

5 Sites of Production

5.1 Kidney

5.1.1 Organ Aspects

The important role of the kidney in erythropoietin production was first demonstrated by *Jacobson* et al. (1957a), who found that bilateral nephrectomy in rats and rabbits prevented the gross rise in plasma erythropoietin following cobalt treatment or hemorrhage. Similar observations were made in anephric mice (*Jacobson* et al. 1959), dogs (*Naets* 1958; *Mirand* et al. 1968) and, less significantly, baboons (*Mirand* et al. 1969). Anephric patients were also found to produce relatively little erythropoietin. This has been shown using in vivo (*Naets* and *Wittek* 1968; *Caro* et al. 1979) and in vitro bioassays (*Radtke* et al. 1978b) and radioimmunoassays (*Garcia* et al. 1979; *Sherwood* and *Goldwasser* 1979; *Cotes* 1982). Furthermore, the assumption that the anemia of patients with chronic renal failure is partly due to inappropriate erythropoietin production (*Penington* 1972) has been verified by radioimmunological measurements (*Zaroulis* et al. 1981; *Cotes* 1982; *McGonigle* et al. 1984a,b; *Chandra* et al. 1985).

Isolated perfused kidneys can produce erythropoietin. This result has been confirmed in blood- or serum-perfused kidneys from various species, including rabbit (*Kuratowska* et al. 1961; *Reissmann* and *Nomura* 1962), dog (*Fisher* and *Birdwell* 1961; *Zangheri* et al. 1963), and baboon (*Murphy* et al. 1971a). *Erslev* (1974) has reported erythropoietin production by rabbit kidneys perfused with serum-free medium. Erythropoietin formation was abolished when puromycin was added to the perfusion medium, which indicates that ribosomal protein synthesis of some kind was involved. Significantly more erythropoietin was produced in hypoxic than in normoxic perfused kidneys (*Kuratowska* et al. 1961; *Reissmann* and *Nomura* 1962; *Zangheri* et al. 1963; *Fisher* and *Langston* 1968). This fact indicates that the kidney is also involved in the mechanism of oxygen sensing which controls the production of erythropoietin. Nevertheless, isolated perfused kidneys produced considerably more erythropoietin if the kidney donor animals were previously exposed to short-term hypobaric hypoxia (*Malgor* and *Fisher* 1970; *Erslev* 1975). *Erslev* (1975) has assumed two phases of erythropoietin formation: one of oxygen sensing (or programming) and the other of hormone synthesis. Although the mechanisms underlying the programming phase are not completely understood, this point is noteworthy because programmed kidneys from animals exposed to hypoxia have been frequently used for the in vitro study of erythropoietin production (*Paulo* et al. 1973a; *Gross* et al. 1976a; *Foley* et al. 1978; *Jelkmann* et al. 1979a; *Radtke* et al. 1980).

There is some evidence for the capacity of isolated kidney slices to produce erythropoietin. While *Erslev* (1962) failed to detect erythropoietin in the incubation medium of hypoxic rat kidney slices, *Chowdhury* and *Datta* (1973) have reported that cobalt stimulated the synthesis by sheep kidney medullary slices of heat- and acid-stable protein that was supposedly identical to erythropoietin. However, this conclusion should be accepted with caution, because it was based on immune precipitation reactions with impure antiserum to erythropoietin and on a single measurement of bioactivity in the starved rat assay for erythropoietin (*Bandyopadhyay* and *Datta* 1981). Studies from the same laboratory indicate that the synthesis of glycoproteins by kidney medulla slices was stimulated by the addition of thyroxine (*Banerjee* et al. 1976), testosterone, and prostaglandins of the A- or E-type (*Ghosal* et al. 1980).

The possibility of extracting erythropoietin from the kidney was long a matter of controversy (*Penington* 1963; *Gordon* et al. 1967a; *Muirhead* et al. 1968; *Wong* et al. 1968; *Erslev* et al. 1971; *Peschle* and *Condorelli* 1975; *Gordon* and *Kaplan* 1977). Recent attempts to extract native erythropoietin from the kidney were eventually successful. Significant amounts of the hormone were obtained from kidneys of rodents after exposure to carbon monoxide (*Scaro* et al. 1975a) or hypobaric hypoxia (*Fried* et al. 1981; *Jelkmann* and *Bauer* 1981), bleeding (*Caro* et al. 1982), cobalt treatment (*Katsuoka* et al. 1983a), and intrarenal injection of nickel subsulfide (*Hopfer* et al. 1984). In addition, erythropoietin was extracted from kidneys flushed free of blood (Fig. 3). This result indicates that at least part of the extractable hormone was contained in the kidney itself and not in trapped plasma (*Jelkmann* and *Bauer* 1981). Relatively little erythropoietin was found in the kidney of normoxic animals (*Sherwood* and *Goldwasser* 1978).

In the above studies, erythropoietin was assayed in the supernatant fluids of kidney homogenates after centrifugation at 2000–16 000 g. This indicates that the hormone was present in a fraction composed of microsomes and intra- and extracellular fluid. Data by *Jelkmann* et al. (1983a) show that at least 70% of the in vivo active erythropoietin was in

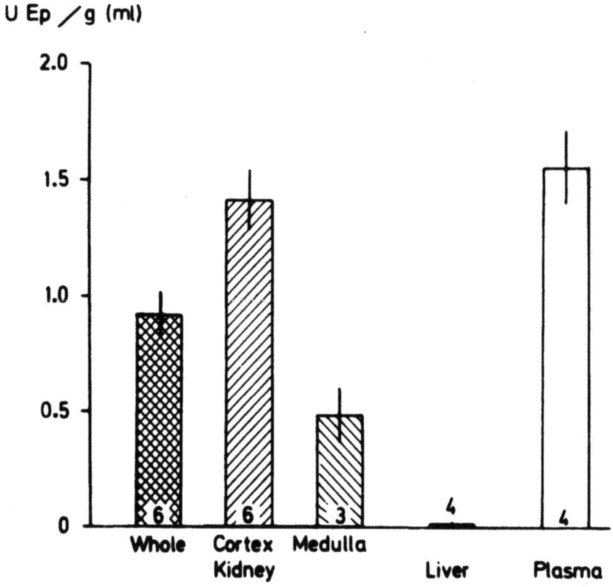

Fig. 3. Erythropoietin (*Ep*) activity in extracts from hypoxic rat kidneys and livers flushed free of blood. Hypoxia was induced by exposure to low pressure at 0.42 atm for 6 h. (*Jelkmann* and *Bauer* 1981)

the extracellular fluid of kidneys flushed free of blood. As yet, erythropoietin has not been extracted unequivocally from renal cells. The possibility still exists that the hormone is extracellularly activated or secreted into the interstitial space immediately after its terminal glycosylation is completed.

Earlier attempts to extract erythropoietin directly from the kidney were rarely successful (cf. *Jelkmann* and *Bauer* 1981). The failure to extract erythropoietin reproducibly from kidney tissue was explained on the basis of the so-called erythrogenin hypothesis. According to this hypothesis, hypoxic kidneys release a lysosomal enzyme (erythrogenin), which in turn liberates erythropoietin from an α-globulin (erythropoietinogen) in plasma (*Gordon* et al. 1967a). Several investigators indeed obtained an activity from the lysosomal fraction of hypoxic kidney homogenates which generated erythropoietin when incubated with serum from normal animals but had no significant intrinsic erythropoietic activity. Most of these studies have been summarized by *Gordon* and *Kaplan* (1977) and therefore are not enumerated here. However, it should be noted that the exact biochemical nature of erythrogenin remains to be clarified, although attempts to purify it (*Wong* et al. 1968) and to obtain antiserum against it (*McDonald* et al. 1971) were partially successful. Other investigators related to lack of erythropoietin activity in kidney extracts to the presence of nonspecific lipid inhibitors (*Erslev* et al. 1971). The capacity of renal cells to synthesize erythropoietin seems to be proved since mRNA coding for the hormone was isolated from anemic baboon kidney and translated into erythropoietin in *Xenopus laevis* oocytes (*Farber* and *Zanjani* 1983).

5.1.2 Cellular Aspects

The cells responsible for erythropoietin production are most probably localized in the kidney cortex. Significantly more erythropoietin was extracted from the cortex than from the medulla (Fig. 3; *Muirhead* et al. 1968; *Jelkmann* and *Bauer* 1981; *Fried* et al. 1982). Furthermore, hormone remnants extracted from the medulla may be derived from tubular fluid. *Murphy* et al. (1977) have demonstrated erythropoietin activity in peripheral lymph from the kidney of ewes following renal artery constriction, unilateral hydronephrosis, or phenylhydrazine-induced hemolytic anemia. Lymph from either popliteal or prescapular nodes was inactive. These findings are important because renal lymph is almost entirely produced in the kidney cortex. The exact route of secretion of erythropoietin is not known. Measurements of renal arteriovenous hormone differences indicate that at least some of the newly synthesized erythropoietin arrives at the renal vein (*Lange* and *Gallagher* 1962; *Abbrecht* and *Malvin* 1966).

Because renal venous blood is relatively well oxygenated, it is often assumed that all cells in the kidney cortex receive plentiful oxygen. However, microelectrode measurements revealed areas of very low oxygen tension in the kidney cortex of dog (*Aperia* et al. 1968) and rat (*Leichtweiss* et al. 1969). *Aperia* et al. (1968) have more directly studied the relation between erythropoietin production and tissue oxygen tension in the kidneys of anemic dogs. The oxygen tension fell markedly in the kidney cortex during anemia, while the normally low oxygen tension in the medulla tended to increase. These results make a cortical oxygen-sensing mechanism most likely.

The site of erythropoietin production in the kidney cortex has not been clearly identified. Glomerular and tubular cells have been mainly proposed as the site of synthesis.

Immunofluorescence histochemical studies have shown that antibody to erythropoietin bound selectively to glomerular cells of anemic sheep (*Fisher* et al. 1965a; *Frenkel* et al. 1968) and human kidneys (*Busuttil* et al. 1971). *Busuttil* et al. (1972) have also reported a cytological localization of erythropoietin in the glomerular tuft of the hypoxia-exposed dog kidney. Kidney sections from normal dogs and a normal dog injected with erythropoietin failed to show fluorescence staining for erythropoietin. In the hypoxic baby rat, immunofluorescence staining for erythropoietin was seen in the liver but not in the kidney until about 3 weeks of age, which correlated with the postnatal liver to kidney shift in the site of erythropoietin production (*Gruber* et al. 1977a). Unfortunately, antisera to impure erythropoietin were used in all of these investigations and definite proof for a glomerular origin of erythropoietin must await immunofluorescence studies with monospecific antierythropoietin. On the other hand, erythropoietic activity was demonstrated in extracts from isolated renal glomeruli of rats exposed to low pressure (*Jelkmann* et al. 1983a; *Caro* and *Erslev* 1984) or carbon monoxide (*Jelkmann* et al. 1984). From chromatographic studies and from the fact that the extractable erythropoietin displayed poor in vivo activity, it has been concluded that glomeruli contain immature erythropoietin which lacks terminal sugar residues (*Jelkmann* et al. 1983a). In addition, *Caro* and *Erslev* (1984) have demonstrated immunoreactive erythropoietin in glomerular extracts. Nevertheless, these authors have favored a tubular origin of erythropoietin, because about three times more erythropoietin was extracted with the tubular mass of the kidney. This reasoning is not completely convincing, however, mainly because the kidney consists mostly of tubular cells, while the glomeruli make up less than 3% of the renal mass. Thus in relation to the cellular mass, significantly more erythropoietin can be extracted from glomerular than from tubular tissue.

The glomerular cells fluorescently stained for erythropoietin were originally considered to belong to the visceral epithelium (*Busuttil* et al. 1971). However, *McCully* et al. (1982) have recently reexamined these cells and concluded that their features would fit glomerular mesangial cells as well. Interestingly, mesangial hyperplasia was noted in polycythemic patients with cyanotic congenital heart disease (*Spear* 1964) and in rabbits exposed to chronic hypoxia (*Spear* and *Kihara* 1972). *Murphy* et al. (1971b) studied endocrine functions and histological graft alterations in a number of patients after renal allotransplantation. Importantly, erythropoietin production was stimulated in all patients who exhibited changes in glomerular structure with increased fibrosis of the mesangium. The combination of mesangial hyperplasia, glomerulomegaly, and increased erythropoietin production also developed in rodents after unilateral intrarenal injection of nickel subsulfide (*McCully* et al. 1982; *Sunderman* et al. 1982). Furthermore, the concept of mesangial origin of erythropoietin is supported by cell culture studies (see Sect. 8.1). Mesangial cells in culture produced erythropoietin in an oxygen-dependent manner (*Kurtz* et al. 1982b, 1983a; *Jelkmann* et al. 1986). Thus mesangial cells seem to have potential for both erythropoietin synthesis and oxygen sensing.

Several questions arise when the glomerular mesangium is regarded as the intrarenal site of erythropoietin production. One of these concerns the role of the extraglomerular mesangium. Formed by the lacis or Goormaghtigh cells, it is cytologically very similar to, and actually continues

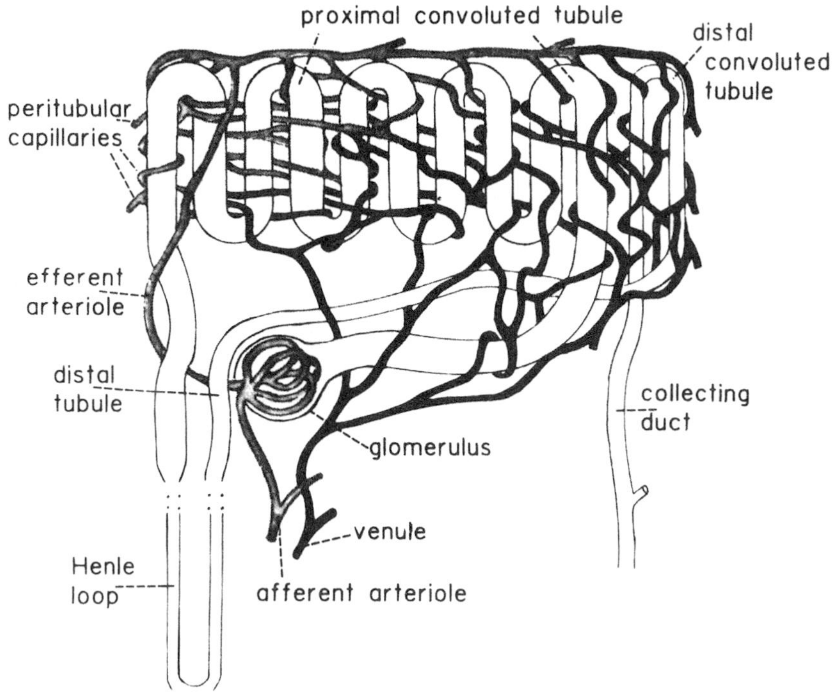

Fig. 4. Schematic presentation of the vascular organization and a nephron in the outer rat kidney cortex. The blackening of the peritubular capillary network indicates the primary site of deoxygenation of blood. (After *Steinhausen* et al. 1979)

with, the intraglomerular mesangium (cf. *Barajas* 1981). In earlier studies, a relationship was proposed between hypoxia-induced changes in the granulation of juxtaglomerular cells and the synthesis of erythropoietin (*Hirashima* and *Takaku* 1962; *Demopoulos* et al. 1965). However, these morphological findings are more readily explicable on the basis of altered renin production (*Donati* et al. 1968; *Varkarakis* et al. 1976). In addition, renal hypoxia involves alterations not only in the ultrastructure of juxtaglomerular cells but also of glomerular and proximal tubule cells (*Dunn* et al. 1976). Positive immunofluorescence staining for erythropoietin was never observed in the juxtaglomerular apparatus of hypoxic kidneys (*Busuttil* et al. 1971, 1972).

A puzzling aspect of the suggested mesangial origin of erythropoietin relates to the low intraglomerular extraction of oxygen from blood. Mesangial cells are embedded between the glomerular capillaries and probably exposed to arterialized blood. As outlined in Fig. 4, significant deoxygenation of blood takes place only in the peritubular capillary network. Hence it must be doubted whether mesangial cells experience a decrease in oxygen tension at a lowered oxygen capacity of the blood, i.e., in anemia. Of course, the possibility still exists that extraglomerular cells distinct from the erythropoietin-producing cells are involved in the mechanism of oxygen sensing. The mesangium is in close contact with the glomerular vascular pole and the distal tubule. Stimuli from the macula densa may reach glomerular cells via gap junctions (*Barajas* 1981). A tubuloglomerular feedback mechanism has been established in the control of renin production, and this mechanism seems to be prostaglandin dependent (*Olson* et al. 1980). It is evident that a similar mechanism may function in the control of erythropoietin production, although this is very speculative. The role of the renal prostaglandin system in the regulation of erythropoietin production will be discussed below (see Sect. 6.3.1). It is reasonable to assume that prostaglandins act as local hormones which mediate the effect of tissue hypoxia.

In the following, theoretical and experimental evidence will be considered which suggests that tubular cells are the site of erythropoietin synthesis. As first emphasized by *Penington* (1963) and outlined above (Fig. 4), the tubular system would seem to be suited for sensing decreases in both blood oxygen tension and content. The application of drugs which affect tubular functions resulted in depressed plasma erythropoietin titers in anemic and cobalt-treated animals (*Reissmann* et al. 1960; *Fisher* et al. 1963). On the other hand, the formation of erythropoietin was not impaired in hypoxia-exposed rats bearing congenital hydronephrosis with extensive destruction of renomedullary tissue (*Lozzio* et al. 1971; *Sušić* et al. 1979). When glomeruli were separated from hypoxic rat kidneys and extracts of the remainder assayed for erythropoietin, activity was demonstrated in the nonglomerular fraction (*Jelkmann* et al. 1983a; *Caro* and *Erslev* 1984). *Caro* and *Erslev* (1984) have assumed that the extractable erythropoietin was contained in tubules. In preliminary studies, *Caro* et al. (1984) furthermore observed in vitro production of erythropoietin by an established proximal tubule cell line (see Sect. 8.1). In addition, the concept of a tubular origin of erythropoietin seems to be supported by the finding of erythropoietin production by renal carcinoma cells (see Sect. 8.2), as renal carcinomas are thought to be derived from tubular cells (cf. *Caro* and *Erslev* 1984).

In summary, the site of origin of erythropoietin in the kidney is still a matter of controversy. Glomerular mesangial cells and proximal tubule cells must be regarded as the primary candidates for producing the hormone (Table 2). Understandably, the author favors the idea of a glomerular origin, since this concept is supported by some findings of his own research group.

Table 2. Suggested sites of origin of erythropoietin (Ep) in the kidney: comparison of evidence

Evidence	Glomeruli	Tubules	Juxta-glomerular apparatus
Morphological alterations following hypoxia	+	+	+
Immunofluorescently stained for Ep	+ (Epithelial cells?)	–	–
Extractable Ep	+	+	Not reported
Miscellaneous in vivo observations	Mesangial hyperplasia associated with Ep production	Ep production by renal carcinomas	–
Apparent aptitude in blood O_2 content sensing	Uncertain	+	Uncertain
Ep production in cell culture	+ (Mesangial cells)	+ (Proximal tubule cells)	Not reported

5.2 Extrarenal Sites

The kidney is certainly not the only source of erythropoietin. It has been estimated that about 5%–15% of plasma erythropoietin derives extrarenally in adult hypoxic rodents (*Jacobson* et al. 1959; *Erslev* et al. 1980a). Appropriate erythropoiesis is not maintained in the renoprival state in adult mammalian species. In the fetal stage, however, erythropoietin is mainly produced at extrarenal sites (*Lucarelli* et al. 1968; *Zanjani* et al. 1974), most likely in the liver (*Peschle* et al. 1976).

In the present section, a short survey is undertaken of extrarenal tissues that have been implicated in erythropoietin production. Other aspects of extrarenal erythropoietin have been described in detail elsewhere, i.e., control of production (*Fried* and *Anagnostou* 1977), cell culture findings (*Zucali* and *Mirand* 1978), clinical relevance (*Fried* and *Anagnostou* 1977; *Fisher* 1979), and role in the regulation of fetal erythropoiesis (*Zanjani* et al. 1977).

The liver is the only extrarenal organ that has been proved to have the capacity for releasing erythropoietin into the blood. Partial hepatectomy prevented an increase in plasma erythropoietin in adult nephrectomized animals exposed to hypoxia (*Katz* et al. 1968; *Fried* 1972). Attempts to extract erythropoietin from hypoxic livers have been unsuccessful (*Jelkmann* and *Bauer* 1981; *Fried* et al. 1982; *Caro* et al. 1983). Erythropoietin was demonstrated in the medium of isolated perfused livers (*Reissmann* and *Nomura* 1962; *Burke* and *Morse* 1962) and in fetal liver cell cultures (*Congote* and *Solomon* 1974; *Zucali* et al. 1975). From carbon particle ingestion studies (*Peschle* et al. 1976; *Gruber* et al. 1977b) and from immunofluorescence histochemical findings (*Gruber* et al. 1977a,b), it has been proposed that liver Kupffer cells produce the hormone. *Paul* et al. (1984) have recently reported erythropoietin production by Kupffer cell cultures from adolescent rat livers.

Apart from the liver, the spleen has mainly been implicated as an extrarenal source of erythropoietin (*Kaplan* et al. 1973; *Peschle* et al. 1976; *Rich* et al. 1982). However, the spleen probably contributes little to circulating erythropoietin. Significant amounts of the hormone were not released by isolated hypoxemic perfused spleens (*Kuratowska* et al. 1961), nor did splenectomy result in reduced erythropoietin production in experimental animals (*Peschle* et al. 1973a; *Bozzini* et al. 1976). Spleen and bone marrow cells produced an erythropoietic activity in long-term cultures (*Rich* et al. 1982). Macrophages have been suggested as its producers. Thus macrophages could act as a local source of erythropoietin in erythropoietic tissues.

The submandibular gland has been regarded as another site of erythropoietin production. Its removal lowered the plasma erythropoietin titer in anephric hypoxic rodents (*Zangheri* et al. 1973, 1977). In addition, positive immunofluorescence staining for erythropoietin was found in the secretory tubules of hypoxic rat and mouse submandibular glands (*Fava-de-Moraes* et al. 1979).

The carotid bodies were earlier thought to secrete an erythropoiesis-stimulating activity, because carotid-body-ablated cats showed reduced reticulocyte counts in response to phlebotomy (*Tramezzani* et al. 1971). The authors also detected erythropoietic activity in carotid body extracts. Yet these findings could not be reproduced in cats exposed to hypoxia (*Hansen* et al. 1973). By contrast, the removal of the carotid bodies caused an increase in erythropoietin production in cats (*Hansen* et al. 1973), rats (*Beynon* and *Balfour* 1973), and rabbits (*Paulo* et al. 1973b) exposed to hypobaric hypoxia.

The variety of organs that can develop potentials for generating erythropoietin is even more numerous if secretory tumors are included. Apart from renal tumors and hepatomas, at least uterine fibromyomas and cerebellar hemangioblastomas incidentally produce activities that behave like erythropoietin (cf. *Hammond* and *Winnick* 1974; *Morse* 1979). Furthermore, erythropoietin-synthesizing cell lines were cultured from a human testicular germ cell tumor (*Ascensao* et al. 1983) and from several virus-induced murine erythroleukemia cell lines (*Tambourin* et al. 1983; *Choppin* et al. 1984, 1985).

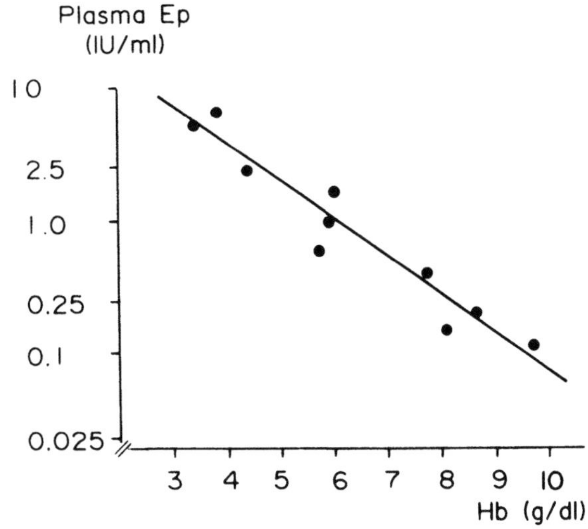

Fig. 5. Semilogarithmic presentation of the relation between blood hemoglobin (*Hb*) concentration and plasma erythropoietin (*Ep*) titer in anemic rats. Anemia was induced by isovolemic exchange transfusion with rat plasma. Blood for the determination of Hb and Ep was taken 17 h after the exchange. (Own unpublished observations)

6 Hypoxic Stimulation of Erythropoietin Production

6.1 Determinants of Tissue Oxygen Tension

Tissue hypoxia is the main stimulus for erythropoietin production. As already noted by *Fried* et al. (1957), the synthesis of the hormone depends basically on the ratio of oxygen supply to demand, with renal cell oxygen tension being the most important parameter.

6.1.1 Blood Oxygen Capacity

Oxygen capacity and volume of blood are certainly the main physiological determinants of tissue oxygen supply. Remember that, teleologically, the action of erythropoietin is to stimulate erythropoiesis after a blood loss. It has been reported that the loss of just 75 ml blood caused a rise in plasma erythropoietin in humans (*Miller* et al. 1982). Urinary erythropoietin excretion and hematocrit are inversely correlated in anemic humans (*Adamson* 1968; *Alexanian* 1973). A similar relationship has been demonstrated in anemic patients with nonrenal diseases between plasma erythropoietin and either the hematocrit (*Van Dyke* et al. 1961; *Erslev* et al. 1980b) or the concentration of hemoglobin in blood (*Alexanian* 1973; *De Klerk* et al. 1981a). Figure 5 illustrates the relationship between the plasma erythropoietin titer and the blood hemoglobin concentration in anemic rats.

Conversely, an increase in red cell mass should suppress erythropoietin production. Indeed, low urinary erythropoietin excretion (*Adamson* 1968) and low plasma titers of the hormone were observed in patients with polycythemia vera (*Erslev* et al. 1979; *De Klerk* et al. 1981b; *Zaroulis* et al. 1981; *Garcia* et al. 1982; *Erslev* and *Caro* 1984a), except in the study by *Birgegård* et al. (1982). Importantly, erythropoietin titers were not increased at even the most extreme hemocrits, indicating that the oxygen tension at the site of control of erythropoietin production is not lowered at high blood viscosities (*Erslev* and *Caro* 1984b). Likewise, hypertransfusion-induced polycythemia lowered erythropoietin production in experimental animals (*Jacobson* et al. 1957b; *Kilbridge* et al. 1969; *Nečas* and *Neuwirt* 1970; *Nečas* et al. 1972).

In the study by *Kilbridge* et al. (1969), the production of erythropoietin was also reduced in hypoxic mice after hypertransfusion with methemoglobin containing erythrocytes that were incapable of oxygen transport. While this finding could indicate a feedback regulation of erythropoietin production based on red cell concentration, further experimental evidence for such a mechanism is not available. However, another example of control of erythropoietin production through factors other than the oxygen supply to demand ratio has been reported by *Alippi* et al. (1983), who found that hypoxic hypoxia caused lower plasma erythropoietin titers in hypertransfused mice than in posthypoxic mice at comparable degrees of polycythemia.

6.1.2 Blood Oxygen Affinity and pH

A change in blood oxygen affinity at normal arterial oxygen tension mainly affects the peripheral oxygen delivery. Patients with high-oxygen-affinity mutant hemoglobins usually have increased hemoglobin concentrations, while low-oxygen-affinity hemoglobins are associated with lowered hemoglobin concentrations (cf. *Bellingham* 1974; *Adamson* and *Finch* 1975). *Napier* (1980) compared the serum erythropoietin titer in an anemic patient before and after transfusion with fresh and with high-oxygen-affinity blood stored in acid citrate dextrose. The author observed a greater fall in plasma erythropoietin after transfusion of fresh blood. In addition, plasma erythropoietin titers increased in rats after exchange transfusion with high-oxygen-affinity blood (*Miller* et al. 1983; *Lechermann* and *Jelkmann* 1985).

Very high plasma titers of erythropoietin were observed in animals after exposure to carbon monoxide (*Syvertsen* and *Harris* 1973; *Scaro* et al. 1975a,b; *Jelkmann* et al. 1984). Carbon monoxide hypoxia results primarily from the decrease in blood oxygen capacity due to the formation of carboxyhemoglobin. However, the effect of carbon monoxide is partly due to the increased oxygen affinity of carboxyhemoglobin (*Jelkmann* et al. 1984).

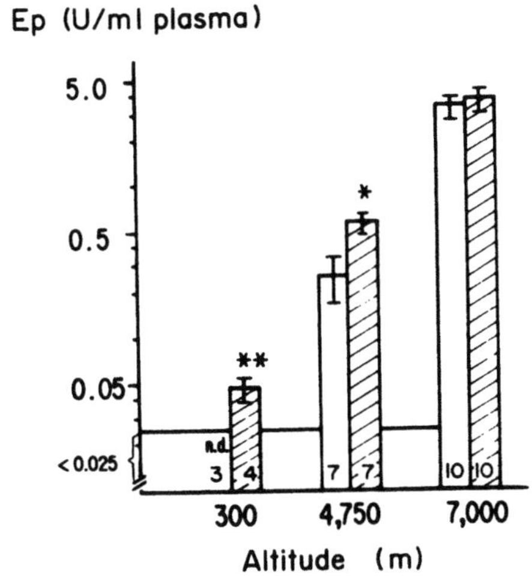

Fig. 6. Effects of blood O_2 affinity: plasma erythropoietin (*Ep*) titers in rats in which altitude exposure was simulated in a hypobaric chamber and which had either normal (*open bars*) or high (*lined bars*) O_2 affinity blood (P_{50} values 42 or 29 Torr O_2 at pH 7.4 and 37°C). Blood samples were taken after 16 h maintenance at the respective altitudes. Ep was not detectable (*n.d.*) in normal O_2 affinity rats at 300 m. Symbols ** and * indicate differences at $p < 0.05$ and $p < 0.1$ respectively, when the two O_2 affinity groups are compared. (*Lechermann* and *Jelkmann* 1985)

Under the condition of hypobaric hypoxia, the position of the hemoglobin-oxygen equilibrium curve influences blood oxygen loading and unloading. Most studies have focused on the effect of oxygen unloading, i.e., it was concluded that a decreased blood oxygen affinity would improve tissue oxygenation. For example, the marked influence of pH on erythropoietin production has been related to changes in oxygen affinity by virtue of the Bohr effect. The erythropoietin response to hypoxia is enhanced in alkalosis and reduced in acidosis, especially in hypercapnia (*Faura* et al. 1969; *Miller* et al. 1973; *Wolf-Priessnitz* et al. 1978; *Zucali* et al. 1978; *Miller* and *Howard* 1979; *Cohen* et al. 1981). Investigators have also emphasized the effect of pH on the respiratory rate and thus the arterial oxygen tension (*Streeter* et al. 1975; *Wolf-Priessnitz* et al. 1978; *Cohen* et al. 1981). However, this explanation is not applicable to the suppressed erythropoietin production in hypercapnic anemic (*Miller* and *Howard* 1979) or cobalt-treated animals (*Miller* et al. 1974). Thus it remains to be determined whether the cellular mechanisms of erythropoietin synthesis are directly sensitive to pH changes.

Hebbel et al. (1978) have reported lower than normal plasma and urinary erythropoietin in two human subjects with a high oxygen affinity hemoglobin mutant during exposure to 3100 m altitude. However, the results of this study are difficult to interpret because the arterial oxygen saturation still exceeds 90% in humans at 3000 m altitude and because the blood hemoglobin concentration in the two high-affinity subjects was 25% greater than in the controls. *Lechermann* and *Jelkmann* (1985) found no difference in the plasma erythropoietin titers of rats exposed to 7000 m altitude after exchange transfusion with either normal or cyanated high-oxygen-affinity blood (Fig. 6). Thus at sea level of leftward shift of the hemoglobin oxygen dissociation curve attenuates the oxygen delivery to the cells controlling the production of erythropoietin, while at extreme altitude the impaired oxygen unloading is equalized by an improved oxygenation in the lung. Computer simulation analyses of the regulation of erythropoiesis are in accord with this conclusion (*Dunn* et al. 1980).

6.1.3 Arterial Oxygen Tension

Patients with chronic respiratory failure and cyanosis have elevated plasma erythropoietin titers and tend to be polycythemic (*Freedman* and *Penington* 1963). High-altitude residents exhibit increased blood hemoglobin concentration and red cell mass. Following acute exposure to hypobaric hypoxia, plasma and urinary erythropoietin titers in man and animals reach a maximum within 1–3 days and then decline despite continued hypoxia (see Sect. 6.2.1). The increase in plasma erythropoietin is proportional to the severity of the hypoxic stress (*Nečas* and *Neuwirt* 1970; *Živný* et al. 1971; *Abbrecht* and *Littell* 1972; *Beynon* 1977; *Cohen* et al. 1981).

Renal production of erythropoietin in hypoxia probably results from de novo protein synthesis. The application of actinomycin D before the initiation of hypoxia inhibited erythropoietin production in rats (*Giger* 1968; *Schooley* and *Mahlmann* 1972). The drug was ineffective if applied after a brief hypoxic period. On the other hand, puromycin or cycloheximide attenuated erythropoietin production in rats exposed to hypoxia in the short term (*Schooley* and *Mahlmann* 1972). These findings indicate that erythropoietin production involves DNA-dependent RNA synthesis.

6.1.4 Renal Blood Flow

Enhanced erythropoietin production following renal artery constriction has been confirmed in rats, rabbits, and dogs (*Takaku* et al. 1962; *Hansen* 1964; *Fisher* and *Samuels* 1967; *Murphy* et al. 1971c). Negative results have also been reported (*Cooper* and *Nocenti* 1961; *Hansen* 1964; *Živný* et al. 1972a). A flow reduction of 50% below normal was necessary to stimulate erythropoietin production in the dog kidney (*Fisher* and *Samuels* 1967). A random look at the literature yields rare incidences of true polycythemia secondary to renal artery stenosis both in humans (*Tarazi* et al. 1966; *Hudgson* et al. 1967; *Bourgoignie* et al. 1968) and in experimental animals (*Johnson* et al. 1982).

Abbrecht and *Malvin* (1966) have reported elevated plasma erythropoietin titers in dogs in which multiple renal infarcts were produced by injections of plastic microspheres. In addition, erythropoiesis and erythropoietin production were stimulated by ischemia in cases of renal tumors

(*Hammond* and *Winnick* 1974; *Kazal* and *Erslev* 1975b; *Morse* 1979) or hydronephrosis (*Toyama* and *Mitus* 1966; *Mitus* et al. 1968). Vasoconstrictive agents such as angiotensin, noradrenaline, 5-hydroxytryptamine (*Fisher* et al. 1971), serotonin (*Lowy* et al. 1970), or bombesin (*Melchiorri* et al. 1976) also stimulated erythropoietin production.

A decrease in renal blood flow will lower renal oxygen consumption because of a reduction in tubular sodium chloride reabsorption (*Kramer* and *Deetjen* 1960; *Thurau* 1961). It is not clear whether this fact is relevant to the regulation of erythropoietin synthesis, as the intrarenal site of oxygen sensing is uncertain. Studies in salt-loaded and salt-restricted animals have not provided clear evidence for an effect of tubular sodium reabsorption on erythropoietin production (*Donati* et al. 1968; *Abbrecht* et al. 1969; *Varkarakis* et al. 1976) with one exception (*Pavlović-Kentera* et al. 1980). The formation of erythropoietin does not appear to require normal urine production, as the erythropoietin response to anemia or hypoxemia was essentially unaltered in ureter-ligated animals (*Jacobson* et al. 1959; *Reissmann* et al. 1960; *Peschle* et al. 1972a).

6.1.5 Oxygen Consumption

Erythropoiesis and metabolic rate are reduced in hypopituitarism or hypothyroidism. Both effects could be normalized by the application of adrenocorticotropic hormone (ACTH) and thyroxine (*Evans* et al. 1961). *Peschle* et al. (1978) observed lower than normal erythropoietin production in hypoxic rats for 7 months after hypophysectomy. In humans a close relation has been found between red cell mass and basal oxygen consumption (*Muldowney* et al. 1957). Erythropoiesis was stimulated when the metabolic rate was enhanced by the administration of dinitrophenol (*Muldowney* et al. 1957; *Fried* et al. 1957; *Evans* et al. 1961). Increased plasma erythropoietin titers were found in mice injected with thyroxine (*Peschle* et al. 1971a). *Caro* et al. (1981) have provided evidence suggesting that the suppressed erythropoietin response to hypoxia in fasting animals is mainly due to caloric deprivation and can be restored by glucose feeding or treatment with replacement doses of thyroid hormone, T_3. In addition, the lowered erythropoietin production in protein-deprived rats was normalized after they had been fed protein (*Reissmann* 1964; *Anagnostou* et al. 1977) or amino acid diets (*Anagnostou* et al. 1978).

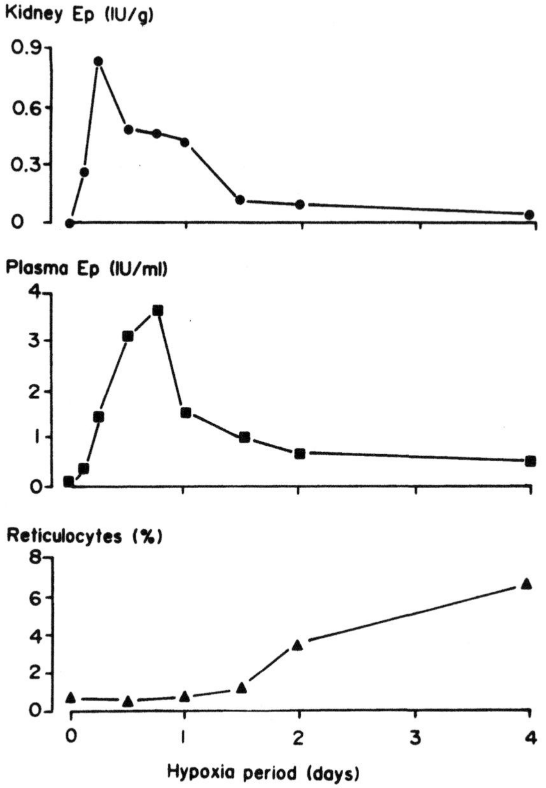

Fig. 7. Temporal pattern of the erythropoietic response to hypobaric hypoxia (0.42 atm) in rats. (After *Jelkmann* 1982)

6.2 Kinetics

6.2.1 Production

Acute exposure to hypoxic hypoxia initiates a marked rise in plasma erythropoietin. Peak levels occurred after 1–3 days in man (*Reynafarje* et al. 1964; *Siri* et al. 1966; *Faura* et al. 1969; *Abbrecht* and *Littell* 1972; *Miller* et al. 1973; *Milledge* and *Cotes* 1985) and after 6-24 h in dogs (*Syvertsen* and *Harris* 1973) and rodents (*Mirand* and *Prentice* 1957; *Stohlman* and *Brecher* 1959; *Fried* et al. 1970; *Abbrecht* and *Littell* 1972; *Schooley* and *Mahlmann* 1975a). Thereafter, plasma erythropoietin declined despite continued exposure to hypoxia and eventually remained moderately elevated above normal. A transient rise in plasma erythropoietin was also seen in acute anemia (*Miller* et al. 1976; *Miller* and *Howard* 1979; *Erslev* et al. 1980a) and carboxyhemoglobinemia (*Jelkmann* et al. 1984). Note that the subsequent fall in the plasma erythropoietin titer occurs before the red blood cell concentration increases significantly (Fig. 7). The initial marked increase in the erythropoietin titer may be necessary to trigger early erythroid progenitor cells into proliferation. A sustained rate of erythropoiesis is maintained in continuous hypoxia, despite the decrease in the plasma erythropoietin titer (*Fried* et al. 1970; *Dunn* et al. 1980).

Several hypotheses have been developed to explain the early decline in the plasma erythropoietin titer, although none of these is really conclusive. The earlier concept of erythropoietin consumption by activated erythropoietic tissues (*Stohlman* and *Brecher* 1959) has received support in some erythropoietin clearance studies (*Russell* and *Keighley* 1972; *Mirand* et al. 1972) but not in others (*Bozzini* 1966; *Naets* and *Wittek* 1969; *Fried* et al. 1970, 1971; *Dunn* et al. 1981). In addition, the finding of a similar rise and fall in the renal erythropoietin titer is suggestive of a change in hormone production (*Jelkmann* 1982; *Caro* and *Erslev* 1984). Nevertheless, there is also evidence that the proliferating erythron exerts a feedback inhibition on erythropoietin production. Plasma erythropoietin titers are usually higher in patients with bone marrow hypoplasia than in comparably anemic patients with active erythropoiesis (*Hammond* et al. 1962; *De Klerk* et al. 1981a). Furthermore, anemic mice with a congenital hemopoietic stem cell defect revealed relatively high plasma erythropoietin titers (*Fried* et al. 1971; *Russell* and *Keighley* 1972). Plasma erythropoietin titers were also enhanced in hypoxic mice with marrow hypoplasia induced by X-irradiation or the injection of 5-fluorouracil (*Barceló* and *Bozzini* 1982).

It is less likely that the early decline of renal erythropoietin production is due to feedback inhibition through the hormone itself. The injection of large doses of erythropoietin in rats did not suppress the endogenous production of the hormone (*Fried* and *Barone-Varelas* 1984).

Other investigators have related the temporal pattern of erythropoietin production to changes in the acid-base status by virtue of an increased ventilation and a lowered blood oxygen affinity at prolonged hypoxia (*Miller* et al. 1973; *Schooley* and *Mahlmann* 1975a; *Wolf-Priessnitz* et al. 1978; *Miller* and *Howard* 1979; *Cohen* et al. 1981). Although an evaluation of the published data reveals rather small changes both in blood oxygen affinity and saturation, ventilatory changes may partly account for the fluctuation in erythropoietin production during hypoxemia.

Dunn et al. (1980, 1981) have concluded from a computer analysis of erythropoiesis that the decreased erythropoietin production could result from the reduced food intake during continuous hypoxic stress. Plasma (*Fried* et al. 1970; *Schooley* and *Mahlmann* 1975a) and kidney (*Jelkmann* 1982; *Caro* and *Erslev* 1984) erythropoietin titers were indeed enhanced when rats were exposed to discontinuous instead of continuous hypoxia. Yet the plasma erythropoietin titer increased similarly upon repeated hypoxia exposure in food-deprived rats (*Fried* et al. 1970; *Jelkmann* et al. 1983b). Although this result makes a simple correlation between food intake and the kinetics of erythropoietin production less likely, the possibility remains that changes in the metabolic rate of the kidney partly account for the fluctuation of erythropoietin production during hypoxia. Actually, rats responded to hypoxia with transient increases both in thyroid function (*Debias* 1966) and the excretion of catecholamines (*Myles* and *Ducker* 1973).

6.2.2 Metabolism

Plasma half-lives of endogenous and exogenous erythropoietin have been studied with variable results. Conceivably, the former provide overestimates, because they were determined in animals previously exposed to hypoxia, whereas the formation of the hormone does not cease immediately after return to normoxia (*Schooley* and *Mahlmann* 1972; *Fried* and *Barone-Varelas* 1984). On the other hand, measurements of the plasma disappearance rate of intravenously injected erythropoietin may have led to underestimates, if the initial fast phase of hormone distribution in the interstitial space was not excluded from the calculation (*Reissmann* et al. 1965). With this in mind, the following half-life data are to be considered: 1.5–3.5 h in rat (*Naets* and *Wittek* 1974; *Dinkelaar* et al. 1981; *Stohlman* and *Howard* 1962; *Reissmann* et al. 1965; *Emmanouel* et al. 1984), 2.2–3.3 h in mouse (*Mirand* et al. 1972; *Abbrecht* and *Littell* 1972; *Dunn* et al. 1981), 8 h in rabbit (*Roh* et al. 1972a), and 9 h in sheep (*Mladenovic* et al. 1985).

About 5%–10% of circulating erythropoietin seems to be cleared by the kidney (*Emmanouel* et al. 1984). In addition to glomerular filtration, secretion of erythropoietin by cells of the proximal convoluted tubule has been assumed, since probenecid reduced the urinary elimination of the hormone (*Skoglund* and *Finne* 1972). Accordingly, the half-life of plasma erythropoietin was prolonged in anephric (*Naets* and *Wittek* 1974; *Dinkelaar* et al. 1981) or ureter-ligated rats (*Emmanouel* et al. 1984). The role of the liver in erythropoietin degradation is less well understood. While asialoerythropoietin was rapidly cleared by isolated perfused livers (*George* et al. 1977), sialoerythropoietin was metabolized by isolated livers in one study (*Roh* et al. 1972b) but not in another (*Dinkelaar* et al. 1981). Apparently, erythropoietin is not inactivated by blood or bone marrow cells, because the blood hormone titers were unaltered after perfusion through isolated rabbit (*Fisher* et al. 1965b) or canine hindlimbs (*Roh* et al. 1972b).

6.3 Role of Hormones

A hormone may act in either of the following ways:

1. It may influence the production of erythropoietin but not be involved in the process of hypoxia per se.
2. It may function as an obligatory intermediate in the chain of events leading to the production of erythropoietin in hypoxia.

In the latter case, the hormone should not only stimulate the elaboration of erythropoietin but should be released itself in an oxygen-dependent manner. In addition, an inactivation of the hormone should abolish the erythropoietin response to hypoxia.

6.3.1 Renal Prostaglandins

Several lines of evidence suggest that the renal prostaglandin (PG) system is involved in the mechanism by which severe hypoxia induces the elaboration of erythropoietin. Prostaglandins of the E- and A-type, as well as prostacyclin (PGI_2) and its stable metabolite, 6-keto-PGE_1, stimulate erythropoiesis in vivo, whereas prostaglandins of the F-type inhibit (*Schooley* and *Mahlmann* 1971a; *Dukes* et al. 1973; *Keighley* and *Cohen* 1978; *Gross* and *Fisher* 1980; *Neal* et al. 1980; *Lewis* et al. 1981; *Nelson* et al. 1983). A recent clinical trial of PGE_2 in patients with anemia of end-stage renal insufficiency resulted in an increment in erythropoietic progenitor cells in blood and an increase in plasma erythropoietin in some of the patients (*Ortega* et al. 1984). In addition, polycythemia and enhanced plasma titers of erythropoietin were noted in patients with Bartter's syndrome (*Jepson* and *McGarry* 1968; *Erkelens* and *Van Eps* 1973), which is characterized by enhanced renal PGE_2 production (*Gill* et al. 1976).

Keigley and *Cohen* (1978) have explained the erythropoietic effect of PGE_1 on the basis of renal hypoxia, as this agent reduced renal blood flow in mice. On the other hand, structure-function studies with synthetic endoperoxide analogs revealed that their erythropoietic potency was not correlated with their vasoconstrictor activity. In addition, only prostaglandins possessing oxygen at the C-9 position were erythropoietically active (*Gross* and *Fisher* 1980). The main renal prostaglandin, PGE_2 (*Gross* et al. 1976a), as well as PGE_1 (*Paulo* et al. 1973a) and the prostaglandin precursor arachidonic acid (*Foley* et al. 1978), stimulated the release of erythropoietin in the programmed isolated perfused dog kidney. Furthermore, PGE_2, PGI_2, and arachidonic acid induced the elaboration of erythropoietin in cultures of renal mesangial cells (*Kurtz* et al. 1985b).

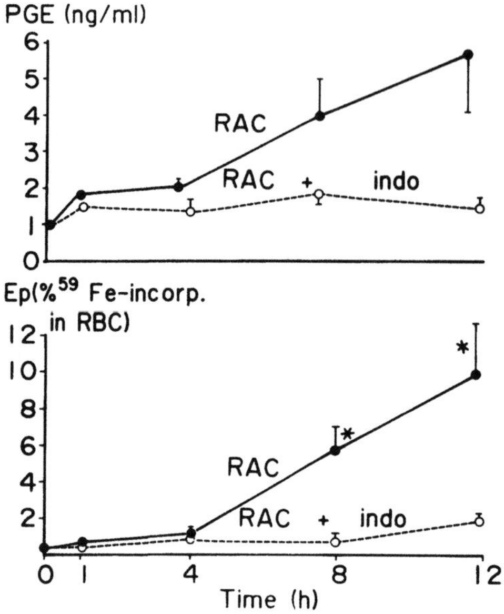

Fig. 8. The effects of indomethacin (*indo*) on serum prostaglandin E (*PGE*) and plasma erythropoietin (*Ep*) titers of dogs following renal artery constriction (*RAC*) to 30% of normal flow. * Indicates significantly ($p < 0.05$) different from the indomethacin-treated group. (Redrawn from *Gross* et al. 1976b)

Severe hypoxia stimulated the synthesis of PGE_2 and/or PGI_2 in various tissues, including the in situ kidney (*Gross* et al. 1976b; *Walker* 1982), isolated hearts and kidneys (*Needleman* et al. 1974; *Kalsner* 1977; *Wennmalm* 1979; *Kirstein* 1979; *Burdowski* et al. 1980), and renal mesangial cell cultures (*Jelkmann* et al. 1985). It is tempting to postulate that the in vivo function of these prostaglandins in turn is to improve the oxygen supply to the tissue both by vasodilation and stimulation of erythropoiesis.

Gross et al. (1976b) have reported a marked increase in both erythropoietin and PGE titers in the blood of dogs after renal artery constriction to 30% of normal flow (Fig. 8). Importantly, neither renal PGE nor erythropoietin were produced in dogs treated with the cyclooxygenase inhibitor drug indomethacin (*Mujovic* and *Fisher* 1974; *Gross* et al. 1976b). Hence the release of prostaglandins appears to be an essential step in the synthesis of erythropoietin during severe hypoxic stress. This view is supported by studies in which the increase in plasma erythropoietin during hypobaric (*Mujovic* and *Fisher* 1975; *Pavlović-Kentera* et al. 1980) or anemic hypoxia (*Jelkmann* et al. 1984) was attenuated by the application of prostaglandin-synthesis-inhibiting drugs. It still remains to be investigated, however, whether the renal prostaglandin system also controls the normal day-to-day production of erythropoietin. It is of interest that isolated canine kidneys responded to hypoxemic perfusion mainly with increased 6-keto-$PGF_{1\alpha}$ and thus probably PGI_2 formation (*Burdowski* et al. 1980), while renal artery constriction caused release of PGE (*Gross* et al. 1976b). Different types of renal cells may contribute to prostaglandin synthesis and thus be involved in the regulation of erythropoietin production. Endothelial cells mainly produce PGI_2, while glomerular mesangial and epithelial cells, tubular cells, and medullary interstitial cells mainly produce PGE_2 (cf. *Hassid* 1982). In cultured renal mesangial cells, hypoxia stimulated the production of PGE_2, 6-keto-$PGF_{1\alpha}$, and $PGF_{2\alpha}$ by about the same degree (*Jelkmann* et al. 1985).

The mechanism of hypoxia-induced prostaglandin production is unknown. Basically, prostaglandin formation is limited by the availability of arachidonic acid, which is liberated from the 2-position of phospholipids by the interaction of phospholipase A_2. Hence the effect of hypoxia could be due to (a) lowered uptake of arachidonic acid into phospholipids at reduced ATP availability, (b) conformational changes in cellular membranes, such as those occurring in the ischemic kidney cortex (*Coleman* et al. 1976), or (c) phospholipase A_2 activation. As earlier suggested by *Markelonis* and *Garbus* (1975), phospholipase A_2 could be stimulated by the release of Ca^{2+} from deteriorating mitochondria in hypoxia. Alternatively, the initial stimulation of phospholipase C, for example by ADP, could result in the formation of the endogenous Ca^{2+} ionophore phosphatidic acid from phosphatidylinositol (cf. *Berridge* 1981). Finally, the synthesis of prostaglandins could be controlled by the cellular oxidation-reduction state. Reduced glutathione, for example, has been shown to stimulate the in vitro formation of PGE_2 (*Lands* et al. 1971; *Takeguchi* et al. 1971).

The mechanism of prostaglandin action on erythropoietin production also remains to be established. PGE_2, PGI_2, and arachidonic acid increased cyclic AMP levels in mesangial cell cultures (*Kurtz* et al. 1985b). It is thought that the activation of cyclic-AMP-dependent protein kinases is an important step in the elaboration of erythropoietin (*Rodgers* et al. 1975c; *Martelo* et al. 1976). Forskolin, a potent activator of adenylate cyclase, induced erythropoietin production in mesangial cell cultures (*Kurtz* et al. 1985b). The same effect was observed, when the dibutyryl derivative of cyclic AMP was directly added to erythropoietin producing renal carcinoma cell cultures (*Hagiwara* et al. 1985).The in vivo application of cyclic AMP produced a marked increase in erythropoiesis (*Schooley* and *Mahlmann* 1971b; *Gidari* et al. 1971; *Bottomley* et al. 1971; *Winkert* et al. 1971; *Rodgers* et al. 1975d; *Keighley* and *Cohen* 1978). Cyclic AMP probably stimulates erythropoiesis by increasing renal production of erythropoietin, because the effect of cyclic AMP was prevented in nephrectomized animals (*Peschle* et al. 1973b) or by the simultaneous administration of antiserum to erythropoietin (*Schooley* and *Mahlmann* 1975b). In addition, the renal cyclic AMP titer increased in experimental animals after exposure to hypoxia (*Rodgers* et al. 1974, 1975b) or following the application of PGE_1 (*Paulo* et al. 1973a).

6.3.2 Renin-Angiotensin

It is mainly the increased granulation of juxtaglomerular cells occurring in hypoxemic (*Demopoulos* et al. 1965; *Oliver* and *Brody* 1965; *Hartcroft* et al. 1969) and anemic animals (*Hirashima* and *Takaku* 1962) that has led investigators to study the relationship between renin and erythropoietin secretion. Attempts to stimulate erythropoietin production in normoxia by the injection of renin have failed. High doses of renin eventually enhanced plasma erythropoietin titers in hypoxia (*Gould* et al. 1973). This effect probably resulted from ischemic hypoxia caused by angiotensin II, for it was prevented by the antiotensin-I-converting enzyme inhibitor SQ 14225 (*Gould* et al. 1980). It is of interest that hypoxia inhibited the activity of angiotensin-converting enzyme (*Stalcup* et al. 1979).

Renin and erythropoietin secretion were clearly dissociated in a variety of experimental and clinical situations. Renal renin content was not altered in hypoxia (*Gould* et al. 1973; *Bridges* and *Taylor* 1975), while plasma renin activity was even lowered (*Donati* et al. 1968; *Murphy* et al. 1971c; *Gould* et al. 1973; *Kotchen* et al. 1973a; *Varkarakis* et al. 1976). Note, though, that hypoxia has proved to enhance the effect of physical exercise on the release of renin (*Milledge* and *Cotes* 1985). However, plasma erythropoietin titers remained low when renin production was stimulated by salt restriction (*Donati* et al. 1968; *Varkarakis* et al. 1976). In kidney-transplated dogs, salt loading reduced renin but not erythropoietin secretion during immunological rejection (*Abbrecht* et al. 1969). These results have been complemented by the clinical observation that the plasma erythropoietin titer is not correlated with renin activity in patients with non-renovascular hypertension (*Bourgoignie* et al. 1968). Finally, a case has been reported of hyporeninemic hypoaldosteronism in combination with anemia and elevated plasma erythropoietin (*Cosgriff* et al. 1978).

A recent concept proposes that erythropoietin could be a derivative of angiotensinogen, because the two glycoproteins have been found to share several chemical (*Gould* et al. 1980) and immunological properties (*Fyhrquist* et al. 1984). Purified human angiotensinogen stimulated HbF synthesis in cultures of the human erythroleukemia cell line K 562 (*Fyhrquist* et al. 1984). *Celio* and *Inagami* (1981) have demonstrated both angiotensin II and renin immunoreactivity in juxtaglomerular granulated cells. Thus there may be intrarenal angiotensinogen synthesis. Nevertheless, it seems less likely that erythropoietin originates from angiotensinogen. Note the poor quantitative correlation between plasma titers of erythropoietin and angiotensinogen during hypoxia. In the experiments by *Gould* et al. (1980), plasma titers of angiotensinogen increased by a factor of 1.2 in rats during 8 h hypoxia exposure, while erythropoietin increased about 100-fold. In addition, comparison of the human erythropoietin amino-acid sequence with the rat angiotensinogen amino-acid sequence revealed no regions of homology (*Jacobs* et al. 1985).

6.3.3 Hypothalamic-Hypophyseal Hormones

Hypophysectomy causes a successive decline in basal oxygen consumption, erythropoietin formation, and red cell production (*Fried* et al. 1957; *Crafts* and *Meineke* 1959; *Evans* et al. 1961). In addition, the pituitary gland is required for a normal erythropoietin response in hypoxia (*Halvorsen* et al. 1968; *Peschle* et al. 1978). Several hormones of the hypothalamic-hypophyseal system stimulate erythropoiesis at physiological or pharmacological doses, including thyroid-stimulating hormone (TSH), ACTH, gonadotropins, growth hormone, and antidiuretic hormone (ADH) (cf. *Fisher* and *Gross* 1977). The glandotropic hormones are thought to affect erythropoiesis indirectly through their target hormones. Growth hormone, thyroxine, and certain steroids stimulate the in vitro proliferation of erythroid progenitor cells (see Sect. 2.2). In vivo, however, these hormones appear to act mainly by increasing the production of erythropoietin, because their erythropoietic effect was abolished after nephrectomy or treatment with antiserum to erythropoietin (*Fisher* et al. 1967; *Peschle* et al. 1971b, 1972b). In addition, ACTH, cortisol, testosterone, triiodothyronine (*Peschle* et al. 1971b), and growth hormone (*Peschle* et al. 1972b) failed to enhance the effect of simultaneously injected erythropoietin in polycythemic mice. On the other hand, the application of growth hormone produced an increase in the plasma erythropoietin titer in hypophysectomized rats (*Peschle* et al. 1972b).

Prolactin was found to stimulate murine erythropoiesis (*Jepson* and *Lowenstein* 1964, 1966a, 1967). The role of prolactin in the regulation of erythropoietin production during hypoxia has not been established. Noteworthily, plasma prolactin titers in man rise after acute exposure to hypobaric hypoxia (*Heyes* et al. 1982).

The release of ADH from the posterior pituitary is probably not relevant to the maintenance of erythropoiesis. Anemia was not observed in experimental animals in which only the posterior pituitary lobe was removed (*Van Dyke* et al. 1952). The in vivo application of ADH enhanced the production of erythropoietin at doses high enough to lower renal blood flow to a degree at which renal hypoxia occurs (*Jepson* et al. 1968).

6.3.4 Thyroxine

The erythropoiesis-stimulating activity of thyroxine has already been mentioned (Sect. 6.1.5). The present paragraph is confined to the relationship between erythropoietin production and thyroid activity during hypoxic stress. The plasma levels of thyroid hormones T_3 and T_4 increased in humans exposed to hypobaric hypoxia (*Surks* et al. 1967; *Kotchen* et al. 1973b; *Rastogi* et al. 1977). Yet this increase was probably due to altered T_3 and T_4 distribution pools rather than to enhanced thyroid activity (*Rastogi* et al. 1977). In addition, basal oxygen consumption increased transiently in humans exposed to hypoxia (*Surks* et al. 1967). Studies in hypoxic rats have revealed an initial increase in thyroid activity (*Debias* 1966) and intrathyroidal inhibition of thyroxine synthesis in continued hypoxia (*Mulvey* and *Macaione* 1969; *Connors* and *Martin* 1982). Thus the corresponding temporal changes in plasma activities of thyroid hormones and of erythropoietin could indicate that these are causally related.

6.3.5 Steroids

Effects of androgenic steroids on erythropoiesis have been studied in detail with a view to their therapeutic value (cf. *Adamson* 1977). Erythropoiesis was clearly stimulated by testosterone and related 5α(H) derivatives. On the other hand, 5β(H) derivatives were erythropoietically inactive, at least in mice (*Fisher* et al. 1977). Androgenic steroids stimulated both kidney production of erythropoietin (*Fried* and *Gurney* 1965) and the proliferation of erythrocytic progenitors (see Sect. 2.2). Sex differences in red blood cell counts have been related in part to the effect of androgens on erythropoietin production (*Adamson* 1977). Male adult humans excrete about three times more bioactive erythropoietin with the urine than females or prepubertal boys (*Alexanian* 1966). However, note the lack of sex difference in serum titers of immunoreactive erythropoietin (*Miller* et al. 1981; *Garcia* et al. 1982; *Cotes* 1982; *Rege* et al. 1982). Testosterone enhanced the effect of hypoxia on the formation of erythropoietin (*Fried* and *Gurney* 1965; *Jepson* and *Lowenstein* 1967; *Gordon* et al. 1968), which explains the higher plasma erythropoietin activity in hypoxic males compared to females (*Wang* and *Fried* 1972). *Gurney* et al. (1980) were able to augment the effect of hypoxia on red cell mass in mice by the long-term application of testosterone.

Isolated perfused kidneys from dogs (*Fisher* and *Langston* 1968) and rabbits (*Gordon* et al. 1968) produced significantly more erythropoietin when the kidney donor animals were pretreated with testosterone. *Malgor* and *Fisher* (1970), utilizing programmed canine kidneys, observed a similar effect when testosterone was added directly to the perfusate. *Paulo* et al. (1974) have compared the effects of various steroid metabolites in this system. 5α-Dihydrotestosterone proved to be the most potent erythropoietic compound. In addition, certain structure-function relationships could be established. For example, the 19-methyl group of testosterone was required for the stimulation of erythropoietin production (*Paulo* et al. 1974).

Estrogenic steroids inhibited erythropoiesis both at the kidney (*Mirand* and *Gordon* 1966) and bone marrow level (*Jepson* and *Lowenstein* 1966b). *Peschle* et al. (1973c) found no difference in the erythropoietin response between male and ovariectomized female rats. However, much lower doses of estradiol were required in ovariectomized than in male rats to reduce the production of erythropoietin. These results indicate that both the inhibition by estrogens and the stimulation by androgens account for the sex difference in erythropoietin production.

The role of glucocorticoids in the regulation of erythropoiesis is less evident. Both stimulatory (*Fisher* 1958; *Peschle* et al. 1971b; *Jepson* and *Manasc* 1974) and inhibitory (*Gordon* et al. 1967b; *Glader* et al. 1968) in vivo effects of glucocorticoids on erythropoiesis have been reported. High doses of prednisolone suppressed the formation of erythropoietin in hypoxia-exposed mice (*Gordon* et al. 1967b).

The application of the mineralocorticoid aldosterone had no significant effect on erythropoiesis in normal or polycythemic mice (*Mann* et al. 1967; *Živný* et al. 1972b). Neither did aldosterone enhance the plasma titer of erythropoietin in normal rats. High doses of aldosterone produced a significant increase in plasma erythropoietin in adrenalectomized rats exposed to hypoxia (*Živný* et al. 1972b). This finding has reen related to the effect of aldosterone on tubular sodium reabsorption, but there is no proof for this explanation.

6.4 Role of the Autonomic Nervous System

Several lines of evidence suggest that the hypothalamus (*Halvorsen* 1966) and efferent sympathetic nerve fibers (*Fink* and *Fisher* 1977c) play a role in the regulation of erythropoiesis. However, the exact reflex mechanisms remain to be clarified. Virtually nothing is known about the sensory and afferent elements. Afferent fibers from renal chemoreceptors have been described (*Recordati* et al. 1978) but their relevance to the production of erythropoietin has not been investigated. Hypoxia stimulation of the carotid chemoreceptor bodies produced no increase in erythropoietin production (*Paulo* et al. 1973b; *Beynon* and *Balfour* 1973; *Hansen* et al. 1973). Electrical stimulation of the hypothalamus caused an increase in the plasma erythropoietin titer in rabbits (*Halvorsen* 1961, 1966) and monkeys (*Mirand* et al. 1964). *Halvorsen* (1964) has reported that lesions in the posterior, middle, and ventral hypothalamus, but not in the anterior hypothalamus, reduced the reticulocyte response to hypoxia in rabbits. Only the effect of the posterior hypothalamic lesion was not normalized after the application of ACTH. *Halvorsen* (1966) has proposed that oxygen-sensitive cells in the hypothalamus may influence erythropoietin production side by side through humoral and nervous mechanisms.

The hypothalamus is important in the control of sympathetic nervous activity. Basically, the sympathetic nervous system can act directly at the target organ and indirectly through blood-borne catecholamines from the adrenal medulla. Transplanted, and thus denervated, kidneys retained the ability to respond to anemia with enhanced production of erythropoietin (*Mirand* et al. 1968; *Varkarakis* et al. 1976), although it remains to be established whether the response of such kidneys is quantitatively normal. Section of the renal nerves caused a decrease in the plasma erythropoietin titer in rabbits exposed to hypoxia in the short term (*Fink* and *Fisher* 1976) but an increase in similarly treated rats (*Beynon* 1977). Pretreatment of rabbits with reserpine also attenuated the production of erythropoietin during a 5-h hypoxic period (*Fink* and *Fisher* 1977c). Cutting of the splanchnic nerves in rats resulted in lowered erythropoietin production if the animals were exposed to hypoxia immediately after the operation (*Beynon* 1977), but was without effect 1 week thereafter (*Finne* and *Skoglund* 1970; *Beynon* 1977). Pretreatment with the β_2-adrenergic antagonist butoxamine resulted in lowered plasma erythropoietin titers in hypoxic rabbits, while the α-adrenergic blocking agent phenoxybenzamine produced the opposite effect (*Fink* et al. 1975; *Fink* and *Fisher* 1977c). In addition, renal production of erythropoietin in normoxic animals increased following the application of single doses of either nonspecific β-adrenergic or selective β_2-adrenergic agonists (*Fink* and *Fisher* 1977a,b; *Gross* and *Fisher* 1978; *Przala* et al. 1979). These effects were abolished by pretreatment with either nonspecific β-adrenergic or selective β_2-adrenergic antagonists (*Fink* and *Fisher* 1977a; *Przala* et al. 1979). The specific β_1-adrenergic agonist dobutamine was erythropoietically ineffective (*Fink* and *Fisher* 1977a). These results indicate a β_2-adrenergic receptor involvement in erythropoietin production. Of the naturally occurring catecholamines, it is mainly epinephrine that has the potential to activate β_2-adrenergic receptors. Thus effects of the sympathetic nervous system on erythropoietin production are more likely to be mediated by blood-borne epinephrine derived from the adrenal medulla than by norepinephrine released from sympathetic nerve terminals. Note, however, that circulating epinephrine can be taken up by sympathetic nerves and then be released like a classic neurotransmitter (*Berecek* and *Brody* 1982).

It is not known whether the sympathetic nervous system is important in the normal day-to-day control of erythropoiesis. Attempts to enhance red cell mass in normoxic animals by the long-term administration of the β_2-adrenergic agonist albuterol failed (*Jelkmann* et al. 1979b). Obviously, β_2-adrenergic activation does not overcome the normal feedback mechanism which controls erythropoiesis on the basis of oxygen supply to the tissues. Under conditions of hypoxia, however, chronic β_2-adrenergic activation enhances erythropoiesis, as demonstrated by the increase in red cell mass in hypoxemic rabbits (*Jelkmann* et al. 1979b) and anemic mice (*Jelkmann* and *Bauer* 1980). Indeed, exposure to hypoxia has been shown to stimulate adrenal medullary activity in a number of species (*Kotchen* et al. 1973a; *Myles* and *Ducker* 1973; *Claustre* and *Peyrin* 1982; *Johnson* et al. 1983). It may thus be inferred that the enhanced release of epinephrine during hypoxic stress augments erythropoiesis at the kidney and, synergistically, at the erythropoietic progenitor cell level (see Sect. 2.2).

The mechanism by which β_2-adrenoceptor activation induces the elaboration of erythropoietin is not completely understood, yet it does not appear to involve changes in total renal blood flow. *Jelkmann* et al. (1979a) demonstrated a significant increase in erythropoietin titers when the β_2-adrenergic agonist albuterol was added to the perfusate of isolated dog kidneys. The effect of albuterol was abolished when the kidney donor dogs were pretreated with the prostaglandin synthesis inhibitor drug indomethacin. These findings have been confirmed by *Radtke* et al. (1980), who used meclofenamate instead of indomethacin. Furthermore, these authors found that β_2-adrenergic activation with albuterol stimulated the production of prostaglandin E_2 in the programmed isolated perfused dog kidney. This effect is exceptional, because cells usually respond to β-adrenoceptor activation with an increase in cyclic AMP synthesis and possibly Ca^{2+} extrusion (*Scheid* and *Fay* 1984). Prostaglandin synthesis, on the other hand, is stimulated more generally by agents which promote Ca^{2+} influx, such as α-adrenergic agonists (*Berridge* 1981).

A cholinergic mechanism has also been implicated in the control of erythropoietin production. Treatment with atropine attenuated the increase in plasma erythropoietin in hypoxia-exposed rabbits (*Paulo* et al. 1972) and in cobalt-treated rats (*Rodgers* et al. 1975a). In addition, erythropoietin production in hypoxic dogs was significantly reduced by pretreatment with methylatropine, which is a peripherally acting cholinolytic agent (*Spannhake* and *Fisher* 1977). However, the idea of parasympathetic involvement in the erythropoietin response to hypoxia remains problematic. Firstly, there is no agreement as to whether systemic hypoxia causes an increase or a decrease in the efferent parasympathetic activity (*Gupta* and *Singh* 1981). Secondly, electron microscopy studies in rat and monkey would seem to indicate that the innervation of vessels and tubules in the renal juxtaglomerular region is exclusively monoaminergic (*Barajas* 1978, 1981).

7 Effects of Cobalt on Erythropoietin Production

The stimulation of erythropoietin production by cobalt is relatively specific and not shared by other divalent metal ions such as Mn^{2+}, Fe^{2+}, and Zn^{2+} (*Goldwasser* et al. 1958; *Thorling* 1973). Ni^{2+} stimulated erythropoietin production only when it was injected intrarenally (*Hopfer* et al. 1979; *McCully* et al. 1982; *Sunderman* et al. 1982; *Hopfer* et al. 1984). *Jensen* and *Thorling* (1965) have suggested that cobalt induces tissue hypoxia, because the injection of cobalt produced a change in the lactate dehydrogenase isoenzyme pattern of rabbit kidneys similar to that seen in anemia. *Rodgers* et al. (1974) found an increase in the concentration of lactate in the blood of cobalt-treated rats. These authors have also reported that lactate injections stimulated erythropoiesis in polycythemic mice. However, *Živný* et al. (1971) failed to demonstrate a direct effect of lactic acid on erythropoietin production. *Thorling* (1973) has doubted that cobalt acts by inhibiting the cellular oxidative pathway, because metabolic inhibitors such as cyanide, fluoroacetate, and monoiodoacetate were without any effect on erythropoietin production (*Nečas* and *Thorling* 1972; *Nečas* and *Neuwirt* 1972).

Miller et al. (1974) have related the effect of cobalt to a pH-dependent increase in blood oxygen affinity caused by hyperventilation. However, this explanation turns out to be insufficient, if one contrasts the small change in oxygen affinity with the marked erythropoietic effect exerted by cobalt. In addition, cobalt stimulated the production of erythropoietin in isolated perfused kidneys (*Fisher* and *Birdwell* 1961; *Fisher* and *Langston* 1968) and in renal mesangial cell cultures (*Kurtz* et al. 1983a).

By contrast with erythropoietin production due to anemic or hypoxic hypoxia, the action of cobalt does not appear to involve renal prostaglandins, for the effect of cobalt was not inhibited by indomethacin (*Jelkmann* et al. 1984). Note, too, that cobaltous ion did not stimulate the synthesis of prostaglandins in kidney medulla slices (*Erman* and *Raz* 1979) or in mesangial cell cultures (*Jelkmann* et al. 1985).

Both cyclic GMP and cyclic AMP have been implicated in the mechanism of cobalt action. The application of cobalt to rats caused an early rise in renal cyclic GMP levels which was accompanied by an increase in the activity of lysosomal enzymes in plasma. Because these effects, along with the erythropoietin response, were inhibited by atropine and enhanced by physostigmine, a cholinergic mechanism has been assumed (*Rodgers* et al. 1975a). The renal cyclic AMP level also rose in cobalt-treated rats (*Rodgers* et al. 1972, 1975b). Cobalt stimulated adenylate cyclase prepared from rat kidneys (*Rodgers* et al. 1975b), while it had no effect on the activity of isolated phosphodiesterase (*Rodgers* et al. 1975c).

In conclusion, the mechanism of cobalt action on renal production of erythropoietin is not completely understood. Although there are basic differences compared to the effect of hypoxia, both mechanisms may eventually involve the activation of cyclic-AMP-dependent protein kinases.

8 Biogenesis in Kidney Cell Cultures

The use of renal cell cultures for the study of erythropoietin production has several appealing aspects. The use of specific cell markers and the evaluation of the ultrastructural characteristics of homogeneous cell lines in culture have enabled investigators to identify the type of erythropoietin-producing cells. In addition, the cellular mechanisms that control erythropoietin production can be studied under defined conditions. Another aspect is the use of human renal cell cultures for the production of erythropoietin for analytical and therapeutic uses. Several of the earlier attempts to produce erythropoietin in renal cell cultures have been summarized by *Ogle* et al. (1978a).

8.1 Normal Renal Cell Cultures

Ozawa (1967) cocultured primary cultures from newborn rabbit kidneys with rabbit bone marrow cells in the same culture tube. When the oxygen concentration in the incubator was reduced from 21% to 3%, heme synthesis by the bone marrow cells increased. Heme synthesis was not stimulated at 1% O_2 or when cobalt was added to the cultures. Only cultures grown from the kidney cortex induced heme synthesis, while medullary cultures were inactive. *McDonald* et al. (1969) reported that culture medium from bovine kidney cells grown at 5% O_2 stimulated erythropoiesis in polycythemic mice. This effect was abolished by incubating the culture medium with antibody to erythropoietin. *Sherwood* et al. (1972) detected erythropoietic activity in the culture medium of rat kidney tissue. This activity was thought to be erythrogenin, because it stimulated erythropoiesis in polycythemic mice only after incubation with rat serum. Kidney cultures from human midterm fetuses (*Congote* and *Solomon* 1974) and from adult rabbits (*Ogle* et al. 1978b) produced an erythropoietic activity that stimulated heme synthesis in fetal mouse liver cell cultures. Kidney cells in culture also produced thrombopoietin but there was no quantitative relationship between the production of the two hormones (*Ogle* et al. 1978c).

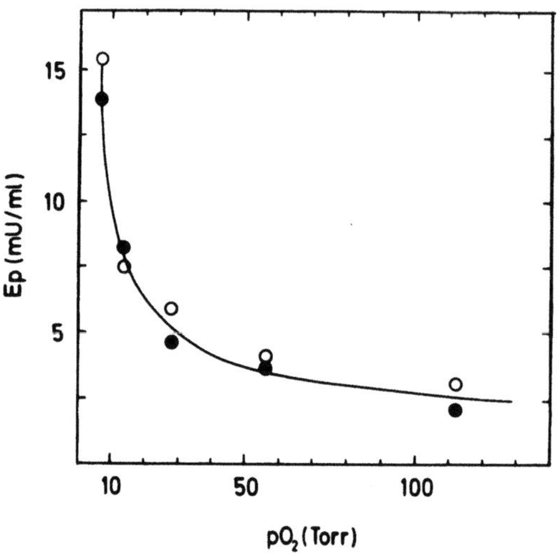

Fig. 9. Relationship between erythropoietin activity (Ep, fetal mouse liver cell assay) in mesangial cell culture medium and the partial pressure of oxygen (pO_2) in the incubation atmosphere. Data represent Ep activity produced by 7×10^5 cells/ml during 3 days of incubation. Results are shown of two independent experiments in which the pO_2 was either increased step by step (○) or decreased (●). (*Kurtz* et al. 1983a)

While in all of the above studies whole kidney fragments were cultured, *Burlington* et al. (1972) studied erythropoietin production in outgrowths from isolated goat kidney glomeruli. Erythropoietin was detected both in long-term primary cultures and in subcultures. The erythropoietic activity in the culture medium was greater when assayed in vitro than in vivo. Somewhat surprisingly, the production of erythropoietin required an interval of at least 32 days between changes of the culture medium without feeding of the cells.

Kurtz et al. (1982b) studied erythropoietin production in outgrowths from isolated rat glomeruli. Erythropoietic activity was detected in primary cultures and in subcultures prepared after 21 days of primary culture, while subcultures prepared after 6 days of primary culture were inactive. The day-21 subcultures were identified as mesangial cells on the basis of several criteria, including cell morphology and immunofluorescence staining of cell markers (*Kurtz* et al. 1982b, 1983a). The mesangial cell-derived activity stimulated the growth of erythroid clusters from fetal liver and adult bone marrow. This effect was inhibited by antibody to erythropoietin. *Kurtz* et al. (1983a) assumed that their mesangial cell cultures produced erythropoietin with an abnormal carbohydrate portion because it was inactive in the polycythemic mouse assay for erythropoietin and was bound to *Ricinus communis* lectin, which is specific for free galactosyl residues. *Jelkmann* et al. (1986) have recently detected in vivo active erythropoietin (polycythemic mouse assay) in mesangial cell cultures grown without HEPES (N-2-hydroxyethylpiperazine-N′-2-ethanesulfonic acid) buffer and suggested that the carbohydrate composition of erythropoietin may be influenced by the cell culture components employed.

The production of erythropoietin activity in mesangial cell cultures was oxygen dependent and increased exponentially when the oxygen tension in the incubator was lowered (Fig. 9). Also, as in vivo, cobalt chloride induced erythropoietin production in mesangial cell cultures (*Kurtz* et al. 1983a). Erythropoietin production was also observed following the addition of prostaglandin E_2 to mesangial cell cultures. The prostaglandin synthetase inhibitor drug indomethacin, on the other hand, attenuated the hypoxia-induced production of erythropoietin (*Jelkmann* et al. 1984; *Kurtz* et al. 1985b). These findings support the concept that prostaglandins are involved in the mechanism by which hypoxia induces the elaboration of erythropoietin (*Fisher* 1980).

Recently, *Caro* et al. (1984) have reported preliminary studies showing erythropoietin production by cultures of an established proximal tubule cell line ($LLCPK_1$) originally derived from normal pig kidney. This in vitro erythropoietin synthesis was also stimulated by hypoxia and cobalt. *Caro* et al. (1984) were able to detect immunoreactive erythropoietin in their cultures. Yet the hormone activity was very low and did not exceed 1 mU/ml culture medium.

8.2 Kidney Tumor Cell Cultures

Elevated plasma and urinary erythropoietin titers in association with erythrocytosis have been noted incidentally in patients with renal tumors. Among such tumors were carcinomas, sarcomas, adenomas, and cysts (*Donati* et al. 1963; *Hammond* and *Winnick* 1974; *Kazal* and *Erslev* 1975b; *Sufrin* et al. 1977). Inappropriate production of erythropoietin was probably caused by local hypoxia in some of these cases (*Hammond* and *Winnick* 1974). However, there is also evidence for autonomous hormone production by renal carcinoma cells. Erythrocytosis and enhanced plasma titers of erythropoietin developed in nude mice receiving lung metastatic transplants from a patient with a kidney carcinoma and elevated plasma erythropoietin (*Toyama* et al. 1979; *Katsuoka* et al. 1983b). *Saito* et al. (1985) extracted poly (A+) RNA from xenotransplanted renal tumor cells that induced production of erythropoietin when injected in *Xenopus* oocytes. *Lee-Huang* (1984) used poly (A+) RNA from human renal carcinomas as a template for the synthesis of erythropoietin cDNA. The cDNA, together with a plasmid vector, induced in *Escherichia coli* the formation of recombinant colonies that expressed erythropoietin gene activity as identified with monoclonal antibody to human erythropoietin (*Lee-Huang* 1984).

Murphy et al. (1970) first detected erythropoietin activity in human renal carcinoma cultures. *Sherwood* and *Goldwasser* (1976) maintained three renal carcinomas in reportedly monolayer cultures for several weeks. All three cell lines secreted initially an erythropoietic activity that was identified as asialoerythropoietin on the basis of bioassay and chromatographic studies. Erythropoietin production was also accomplished in cultures of human renal carcinoma cells after previous transplantation into nude mice (*Katsuoka* et al. 1983b; *Hagiwara* et al. 1984a; *Saito* et al. 1985; *Okabe* et al. 1985). Again, this erythropoietin was much less active when assayed in polycythemic mice than in fetal mouse liver cultures (*Katsuoka* et al. 1983b; *Hagiwara* et al. 1984a) or by radioimmunoassay (*Hagiwara* et al. 1984a). *Hagiwara* et al. (1984b) observed an exponential increase in erythropoietin production when renal carcinoma cell cultures reached confluency. Recent findings by *Hagiwara* et al. (1984c) indicate that endogenous prostaglandins support erythropoietin production by renal carcinoma cultures, because meclofenamate, an inhibitor of prostaglandin synthesis, lowered erythropoietin titers in the cultures.

An erythropoietin-producing continuous carcinoma cell line, designated GKA, has been derived from the kidney of a polycythemic patient with renal adenocarcinoma (*Sytkowski* et al. 1983). The primary uncloned cell line produced erythropoietin-like activity during serial passages for over 3 years. In addition, the authors prepared more homogeneous clones from the original GKA cell line (*Sytkowski* et al. 1984). Some of the sublines secreted higher erythropoietic activity than the uncloned cell line, when assayed by in vitro bioassay. No activity was demonstrable by radioimmunoassay for erythropoietin. There are reportedly two explanations for this descrepancy. One is that the cells produced an erythropoietin derivative with altered immunological properties. The other is that they produced an erythropoietic factor distinct from erythropoietin (*Sytkowski* et al. 1984).

9 Summary and Current Problems

Hemopoiesis is maintained by the proliferative activity of pluripotent stem cells. It remains to be clarified how myelopoietic pluripotent stem cells are committed to differentiate into a distinct blood cell line. Beginning at the progenitor stage (i.e., the BFU-Es in the erythroid line), cell proliferation is controlled by specific glycoprotein factors such as erythropoietin. Erythropoietin stimulates erythropoiesis mainly by amplifying the pool of progenitor cells. The CFU-E is considered the main target cell of erythropoietin (Fig. 1). Reportedly, the hormone binds to a protein receptor on the cell surface and acts via an intracellular second messenger. This messenger is most likely a cytoplasmic protein, but other factors such as cyclic nucleotides have also been considered. Intracellular effects induced by erythropoietin binding include the synthesis of RNAs and DNA, and eventually cell division. Several other hormonal agents may support the effect of erythropoietin. These include androgenic steroids, growth hormone, insuline-like growth factors, thyroxine, and adrenomimetic agents (Fig. 1).

Erythropoietin purified from human urine has been identified as an acidic glycoprotein with a molecular weight of 34 000 daltons. The carbohydrate component (about 30%) is not necessary for the in vitro action of erythropoietin, but is necessary for its in vivo survival (plasma half-life 2–11 h depending on species). Renal mRNA encoding for human erythropoietin has been prepared and used as the template for the synthesis of erythropoietin cDNA. Previously, the complete nucleotide and the deduced amino-acid sequence have been described for human fetal liver erythropoietin cDNA.

Current assays of erythropoietin include in vivo and in vitro bioassays and immunoassays. Though the use of radioimmunological techniques is desirable, their general applicability is still hampered by the lack of monospecific antibody and of pure antigen.

Attempts to produce the hormone in vitro have been partially successful. Erythropoietin-like activity has been demonstrated in mesangial cell cultures and tubule cultures from normal kidneys, and in kidney carcinoma cell cultures. The main problem with in vitro studies of erythropoietin production relates to difficulties in obtaining hormone with a normal carbohydrate portion. In general, cell culture products display poor erythropoietin activity when tested by in vivo bioassay. The lack of in vivo activity is probably due to an abnormal assembly of erythropoietin glycans by cultured cells.

The plasma titer of erythropoietin can rise by several orders of magnitude above normal (10–20 mU/ml) within a few hours of hypoxic stress. Subsequently, the plasma erythropoietin level decreases with continuing hypoxia. Most of the circulating hormone is derived from the kidney. There is minute storage of erythropoietin in the normal kidney, yet significant amounts can be extracted from the kidneys of animals previously rendered hypoxic. The intrarenal site of erythropoietin synthesis is still not exactly known, although several lines of evidence point to the glomerular mesangium. This evidence includes the immunofluorescence localization of erythropoietin in the glomerular tuft of hypoxic kidneys, the extraction of erythropoietin from isolated glomeruli, and the ability of mesangial cells to produce erythropoietin in cell culture. Alternatively, some authors favor a tubular origin of the hormone, mainly because some renal carcinomas produce erythropoietin in vivo and in vitro.

The formation of erythropoietin is initiated by systemic or renal hypoxia of different origin, e.g., anemia, hypoxemia, or renal ischemia. Renal cell oxygen tension appears to be the primary determinant. The kidney is well suited to control the production of an oxygen-dependent hormone because, by contrast with several other organs, a reduction in renal oxygen supply is not readily masked by an increase in blood flow (cf. *Kiil* 1971). Renal prostaglandins are thought to function as local mediators of the effect of hypoxia on the synthesis of erythropoietin (Fig. 10). Cellular hypoxia is a stimulus to prostaglandin formation. Prostaglandin of the E-type and prostacyclin stimulate the production of erythropoietin, while inhibitors of prostaglandin synthesis also attenuate erythropoietin formation.

Extrarenal mechanisms exert a supportive, if not regulative, influence on renal production of erythropoietin (Fig. 10). Note the moderate erythropoietin response to renal artery constriction. Note, too, that hypoxemic perfused isolated kidneys produce significantly more erythropoietin if they were programmed by hypoxia in vivo. The hypothalamus is considered to be the primary extrarenal center of control of renal erythropoietin production. The endocrine function of the hypothalamic-hypophyseal system is required to maintain basal erythropoiesis and to elicit a normal erythropoietin response to hypoxic stimuli. In addition, the sympathetic nervous system is activated during hypoxic stress, and renal β_2-adrenergic receptor activation augments the production of erythropoietin. This effect, too, is prostaglandin dependent.

Calcium, prostanoids, cyclic AMP, and cyclic GMP have been implicated as positive messengers in erythropoietin production. It still remains to be elucidated how these effectors are coupled, whether they act in parallel or in series, and whether different types of renal cells are involved. Present data suggest the following sequence of events: (a) hypoxia initiates an intracellular mobilization of Ca^{2+}; (b) Ca^{2+}-dependent phospholipase A_2 is activated, thus increasing prostaglandin formation; (c) prostaglandin E_2 and prostacyclin stimulate adenylate cyclase in adjacent cells, thereby expanding the effect of hypoxia; and (d) cyclic-AMP-dependent phosphorylation of nuclear proteins evokes erythropoietin gene transcription.

Unfortunately, the evidence for this concept is not yet complete. Thus it is puzzling that several of the reported stimuli to erythropoietin production are known to lower the cytosolic Ca^{2+} concentration elsewhere. Examples of this kind are α-adrenergic blockade, β_2-adrenergic activation, and the action of cobalt. *Fisher* (1983) has also mentioned the observation that the calcium-entry-blocking agent verapamil enhanced erythropoietin production in hypoxia-exposed rats. Hence a clue to a better understanding of the regulation of erythropoietin production might be

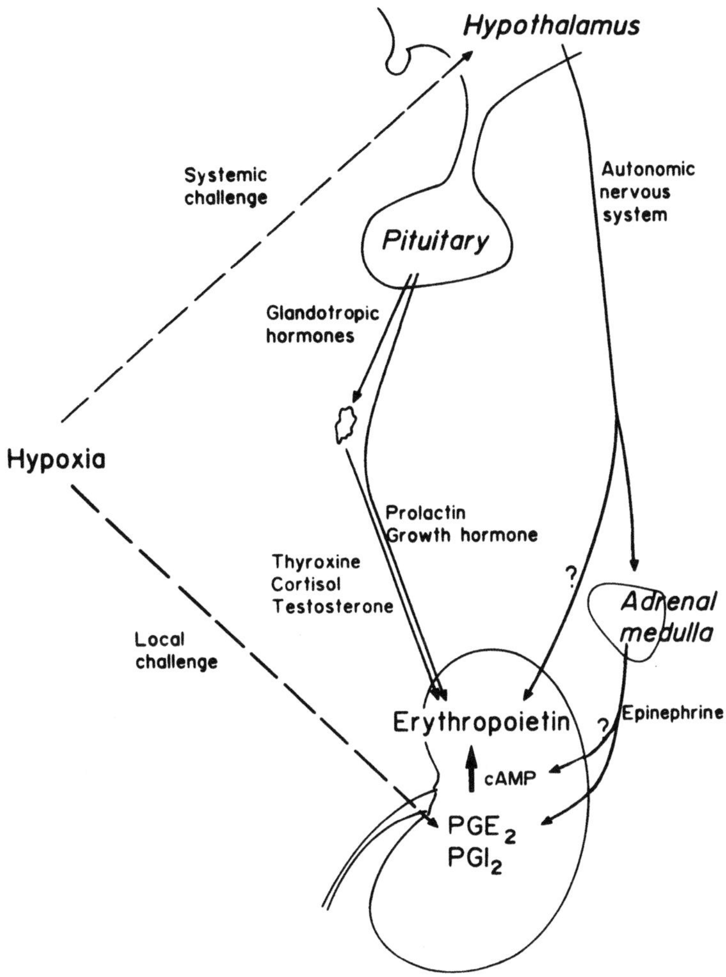

Fig. 10. Schematic representation of local and systemic factors that appear relevant to the regulation of renal production of erythropoietin. The renal prostaglandin system may be involved directly in the action of hypoxia. The sympathetic nervous system is thought to augment erythropoietin production through renal β_2-adrenoceptor activation. Whether the other hormonal agents mediate the effect of hypoxia per se is uncertain but they may have a supportive action, perhaps through increasing cellular metabolism. (Based on findings by several groups of investigators, including *Crafts* and *Meineke* 1959; *Halvorsen* 1966; *Fink* and *Fisher* 1977c; *Fisher* et al. 1978; *Jelkmann* et al. 1979a, and *Kurtz* et al. 1985b)

sought in the effects of hypoxia on the handling of Ca^{2+} by the respective cells. In addition, little is known about the specific mechanisms which control the subsequent steps in erythropoietin production, including ribosomal polypeptide synthesis, glycosylation of the nascent protein and, eventually, its secretion.

Acknowledgments. My thanks are due to Drs. Christopher D.R. Dunn, James W. Fisher, and Christoph Weiss for reviewing an earlier version of this manuscript and providing valuable suggestions. Dr. Allan J. Erslev also made helpful comments. Of course, there is no implication that those listed necessarily share all of the opinions in this review.

I am also grateful to Mrs. Gisela Thaler for her expert secretarial help in the preparation of the manuscript.

References

Abbrecht PH, Littell JK (1972) Plasma erythropoietin in men and mice during acclimatization to different altitudes. J Appl Physiol 32:54–58

Abbrecht PH, Malvin RL (1966) Renal production of erythropoietin in the dog. Am J Physiol 210:237–242

Abbrecht PH, Vander AJ, Turcotte JG (1969) Effects of saline loading on the renin, erythropoietin, and blood pressure responses to canine renal allotransplantation. Circ Res 25:99–105

Abramson S, Miller RG, Phillips RA (1977) The identification in adult bone marrow of pluripotent and restricted stem cells of the myeloid and lymphoid systems. J Exp Med 145:1567–1579

Adamson JW (1968) The erythropoietin/hematocrit relationship in normal and polycythemic man: implications of marrow regulation. Blood 32:597–609

Adamson JW (1977) Steroid metabolites and hemopoiesis. In: Fisher JW (ed) Kidney hormones, vol II, erythropoietin. Academic, London, pp 437–462

Adamson JW, Finch CA (1975) Hemoglobin function, oxygen affinity, and erythropoietin. Annu Rev Physiol 37:351–369

Adamson JW, Torok-Storb B, Lin N (1978) Analysis of erythropoiesis by erythroid colony formation in culture. Blood Cells 4:89–103

Alexanian R (1966) Urinary excretion of erythropoietin in normal men and women. Blood 28:344–353

Alexanian R (1973) Erythropoietin excretion in bone marrow failure and hemolytic anemia. J Lab Clin Med 82:438–445

Alippi RM, Barceló AC, Bozzini CE (1983) Erythropoietic response to hypoxia in mice with polycythemia induced by hypoxia or transfusion. Exp Hematol 11:122–128

Anagnostou A, Schade S, Ashkinaz M, Barone J, Fried W (1977) Effect of protein deprivation on erythropoiesis. Blood 50:1093–1097

Anagnostou A, Schade SG, Fried W (1978) Stimulation of erythropoietin secretion by single amino acids. Proc Soc Exp Biol Med 159:139–141

Annable L, Cotes PM, Mussett MV (1972) The second international reference preparation of erythropoietin, human, urinary, for bioassay. Bull WHO 47:99–112

Aperia AC, Liebow AA, Roberts LE (1968) Renal adaptation to anemia. Circ Res 22: 489–500

Ascensao JL, Gaylis F, Bronson D, Fraley EE, Zanjani ED (1983) Erythropoietin production by a human testicular germ cell line. Blood 62:1132–1134

Axelrad AA, McLeod DL, Shreeve MM, Heath DS (1974) Properties of cells that produce erythrocytic colonies in vitro. In: Robinson WA (ed) Hemopoiesis in culture. US Government Printing Office, Washington DC, pp 226–234

Aye MT (1977) Erythroid colony formation in cultures of human marrow: effect of leukocyte conditioned medium. J Cell Physiol 91:69–78

Bandyopadhyay R, Datta AG (1981) Further studies on the biosynthesis of erythropoietin. Indian J Biochem Biophys 18:241–244

Banerjee RK, Moulik S, Ghosal J, Datta AG (1976) Effect of thyroxine on the vitro formation of erythropoietin in kidney medulla slices. Indian J Biochem Biophys 13:145–149

Barajas L (1978) Innervation of the renal cortex. Fed Proc 37:1192–1201

Barajas L (1981) The juxtaglomerular apparatus: anatomical considerations in feedback control of glomerular filtration rate. Fed Proc 40:78–86

Barceló AC, Bozzini CE (1982) Erythropoietin formation during hypoxia in mice with impaired responsiveness to erythropoietin induced by irradiation or 5-fluorouracil injection. Experientia 38:504–505

Bartley ET, Powsner ER (1974) Erythropoietin in vitro. A culture plate method for marrow. Am J Clin Pathol 61:69–73

Beckman B, Fisher JW (1979) Changes in beta-2 adrenergic receptor sensitivity with maturation of erythroid progenitor cells. Experientia 35:1671–1672

Bellingham AJ (1974) The red cell in adaptation to anaemic hypoxia. Clin Haematol 3:577–594

Berecek KH, Brody MJ (1982) Evidence for a neurotransmitter role for epinephrine derived from the adrenal medulla. Am J Physiol 242:H593–H601

Berridge MJ (1981) Phosphatidylinositol hydrolysis: a multifunctional transducing mechanism. Mol Cell Endocrinol 24:115–140

Beynon G (1977) The influence of the autonomic nervous system in the control of erythropoietin secretion in the hypoxic rat. J Physiol 266:347–360

Beynon GW, Balfour WE (1973) The carotid body and erythropoiesis in the rat. Nature (New Biol) 243:61–62

Birgegård G, Miller O, Caro J, Erslev A (1982) Serum erythropoietin levels by radioimmunoassay in polycythaemia. Scand J Haematol 29:161–167

Bonsdorff E, Jalavisto E (1948) A humoral mechanism in anoxic erythrocytosis. Acta Physiol Scand 16:150–170

Bottomley SS, Whitcomb WH, Smithee GA, Moore MZ (1971) Effect of cyclic adenosine-3′, 5′-monophosphate on bone marrow Δ-aminolevulinic acid synthetase and erythrocyte iron uptake. J Lab Clin Med 77:793–801

Bourgoignie JJ, Gallagher NI, Perry HM, Kurz L, Warnecke MA, Donati RM (1968) Renin and erythropoietin in normotensive and in hypertensive patients. J Lab Clin Med 71:523–536

Bozzini CE (1966) Influence of the erythroid activity of the bone marrow on the plasma disappearance of injected erythropoietin in dogs. Nature 209:1140–1141

Bozzini CE, Ugarte CAA, Martinez MA, Soriano G, Alippi RM, Giglio MJ (1976) Production of and response to erythropoietin in the splenectomized mouse. Exp Hematol 4:114–120

Brandan NC, Cotes PM, Espada J (1981) In vitro assay of erythropoietin in fetal mouse liver cultures. II. Effects of human transferrin bound iron and of serum on the stimulation of incorporation of ^{3}H-thymidine into DNA. Br J Haematol 47:469–478

Braun JL, Greenberg HM, Robinson SH (1979) Hemopoietic progenitor cells in short-term culture: effects of erythropoietin and endotoxin contamination of erythropoietin preparations. Exp Hematol 7:189–196

Brennan JK, Lichtman MA, DiPersio JF, Abboud CN (1980) Chemical mediators of granulopoiesis: a review. Exp Hematol 8:441–464

Bridges JB, Taylor IC (1975) Changes in the kidney and blood following hypoxic hypoxia in rats. Ir J Med Sci 144:274–285

Brown JE, Adamson JW (1977a) Modulation of in vitro erythropoiesis. The influence of β-adrenergic agonists on erythroid colony formation. J Clin Invest 60:70–77

Brown JE, Adamson JW (1977b) Modulation of in vitro erythropoiesis: enhancement of erythroid colony growth by cyclic nucleotides. Cell Tissue Kinet 10:289–298

Burdowski AJ, Brookins J, Kadowski PJ, Jubiz W, Salmon J, Moncada S, Fisher JW (1980) Relationships between prostanoids and renin on erythropoietin (Ep) production following hypoxic stimulation. Exp Hematol 8 (Suppl 8):310–311

Burke WT, Morse BS (1962) Studies on the production and metabolism of erythropoietin in rat liver and kidney. In: Jacobson LO, Doyle M (eds) Erythropoiesis. Grune and Stratton, New York, pp 111–119
Burlington H, Cronkite EP, Reincke U, Zanjani ED (1972) Erythropoietin production in cultures of goat renal glomeruli. Proc Natl Acad Sci USA 69:3547–3550
Busuttil RW, Roh BL, Fisher JW (1971) The cytological localization of erythropoietin in the human kidney using the fluorescent antibody technique. Proc Soc Exp Biol Med 137:327–330
Busuttil RW, Roh BL, Fisher JW (1972) Localization of erythropoietin in the glomerulus of the hypoxic dog kidney using a fluorescent antibody technique. Acta Haematol 47:238–242
Byron JW (1975) Manipulation of the cell cycle of the hemopoietic stem cell. Exp Hematol 3:44–53
Carnot P, Deflandre C (1906) Sur l'activité hémopoiétique des différents organes au cours de la régénération du sang. CR Searces Acad Sci 143:432–435
Caro J, Erslev AJ (1984) Biologic and immunologic erythropoietin in extracts from hypoxic whole rat kidneys and in their glomerular and tubular fractions. J Lab Clin Med 103:922–931
Caro J, Brown S, Miller O, Murray T, Erslev AJ (1979) Erythropoietin levels in uremic nephric and anephric patients. J Lab Clin Med 93:449–458
Caro J, Silver R, Erslev AJ, Miller OP, Birgegård G (1981) Erythropoietin production in fasted rats. J Lab Clin Med 98:860–868
Caro J, Erslev AJ, Silver R, Miller O, Birgegård G (1982) Erythropoietin production in response to anemia or hypoxia in the newborn rat. Blood 60:984–988
Caro J, Zon LI, Silver R, Miller O, Erslev AJ (1983) Erythropoietin in liver tissue extracts and in liver perfusates from hypoxic rats. Am J Physiol 244:E431–E434
Caro J, Hickey J, Erslev AJ (1984) Erythropoietin production by an established kidney proximal tubule cell line ($LLCPK_1$). Exp Hematol 12:357A
Celio MR, Inagami T (1981) Angiotensin II immunoreactivity coexists with renin in the juxtaglomerular granular cells of the kidney. Proc Natl Acad Sci USA 78: 3897–3900
Chandra M, Miller ME, Garcia JF, Mossey RT, McVicar M (1985) Serum immunoreactive erythropoietin levels in patients with polycystic kidney disease as compared with other hemodialysis patients. Nephron 39:26–29
Chang C-S, Goldwasser E (1973) On the mechanism of erythropoietin-induced differentiation. XII. A cytoplasmatic protein mediating induced nuclear RNA synthesis. Dev Biol 34:246–254
Chang SC-S, Sikkema D, Goldwasser E (1974) Evidence for an erythropoietin receptor protein on rat bone marrow cells. Biochem Biophys Res Commun 57:399–405
Chiuini F, Della Torre G, Fanò G, Viti A (1979) Early increase of cyclic adenosine monophosphate level induced by erythropoietin on rabbit bone marrow cell suspensions. Acta Haematol 61:251–257
Choppin J, Lacombe C, Casadevall N, Muller O, Tambourin P, Varet B (1984) Characterization of erythropoietin produced by IW32 murine erythroleukemia cells. Blood 64:341–347
Choppin J, Casadevall N, Lacombe C, Wendling F, Goldwasser E, Berger R, Tambourin P, Varet B (1985) Production of erythropoietin by cloned malignant murine erythroid cells. Exp Hematol 13:610–615
Chowdhury RR, Datta AG (1973) Studies on the in vitro formation of erythropoietin in sheep kidney medulla and the effect of cobalt thereon. Biochem Biophys Res Commun 52:1329–1337
Chui DHK, Djaldetti M, Marks PA, Rifkind RA (1971) Erythropoietin effects on fetal mouse erythroid cells. I. Cell population and hemoglobin synthesis. J Cell Biol 51: 585–595

Clarke BJ, Housman D (1977) Characterization of an erythroid precursor cell of high proliferative capacity in normal human peripheral blood. Proc Natl Acad Sci USA 74:1105–1109

Claustre J, Peyrin L (1982) Free and conjugated catecholamines and metabolites in cat urine after hypoxia. J Appl Physiol 52:304–308

Cline MJ, Golde DW (1979) Cellular interactions in haematopoiesis. Nature 277:177–181

Cohen RA, Miller ME, Garcia JF, Moccia G, Cronkite EP (1981) Regulatory mechanism of erythropoietin production: effects of hypoxemia and hypercarbia. Exp Hematol 9:513–521

Cohen RA, Clemons G, Ebbe S (1985) Correlation between bioassay and radioimmunoassay for erythropoietin in human serum and urine concentrates. Proc Soc Exp Biol Med 179:296–299

Cole RJ, Paul J (1966) The effects of erythropoietin on haem synthesis in mouse yolk sac and cultured foetal liver cells. J Embryol Exp Morphol 15:245–260

Coleman SE Duggan J, Hackett RL (1976) Plasma membrane changes in freeze-fractured rat kidney cortex following renal ischemia. Lab Invest 35:63–70

Congote LF (1983) Isolation of two biologically active peptides, erythrotropin I and erythrotropin II from fetal calf intestine. Biochem Biophys Res Commun 115: 477–483

Congote LF (1984a) Extraction from fetal bovine serum of erythrotropin, an erythroid cell-stimulating factor. Anal Biochem 140:428–433

Congote LF (1984b) High-performance liquid chromatographic separation of serum erythrotropin and erythropoietin. J Chromatogr 310:396–400

Congote LF (1985) Similarities in structure and function of bovine serum erythrotropin and human insulin-like growth factor II. two fetal erythroid cell stimulating factors. Biochem Biophys Res Commun 126:653–659

Congote LF, Solomon S (1974) On the site of origin of erythropoietic factors in human fetal tissues. Endocr Res Commun 1:495–504

Connors JM, Martin LG (1982) Altitude-induced changes in plasma thyroxine, 3, 5, 3′-triiodothyronine, and thyrotropin in rats. J Appl Physiol 53:313–315

Cooper GW, Nocenti MR (1961) Unilateral renal ischemia and erythropoietin. Proc Soc Exp Biol Med 108:546–549

Cosgriff TM, Olin DB, Nash DA (1978) Hyporeninemic hypoaldosteronism – case report and observation of dissociated renin and erythropoietin activity. J Chronic Dis 31:195–200

Cotes PM (1982) Immunoreactive erythropoietin in serum. I. Evidence for the validity of the assay method and the physiological relevance of estimates. Br J Haematol 50:427–438

Cotes PM, Bangham DR (1961) Bioassay of erythropoietin in mice made polycythemic by exposure to air at reduced pressure. Nature 191:1065–1067

Cotes PM, Brozovic B (1982) Diurnal variation of serum immunoreactive erythropoietin in a normal subject. Clin Endocrinol 17:419–422

Crafts RC, Meineke HA (1959) The anemia of hypophysectomized animals. Ann NY Acad Sci 77:501–517

Dainiak N, Hoffman R, Maffei LA, Forget BG (1978) Potentiation of human erythropoiesis in vitro by thyroid hormone. Nature 272:260–262

Dainiak N, Davies G, Kalmanti M, Lawler J, Kulkarni V (1983) Platelet-derived growth factor promotes proliferation of erythropoietic progenitor cells in vitro. J Clin Invest 71:1206–1214

Datta MC, Dukes PP (1982) Comparison of the effects of crude and purified erythropoietin on the synthesis of hemoglobin in rat bone marrow cell cultures. Exp Hematol 10:689–697

Daughaday WH, Ward AP, Goldberg AC, Trivedi B, Kapadia M (1982) Characterization of somatomedin binding in human serum by ultracentrifugation and gel filtration. J Clin Endocrinol Metab 55:916–921

Debias DA (1966) Thyroid-adrenal relationship in altitude tolerance. Fed Proc 25: 1227–1229

De Klerk G, Hart AAM, Kruiswijk C, Goudsmit R (1978) Modified method of erythropoietin (ESF) bioassay in vitro using mouse fetal liver cells. II. Measurement of ESF in human serum. Blood 52:569–577

De Klerk G, Vet RJWM, Rosengarten PCJ, Goudsmit R (1980) Comparison of the hemagglutination inhibition assay kit for erythropoietin (ESF) with the fetal mouse liver cell bioassay in vitro. Blood 55:955–959

De Klerk G, Rosengarten PCJ, Vet RJWM, Goudsmit R (1981a) Serum erythropoietin (ESF) titers in anemia. Blood 58:1164–1170

De Klerk G, Rosengarten PCJ, Vet RJWM, Goudsmit R (1981b) Serum erythropoietin (ESF) titers in polycythemia. Blood 58:1171–1174

Demopoulos HB, Highman B, Altland PD, Gerving MA, Kaley G (1965) Effects of high altitude on granular juxtaglomerular cells and their possible role in erythropoietin production. Am J Pathol 46:497–509

Denton MJ, Arnstein HRV (1973) Characterization of developing adult mammalian erythroid cells separated by velocity sedimentation. Br J Haematol 24:7–17

Dessypris EN, Krantz SB (1984) Effect of pure erythropoietin on DNA-synthesis by human marrow day 15 erythroid burst forming units in short-term liquid culture. Br J Haematol 56:295–306

Dinkelaar RB, Engels EY, Hart AAM, Schoemaker LP, Bosch E, Chamuleau RAFM (1981) Metabolic studies on erythropoietin (Ep): II. The role of liver and kidney in the metabolism of Ep. Exp Hematol 9:796–803

Donati RM, McCarthy JM, Lange RD, Gallagher NI (1963) Erythrocythemia and neoplastic tumors. Ann Intern Med 58:47–55

Donati RM, Bourgoignie JJ, Kuhn C, Gallagher NI, Perry HM (1968) Dissociation of circulating renin and erythropoietin in rats. Circ Res 22:91–95

Dorado M, Langton AA, Brandan NC, Espada J (1972) Electrophoretic behavior of erythropoietin in polyacrylamide gel. Biochem Med 6:238–245

Dorado M, Espada J, Langton AA, Brandan NC (1974) Molecular weight estimation of human erythropoietin by SDS-polyacrylamide gel electrophoresis. Biochem Med 10:1–7

Dordal MS, Wang FF, Goldwasser E (1985) The role of carbohydrate in erythropoietin action. Endocrinology 116:2293–2299

Draganac PS, Andrews RB, Trent DJ, Saito T, Farkas WR, Lange RD (1983) Rapid preparation of human urinary erythropoietin by high performance liquid chromatography. Exp Hematol 11 (Suppl 14):58A

Dukes PP (1968) In vitro studies on DNA synthesis of bone marrow cells stimulated by erythropoietin. Ann NY Acad Sci 149:437–444

Dukes PP, Einbinder M (1975) Response of marrow cell cultures to glass bound erythropoietin. Blood 16:1024A

Dukes PP, Takaku F, Goldwasser E (1964) In vitro studies on the effect of erythropoietin on glucosamine-1-^{14}C incorporation into rat bone marrow cells. Endocrinology 74:960–972

Dukes PP, Hammond D, Shore NA, Ortega JA (1970) Erythropoietin: a complex with different in vivo and in vitro activities. J Lab Clin Med 76:439–444

Dukes PP, Shore NA, Hammond D, Ortega JA, Datta MC (1973) Enhancement of erythropoiesis by prostaglandins. J Lab Clin Med 82:704–712

Dunn CDR (1983) Current concepts in erythropoiesis. Wiley, Chichester

Dunn CDR, Gibson L (1983) Erythropoietin bioassays using fetal mouse liver cells: validations and technical improvements. Exp Hematol 11:590–600

Dunn CDR, Lange RD (1980) Erythropoietin: assay and characterization. In: Roath S (ed) Topical reviews in haematology. Wright, Bristol, pp 1–32

Dunn CDR, Napier JAF (1975) An evaluation of factors affecting the in vitro bioassay for erythropoietin. Exp Hematol 3:362–374

Dunn CDR, Napier JAF (1978) Technical comments on the bioassay of erythropoietin. Exp Hematol 6:577–584
Dunn CDR, Trent D (1981) The effect of parathyroid hormone on erythropoiesis in serum-free cultures of fetal mouse liver cells. Proc Soc Exp Biol Med 166:556–561
Dunn CDR, Jarvis JH, Greenman JM (1975) A quantitative bioassay for erythropoietin using mouse fetal liver cells. Exp Hematol 3:65–78
Dunn CDR, Jarvis JH, Napier JAF (1976) Changes in erythropoiesis and renal ultrastructure during exposure of mice to hypoxia. Exp Hematol 4:365–381
Dunn CDR, Smith LN, Leonard JI, Andrews RB, Lange RD (1980) Animal and computer investigations into the murine erythroid response to chronic hypoxia. Exp Hematol 8 (Suppl 8):259–279
Dunn CDR, Leonhard JI, Kimzey SL (1981) Interactions of animal and computer models in investigations of the "anemia" of space flight. Aviat Space Environ Med 52:683–690
Eaves AC, Eaves CJ (1984) Erythropoiesis in culture. Clin Haematol 13:371–391
Eliason JF, Van Zant G (1983) Erythropoietin-dependent and erythropoietin-independent enhancement of colony formation by immature erythroid progenitors (BFUe). Cell Tissue Kinet 16:65–76
Emmanouel DS, Goldwasser E, Katz AI (1984) Metabolism of pure human erythropoietin in the rat. Am J Physiol 247:F168–F176
Erkelens DW, Van Eps LWS (1973) Bartter's syndrome and erythrocytosis. Am J Med 55:711–719
Erman A, Raz A (1979) Effects of bivalent cations on prostaglandin biosynthesis and phospholipase A_2 activation in rabbit kidney medulla slices. Biochem J 182:821–825
Erslev AJ (1962) Erythropoietin in vitro. In: Jacobson LO, Doyle M (eds) Erythropoiesis. Grune and Stratton, New York, pp 275–285
Erslev AJ (1974) In vitro production of erythropoietin by kidneys perfused with a serum-free solution. Blood 44:77–85
Erslev AJ (1975) Renal biogenesis of erythropoietin. Am J Med 58:25–30
Erslev AJ (1980) Blood and mountains. In: Wintrobe MM (ed) Blood, pure and eloquent. McGraw Hill, New York, pp 257–280
Erslev AJ, Caro J (1984a) Pure erythrocytosis classified according to erythropoietin titers. Am J Med 76:57–61
Erslev AJ, Caro J (1984b) Secondary polycythemia: a boon or a burden? Blood Cells 10:177–191
Erslev AJ, Kazal LA, Miller OP (1971) The action and neutralization of a renal lipid inhibitor of erythropoietin. Proc Soc Exp Biol Med 138:1025–1029
Erslev AJ, Caro J, Kansu E, Miller O, Cobbs E (1979) Plasma erythropoietin in polycythemia. Am J Med 66:243–247
Erslev AJ, Caro J, Kansu E, Silver R (1980a) Renal and extrarenal erythropoietin production in anaemic rats. Br J Haematol 45:65–72
Erslev AJ, Caro J, Miller O, Silver R (1980b) Plasma erythropoietin in health and disease. Ann Clin Lab Sci 10:250–257
Espada J (1977) Chemistry and purification of erythropoietin. In: Fisher JW (ed) Kidney hormones, vol II, erythropoietin. Academic, London, pp 37–71
Espada J, Brandan NC, Dorado M (1973) Effect of chemical and enzymatic agents on the biological activity of erythropoietin. Acta Physiol Lat Am 23:193–201
Espada J, Brandan NC, Dorado M (1974) Molecular heterogeneity of human erythropoietin. Biochim Biophys Acta 359:369–378
Essers U, Heilmann E (1981) Nachweis von Erythropoietin: Vergleich eines biologischen Verfahrens mit einem Haemagglutinations-Hemmtest. Klin Wochenschr 59: 1217–1222
Evans ES, Rosenberg LL, Simpson ME (1961) Erythropoietic response to calorigenic hormones. Endocrinology 68:517–532

Fagg B (1981) Is erythropoietin the only factor which regulates late erythroid differentiation? Nature 289:184–186
Farber NM, Zanjani ED (1983) Translation of mRNA from anemic baboon kidney into biologically active erythropoietin. Exp Hematol 11 (Suppl 14):57A
Faura J, Ramos J, Reynafarje C, English E, Finne P, Finch CA (1969) Effect of altitude on erythropoiesis. Blood 33:668–676
Fava-de-Moraes F, Zangheri EO, Doine AI (1979) Immunohistochemical localization of erythropoietin in the rat and mouse submandibular gland. Histochem J 11:97–102
Fehér I, Gidáli J (1974) Prostaglandin E_2 as stimulator of haemopoietic stem cell proliferation. Nature 247:550–551
Fink GD, Fisher JW (1976) Erythropoietin production after renal denervation or beta-adrenergic blockade. Am J Physiol 230:508–513
Fink GD, Fisher JW (1977a) Stimulation of erythropoiesis by *beta* adrenergic agonists. I. Characterization of activity in polycythemic mice. J Pharmacol Exp Ther 202:192–198
Fink GD, Fisher JW (1977b) Stimulation of erythropoiesis by *beta* adrenergic agonists. II. Mechanism of action. J Pharmacol Exp Ther 202:199–208
Fink GD, Fisher JW (1977c) Role of the sympathetic nervous system in the control of erythropoietin production. In: Fisher JW (ed) Kidney hormones, vol II, erythropoietin. Academic, London, pp 387–413
Fink GD, Paulo LG, Fisher JW (1975) Effects of beta-adrenergic blocking agents on erythropoietin production in rabbits exposed to hypoxia. J Pharmacol Exp Ther 193:176–181
Finne PH, Skoglund RW (1970) Erythropoietin production in the rat following splanchnic neurectomy. J Lab Clin Med 76:103–106
Firkin FC, Russell SH (1983) Influence of human serum components on measurement of erythropoietin biological activity in vitro. Scand J Haematol 31:349–358
Fisher JW (1958) Increase in circulating red cell volume of normal rats after treatment with hydrocortisone or corticosterone. Proc Soc Exp Biol Med 97:502–505
Fisher JW (1972) Erythropoietin: pharmacology, biogenesis and control of production. Pharmacol Rev 24:459–508
Fisher JW (ed) (1977) Kidney hormones, vol II, erythropoietin. Academic, London
Fisher JW (1979) Extrarenal erythropoietin production. J Lab Clin Med 93:695–699
Fisher JW (1980) Prostaglandins and kidney erythropoietin production. Nephron 25:53–56
Fisher JW (1983) Control of erythropoietin production. Proc Soc Exp Biol Med 173:289–305
Fisher JW, Birdwell BJ (1961) The production of an erythropoietic factor by the in situ perfused kidney. Acta Haematol 26:224–232
Fisher JW, Gross DM (1977) Hormonal influences on erythropoiesis: anterior pituitary, adrenocortical, thyroid, growth and other hormones. In: Fisher JW (ed) Kidney hormones, vol II, erythropoietin. Academic, London, pp 415–435
Fisher JW, Langston JW (1968) Effects of testosterone, cobalt and hypoxia on erythropoietin production in the isolated perfused dog kidney. Ann NY Acad Sci 149:75–87
Fisher JW, Samuels AI (1967) Relationship between renal blood flow and erythropoietin production in dogs. Proc Soc Exp Biol Med 125:482–485
Fisher JW, Knight DB, Couch C (1963) Influence of several diuretic drugs on erythropoietin formation. J Pharmacol Exp Ther 141:113–121
Fisher JW, Taylor G, Porteous DD (1965a) Localization of erythropoietin in glomeruli of sheep kidney by fluorescent antibody technique. Nature 205:611–612
Fisher JW, Lajtha LG, Buttoo AS, Porteous DD (1965b) Direct effects of erythropoietin on the bone marrow of the isolated perfused hind limbs of rabbits. Br J Haematol 11:342–349

Fisher JW, Roh BL, Halvorsen S (1967) Inhibition of erythropoietic effects of hormones by erythropoietin antisera in mildly plethoric mice. Proc Soc Exp Biol Med 126:97–100
Fisher JW, Roh BL, Malgor LA, Noveck RJ (1971) Effects of vasoactive agents on erythropoietin production. Isr J Med Sci 7:977–985
Fisher JW, Adamson JW, Camiscoli JF, Fried W, Gordon AS, Schooley J, Zanjani E (1977) Cooperative erythropoietic assay of several steroid metabolites in polycythemic mice. Steroids 30:833–845
Fisher JW, Gross DM, Foley JE, Nelson PK, Rodgers GM, George WJ, Jubiz W (1978) A concept for the control of kidney production of erythropoietin involving prostaglandins and cyclic nucleotides. Contrib Nephrol 13:37–59
Fogh J (1966) A sensitive erythropoietin assay on mice exposed to CO-hypoxia. Scand J Clin Lab Invest 18:33–44
Foley JW, Gross DM, Nelson PK, Fisher JW (1978) The effects of arachidonic acid on erythropoietin production in exhypoxic polycythemic mice and the isolated perfused canine kidney. J Pharmacol Exp Ther 207:402–409
Fredrickson TN, Smith KA, Cornell CJ, Jasmin C, McIntyre OR (1977) The interaction of erythropoietin with fetal liver cells. I. Measurement of proliferation by tritiated thymidine incorporation. Exp Hematol 5:254–265
Freedman BJ, Penington DG (1963) Erythrocytosis in emphysema. Br J Haematol 9: 425–430
Frenkel EP, Suki W, Baum J (1968) Some observations on the localization of erythropoietin. Ann NY Acad Sci 149:292–293
Fried W (1972) The liver as a source of extrarenal erythropoietin production. Blood 40:671–677
Fried W (1975) Erythropoietin and the kidney. Nephron 15:327–349
Fried W, Anagnostou A (1977) Extrarenal erythropoietin production. In: Fisher JW (ed) Kidney hormones, vol II, erythropoietin. Academic, London, pp 231–244
Fried W, Barone-Varelas J (1984) Regulation of the plasma erythropoietin level in hypoxic rats. Exp Hematol 12:706–711
Fried W, Gurney CW (1965) Erythropoietic effect of plasma from mice receiving testosterone. Nature 206:1160–1161
Fried W, Plzak LF, Jacobson LO, Goldwasser E (1957) Studies on erythropoiesis. III. Factors controlling erythropoietin production. Proc Soc Exp Biol Med 94:237–241
Fried W, Johnson C, Heller P (1970) Observations on regulation of erythropoiesis during prolonged periods of hypoxia. Blood 36:607–616
Fried W, Gregory SA, Knospe WH, Trobaugh FE (1971) Regulation of plasma erythropoietin levels in mice with impaired responsiveness to erythropoietin. J Lab Clin Med 78:449–456
Fried W, Barone-Varelas J, Berman M (1981) Detection of high erythropoietin titers in renal extracts of hypoxic rats. J Lab Clin Med 97:82–86
Fried W, Barone-Varelas J, Barone T (1982) The influence of age and sex on erythropoietin titers in the plasma and tissue homogenates of hypoxic rats. Exp Hematol 10:472–477
Fyhrquist F, Rosenlöf K, Grönhagen-Riska C, Hortling L, Tikkanen I (1984) Is renin substrate an erythropoietin precursor? Nature 308:649–652
Gallicchio VS, Murphy MJ (1981) Effects of membrane dialysis and filtration-sterilization on erythropoietin activity. Yale J Biol Med 54:249–254
Gallien-Lartigue O, Goldwasser E (1964) Hemoglobin synthesis in marrow cell culture: the effect of rat plasma on rat cells. Science 145:277–279
Garcia JF (1977) Assays for erythropoietin. In: Fisher JW (ed) Kidney hormones, vol II, erythropoietin. Academic, London, pp 7–35
Garcia JF, Sherwood J, Goldwasser E (1979) Radioimmunoassay of erythropoietin. Blood Cells 5:405–419

Garcia JF, Ebbe SN, Hollander L, Cutting HO, Miller ME, Cronkite EP (1982) Radioimmunoassay of erythropoietin: circulating levels in normal and polycythemic human beings. J Lab Clin Med 99:624–635
Garland JM, Crompton S (1983) A preliminary report: preparations containing Interleukin-3 (IL-3) promote proliferation of multipotential stem cells (CFU_S) in the mouse. Exp Hematol 11:757–761
George WJ, Briggs DW, Rodgers GM, Fisher JW (1977) Metabolism of erythropoietin. In: Fisher JW (ed) Kidney hormones, vol II, erythropoietin. Academic, London, pp 73–109
Ghosal J, Moulik S, Datta AG (1980) Effect of different prostaglandins & methyl testosterone on the formation of erythropoietin by kidney. Indian J Exp Biol 18: 377–381
Gidari AS, Zanjani ED, Gordon AS (1971) Stimulation of erythropoiesis by cyclic adenosine monophosphate. Life Sci 10 (Part II):895–900
Giger K (1968) Wirkung von Actinomycin-D auf die Erythropoietinbildung während Hypoxie bei der Ratte. Klin Wochenschr 46:42–44
Gill JR, Frölich JC, Bowden RE, Taylor AA, Keiser HR, Seyberth HW, Oates JA, Bartter FC (1976) Bartter's syndrome: a disorder characterized by high urinary porstaglandins and a dependence of hyperreninemia on prostaglandin synthesis. Am J Med 61:43–51
Glader BE, Rambach WA, Alt HL (1968) Observations on the effect of testosterone and hydrocortisone on erythropoiesis. Ann NY Acad Sci 149:383–388
Golde DW, Bersch N, Cline MJ (1976) Potentiation of erythropoiesis in vitro by dexamethasone. J Clin Invest 57:57–62
Golde DW, Bersch N, Chopra IJ, Cline MJ (1977a) Thyroid hormones stimulate erythropoiesis in vitro. Br J Haematol 37:173–177
Golde DW, Bersch N, Li CH (1977b) Growth hormone: species-specific stimulation of erythropoiesis in vitro. Science 196:1112–1113
Goldwasser E (1975) Erythropoietin and the differentiation of red blood cells. Fed Proc 34:2285–2292
Goldwasser E (1981) Erythropoietin and red cell differentiation. In: Cunningham D, Goldwasser E, Watson J, Fox CF (eds) Control of cellular division and development, part A. Liss, New York, pp 487–494
Goldwasser E, Inana G (1978) Molecular aspects of the initiation of erythropoiesis. In: Golde DW, Cline MJ, Metcalf D, Fox CF (eds) Hematopoietic cell differentiation. Academic, New York, pp 15–24
Goldwasser E, Kung CKH (1968) Progress in the purification of erythropoietin. Ann NY Acad Sci 149:49–53
Goldwasser E, Kung CKH (1971) Purification of erythropoietin. Proc Natl Acad Sci USA 68:697–698
Goldwasser E, Kung CKH (1972) The molecular weight of sheep plasma erythropoietin. J Biol Chem 247:5159–5160
Goldwasser E, Jacobson LO, Fried W, Plzak LF (1958) Studies on erythropoiesis. V. The effect of cobalt on the production of erythropoietin. Blood 13:55–60
Goldwasser E, White WF, Taylor KB (1962) Further purification of sheep plasma erythropoietin. Biochim Biophys Acta 64:487–496
Goldwasser E, Kung CKH, Eliason J (1974) On the mechanism of erythropoietin-induced differentiation. J Biol Chem 249:4202–4206
Goldwasser E, Eliason JF, Sikkema D (1975) An assay for erythropoietin in vitro at the milliunit level. Endocrinology 97:315–323
Gordon AS (1973) Erythropoietin. Vitam Horm 31:105–174
Gordon AS, Kaplan SM (1977) Erythrogenin (REF). In: Fisher JW (ed) Kidney hormones, vol II, erythropoietin. Academic, London, pp 187-229
Gordon AS, Cooper GW, Zanjani ED (1967a) The kidney and erythropoiesis. Semin Hematol 4:337–358

Gordon AS, Mirand EA, Zanjani ED (1967b) Mechanisms of prednisolone action in erythropoiesis. Endocrinology 81:363–368
Gordon AS, Mirand EA, Wenig J, Katz R, Zanjani ED (1968) Androgen actions on erythropoiesis. Ann NY Acad Sci 149:318–335
Goudsmit R, Krugers Dagneaux PGLC, Krijnen HW (1967) Een immunochemische bepaling van erythropoietine. Folia Med 10:39–45
Gould AB, Goodman SA, Green D (1973) An in vivo effect of renin on erythropoietin formation. Lab Invest 28:719–722
Gould AB, Goodman S, DeWolf R, Onesti G, Swartz C (1980) Interrelation of the renin system and erythropoietin in rats. J Lab Clin Med 96:523–534
Graber SE, Krantz SB (1978) Erythropoietin and the control of red cell production. Annu Rev Med 29:51–66
Graber SE, Carrillo M, Krantz SB (1974) Lack of effect of erythropoietin on cyclic adenosine-3′, 5′-monophosphate levels in rat fetal liver cells. J Lab Clin Med 83: 288–295
Graber SE, Bomboy JD, Salmon WD, Krantz SB (1979) Evidence that endotoxin is the cyclic 3′:5′-GMP-promoting factor in erythropoietin preparations. J Lab Clin Med 93:25–31
Gregory CJ (1976) Erythropoietin sensitivity as a differentiation marker in the hemopoietic system: studies of three erythropoietic colony responses in culture. J Cell Physiol 89:289–302
Gregory CJ, Eaves AC (1977) Human marrow cells capable of erythropoietic differentiation in vitro: definition of three erythroid colony responses. Blood 49:855–864
Gregory CJ, Eaves AC (1978) Three stages of erythropoietic progenitor cell differentiation distinguished by a number of physical and biologic properties. Blood 51:527–537
Gregory CJ, Teppermann AD, McCulloch EA, Till JE (1974) Erythropoietic progenitors capable of colony formation in culture: response of normal and genetically anemic W/W^V mice to manipulations of the erythron. J Cell Physiol 84:1–12
Gross DM, Fisher JW (1978) Effects of terbutaline, a synthetic beta adrenoceptor agonist, on in vivo erythropoietin production. Arch Int Pharmacodyn Ther 236: 192–201
Gross DM, Fisher JW (1980) Erythropoietic effects of PGE_2 and 2 endoperoxide analogs. Experientia 36:458–459
Gross DM, Brookins J, Fink GD, Fisher JW (1976a) Effects of prostaglandins A_2, E_2 and $F_{2\alpha}$ on erythropoietin production. J Pharmacol Exp Ther 198:489–496
Gross DM, Mujovic VM, Jubiz W, Fisher JW (1976b) Enhanced erythropoietin and prostaglandin E production in the dog following renal artery constriction. Proc Soc Exp Biol Med 151:498–501
Gruber DF, Zucali JR, Wleklinski J, LaRussa V, Mirand EA (1977a) Temporal transition in the site of rat erythropoietin production. Exp Hematol 5:399–407
Gruber DF, Zucali JR, Mirand EA (1977b) Identification of erythropoietin producing cells in fetal mouse liver cultures. Exp Hematol 5:392–398
Gupta PD, Singh M (1981) Carotid chemoreceptors and vagi in hypoxic and cyanide-induced tachycardia in the dog. Am J Physiol 240:H874–H880
Gurney CW, Simmons EL, Gaston EO, Cox JC (1980) The erythropoietic effect, in mice, of androgens combined with hypoxia. Exp Hematol 8:192–199
Hågå P, Falkanger B (1979) In vitro assay for erythropoietin: erythroid colony formation in methyl cellulose used for the measurement of erythropoietin in plasma. Blood 53:1172–1181
Hagiwara M, Chen I-Li, McGonigle R, Beckman B, Kasten FH, Fisher JW (1984a) Erythropoietin production in a primary culture of human renal carcinoma cells maintained in nude mice. Blood 63:828–835
Hagiwara M, Chen I-Li, Fisher JW (1984b) Erythropoietin production in long-term cultures of human renal carcinoma cells. The role of cell population density. Exp Cell Res 154:619–624

Hagiwara M, McNamara DB, Chen I-Li, Fisher JW (1984c) Role of endogenous prostaglandin E_2 in erythropoietin production and dome formation by human renal carcinoma cells in culture. J Clin Invest 74:1252–1261

Hagiwara M, Pincus SM, Chen I-Li, Beckman BS, Fisher JW (1985) Effects of dibutyryl adenosine 3′,5′-cyclic monophosphate on erythropoietin production in human renal carcinoma cell cultures. Blood 66:714–717

Halvorsen S (1961) Plasma erythropoietin levels following hypothalamic stimulation in the rabbit. Scand J Clin Lab Invest 13:564–575

Halvorsen S (1964) Effects of hypothalamic lesions on the erythropoietic response to hypoxia in rabbits. Acta Physiol Scand 61:1–19

Halvorsen S (1966) The central nervous system in regulation of erythropoiesis. Acta Haematol 35:65–79

Halvorsen S, Roh BL, Fisher JW (1968) Erythropoietin production in nephrectomized and hypophysectomized animals. Am J Physiol 215:349–352

Hammond D, Winnick S (1974) Paraneoplastic erythrocytosis and ectopic erythropoietins. Ann NY Acad Sci 230:219–227

Hammond GD, Ishikawa A, Keighley G (1962) Relationship between erythropoietin and severity of anemia in hypoplastic and hemolytic states. In: Jacobson LO, Doyle M (eds) Erythropoiesis. Grune and Stratton, New York, pp 351–358

Hansen P (1964) Polycythaemia produced by constriction of the renal artery in a rabbit. Acta Pathol Microbiol Immunol Scand 60:465–472

Hansen AJ, Fogh J, Møllgård K, Sørensen SC (1973) Evidence against erythropoietin production by the carotid body. Resp Physiol 18:101–106

Hara H, Ogawa M (1977) Erythropoietic precursors in murine blood. Exp Hematol 5: 161–165

Harrison PR, Conkie D, Affara N, Paul J (1974) In situ localization of globin messenger RNA formation. J Cell Biol 63:402–413

Hartcroft PM, Bischoff MB, Bucci TJ (1969) Effects of chronic exposure to high altitude on the juxtaglomerular complex and adrenal cortex of dogs, rabbits and rats. Fed Proc 28:1234–1236

Hassid A (1982) Regulation of prostaglandin biosynthesis in cultured cells. Am J Physiol 243:C205–C211

Heath DS, Axelrad AA, McLeod DL, Shreeve MM (1976) Separation of the erythropoietin-responsive progenitors BFU-E and CFU-E in mouse bone marrow by unit gravity sedimentation. Blood 47:777–792

Hebbel RP, Eaton JW, Kronenberg RS, Zanjani ED, Moore LG, Berger EM (1978) Human llamas. Adaptation to altitude in subjects with high hemoglobin oxygen affinity. J Clin Invest 62:593–600

Heim LR (1980) Erythropoietin: what's in a name. Exp Hematol 8:229–230

Heyes MP, Farber MO, Manfredi F, Robertshaw D, Weinberger M, Fineberg N, Robertson G (1982) Acute effects of hypoxia on renal and endocrine function in normal humans. Am J Physiol 243:R265–R270

Hirashima K, Takaku F (1962) Experimental studies on erythropoietin. II. The relationship between juxtaglomerular cells and erythropoietin. Blood 20:1–8

Hoffman R, Dainiak N, Coupal E, Maffei L, Ritchey K, Forget BG (1979) Hormonal influences on globin chain synthesis of fetal cord blood BFU-E in vitro. In: Stamatoyannopoulos G, Nienhuis AW (eds) Cellular and molecular regulation of hemoglobin switching. Grune and Stratton, New York, pp 389–396

Hopfer SM, Sunderman FW, Fredrickson TN, Morse EE (1979) Increased serum erythropoietin activity in rats following intrarenal injection of nickel subsulfide. Res Commun Chem Pathol Pharmacol 23:155–170

Hopfer SM, Sunderman FW, Reid MC, Goldwasser E (1984) Increased immunoreactive erythropoietin in serum and kidney extracts of rats with Ni_3S_2-induced erythrocytosis. Res Commun Chem Pathol Pharmacol 43:299–305

Hudgson P, Pearce JMS, Yeates WK (1967) Renal artery stenosis with hypertension and high haematocrit. Br Med J 1:18–21

Ikegami S, Tamura S, Oide H (1973) Simple purification procedure for human urinary erythropoietin. Proc Soc Exp Biol Med 143:526–527
Iscove NN (1977) The role of erythropoietin in regulation of population size and cell cycling of early and late erythroid precursors in mouse bone marrow. Cell Tissue Kinet 10:323–334
Iscove NN (1978) In: Golde DW, Cline MJ, Metcalf D, Fox CF (eds) Hematopoietic cell differentiation. Academic, New York, pp 37–82
Iscove NN, Guilbert LJ (1978) Erythropoietin-independence of early erythropoiesis and a two-regulator model of proliferative control in the hemopoietic system. In: Murphy MJ (ed) In vitro aspects of erythropoiesis. Springer, New York Heidelberg Berlin, pp 3–7
Iscove NN, Sieber F (1975) Erythroid progenitors in mouse bone marrow detected by macroscopic colony formation in culture. Exp Hematol 3:32–43
Iscove NN, Sieber F, Winterhalter KH (1974) Erythroid colony formation in cultures of mouse and human bone marrow: analysis of the requirement for erythropoietin by gel filgration and affinity chromatography on agarose-concanavalin A. J Cell Physiol 83:309–320
Iscove NN, Guilbert LJ, Weyman C (1980) Complete replacement of serum in primary cultures of erythropoietin-dependent red cell precursors (CFU-E) by albumin, transferrin, iron, unsaturated fatty acid, lecithin and cholesterol. Exp Cell Res 126: 121–126
Jacobs K, Shoemaker C, Rudersdorf R, Neill SD, Kaufman RJ, Mufson A, Seehra J, Jones SS, Hewick R, Fritsch EF, Kawakita M, Shimizu T, Miyake T (1985) Isolation and characterization of genomic and cDNA clones of human erythropoietin. Nature 313:806–810
Jacobson LO, Goldwasser E, Fried W, Plzak L (1957a) Role of the kidney in erythropoiesis. Nature 179:633–634
Jacobson LO, Goldwasser E, Plzak LF, Fried W (1957b) Studies on erythropoiesis. Part IV. Reticulocyte response of hypophysectomized and polycythemic rodents to erythropoietin. Proc Soc Exp Biol Med 94:243–249
Jacobson LO, Marks EK, Gaston EO, Goldwasser E (1959) Studies on erythropoiesis. XI. Reticulocyte response of transfusion-induced polycythemic mice to anemic plasma from nephrectomized mice and to plasma from nephrectomized rats exposed to low oxygen. Blood 14:635–643
Jelkmann W (1982) Temporal pattern of erythropoietin titers in kidney tissue during hypoxic hypoxia. Pfluegers Arch 393:88–91
Jelkmann W, Bauer C (1980) β_2-Adrenergic stimulation of erythropoiesis in busulfan treated mice. Exp Hematol 8:742–748
Jelkmann W, Bauer C (1981) Demonstration of high levels of erythropoietin in rat kidneys following hypoxic hypoxia. Pfluegers Arch 392:34–39
Jelkmann W, Brookins J, Fisher JW (1979a) Indomethacin blockade of albuterol-induced erythropoietin production in isolated perfused dog kidneys. Proc Soc Exp Biol Med 162:65–70
Jelkmann W, Beckman B, Fisher JW (1979b) Enhanced effects of hypoxia on erythropoiesis in rabbits following beta-2 adrenergic activation with albuterol. J Pharmacol Exp Ther 211:99–103
Jelkmann W, Kurtz A, Bauer C (1983a) Extraction of erythropoietin from isolated renal glomeruli of hypoxic rats. Exp Hematol 11:581–588
Jelkmann W, Kurtz A, Bauer C (1983b) Effects of fasting on the hypoxia-induced erythropoietin production in rats. Pfluegers Arch 396:174–175
Jelkmann W, Kurtz A, Seidl J, Bauer C (1984) Mechanisms of the renal glomerular erythropoietin production. In: Grote J, Witzleb E (eds) Atemgaswechsel and O_2-Versorgung der Organe. Akademie der Wissenschaften und der Literatur, Mainz, pp 130–137
Jelkmann W, Kurtz A, Förstermann U, Pfeilschifter J, Bauer C (1985) Hypoxia enhances prostaglandin synthesis in renal mesangial cell cultures. Prostaglandins 30: 109–118

Jelkmann W, Kurtz A, Bauer C (1986) In vitro production of erythropoietin. In: Fisher JW (ed) Kidney hormones, vol III. Academic, London
Jensen K, Thorling EB (1965) The effect of anaemia and cobalt on lactic dehydrogenase isoenzymes in kidney tissues of rabbits and its possible relation to the erythropoietin production. Acta Pathol Microbiol Immunol Scand 63:385–390
Jepson JH, Lowenstein L (1964) Effect of prolactin on erythropoiesis in the mouse. Blood 24:726–738
Jepson JH, Lowenstein L (1966a) Erythropoiesis during pregnancy and lactation in the mouse. II. Role of erythropoietin. Proc Soc Exp Biol Med 121:1077–1081
Jepson JH, Lowenstein L (1966b) Inhibition of the stem-cell action of erythropoietin by estradiol. Proc Soc Exp Biol Med 123:457–460
Jepson JH, Lowenstein L (1967) The effect of testosterone, adrenal steroids and prolactin on erythropoiesis. Acta Haematol 38:292–299
Jepson JH, Manasc B (1974) Enhanced response of erythropoietin sensitive cells to erythropoietin by low dose prednisolone hemisuccinate and its abolition by azathioprine. Biomedicine 20:179–185
Jepson J, McGarry EE (1968) Polycythemia and increased erythropoietin production in a patient with hypertrophy of the juxta-glomerular apparatus. Blood 32:370–375
Jepson JH, McGarry EE, Lowenstein L (1968) Erythropoietin excretion in a hypopituitary patient. Arch Intern Med 122:265–270
Jepson JH, Leonard RA, Gorshein D, Gardner FH, Besa E (1974) Review of the induction of cycling of erythropoietin sensitive cells by androgens and their 5 β-H-metabolites. Biomedicine 20:262–266
Johnson GR, Metcalf D (1977) Pure and mixed erythroid colony formation in vitro stimulated by spleen conditioned medium with no detectable erythropoietin. Proc Natl Acad Sci USA 74:3879–3882
Johnson JA, Kurz KD, Siripaisarnpipat S, Zeigler DW, Payne CG (1982) Fluid volumes and pressor responsiveness in two-kidney rabbits with renal artery stenosis. Am J Physiol 243:H779–H787
Johnson TS, Young JB, Landsberg L (1983) Sympathoadrenal responses to acute and chronic hypoxia in the rat. J Clin Invest 71:1263–1272
Jordan TA, Potter TP, Essary BH (1975) Erythropoietin assay by hemagglutination-inhibition. Am J Med Technol 41:39–46
Kalsner S (1977) The effect of hypoxia on prostaglandin output and on tone in isolated coronary arteries. Can J Physiol Pharmacol 55:882–887
Kaplan SM, Rothmann SA, Gordon AS, Rappaport IA, Camiscoli JF, Peschle C (1973) Extrarenal sites of erythrogenin production. Proc Soc Exp Biol Med 143:310–313
Katsuoka Y, Beckman B, George WJ, Fisher JW (1983a) Increased levels of erythropoietin in kidney extracts of rats treated with cobalt and hypoxia. Am J Physiol 244:F129–F133
Katsuoka Y, McGonigle R, Rege AB, Beckman B, Fisher JW (1983b) Erythropoietin production in human renal carcinoma cells passaged in nude mice and in tissue culture. Gann 74:534–541
Katz R, Cooper GW, Gordon AS, Zanjani ED (1968) Studies on the site of production of erythropoietin. Ann NY Acad Sci 149:120–127
Kazal LA, Erslev AJ (1975a) The measurement of erythropoietin. Ann Clin Lab Sci 5:91–97
Kazal LA, Erslev AJ (1975b) Erythropoietin production in renal tumors. Ann Clin Lab Sci 5:98–109
Keighley G, Cohen NS (1978) Stimulation of erythropoiesis in mice by adenosine 3′, 5′-monophosphate and prostaglandin E_1. J Med 9:129–138
Kiil F (1971) Blood flow and oxygen utilization by the kidney. In: Fisher JW (ed) Kidney hormones. Academic, London, pp 1–30
Kilbridge TM, Fried W, Heller P (1969) The mechanism by which plethora suppresses erythropoiesis. Blood 33:104–113

Kirstein A (1979) Cardiac prostacyclin release: stimulation by hypoxia and various agents. Scand J Haematol S34:105–110

Kolk-Vegter AJ, Kolk AHJ, Napier JAF, Dunn CDR (1975) Some problems concerning the assay of erythropoietin using the haemagglutination inhibition kit. Br J Haematol 30:371–372

Kotchen TA, Hogan RP, Boyd AE, Li T-K, Sing HC, Mason JW (1973a) Renin, noradrenaline and adrenaline responses to simulated altitude. Clin Sci 44:243–251

Kotchen TA, Mougey EH, Hogan RP, Boyd AE, Pennington LL, Mason JW (1973b) Thyroid responses to simulated altitude. J Appl Physiol 34:165–168

Kramer K, Deetjen P (1960) Beziehungen des O_2-Verbrauchs der Niere zu Durchblutung und Glomerulusfiltrat bei Änderung des arteriellen Druckes. Pfluegers Arch 271:782–796

Krantz SB, Goldwasser E (1984) Specific binding of erythropoietin to spleen cells infected with the anemia strain of Friend virus. Proc Natl Acad Sci USA 81:7574–7578

Krantz SB, Jacobson LO (1970) Erythropoietin and the regulation of erythropoiesis. University of Chicago Press, Chicago

Krystal G (1983) Physical and biological characterization of erythroblast enhancing factor (EEF), a late acting erythropoietic stimulator in serum distinct from erythropoietin. Exp Hematol 11:18–31

Krystal G, Eaves AC, Eaves CJ (1981) Determination of normal human serum erythropoietin levels, using mouse bone marrow. J Lab Clin Med 97:158–169

Krystal G, Eaves CJ, Eaves AC (1984) CM Affi-Gel Blue chromatography of human urine: a simple one-step procedure for obtaining erythropoietin suitable for in vitro erythropoietic progenitor assays. Br J Haematol 58:533–546

Kuratowska Z, Lewartowski B, Michalak E (1961) Studies on the production of erythropoietin by isolated perfused organs. Blood 18:527–534

Kurtz A, Jelkmann W, Bauer C (1982a) A new candidate for the regulation of erythropoiesis. Insulin-like growth factor I. FEBS Lett 149:105–108

Kurtz A, Jelkmann W, Bauer C (1982b) Mesangial cells derived from rat glomeruli produce an erythropoiesis stimulating factor in cell culture. FEBS Lett 137:129–132

Kurtz A, Jelkmann W, Sinowatz F, Bauer C (1983a) Renal mesangial cell cultures as a model for study of erythropoietin production. Proc Natl Acad Sci USA 80: 4008–4011

Kurtz A, Jelkmann W, Bauer C (1983b) Insulin stimulates erythroid colony formation independently of erythropoietin. Br J Haematol 53:311–316

Kurtz A, Härtl W, Jelkmann W, Zapf J, Bauer C (1985a) The activity in fetal bovine serum that stimulates erythroid colony formation in fetal mouse livers is insulinlike growth factor I. J Clin Invest 76:1643–1648

Kurtz A, Jelkmann W, Pfeilschifter J, Bauer C (1985b) Role of prostaglandins in hypoxia-stimulated erythropoietin production. Am J Physiol 249:C3–C8

Lafferty MD, Ackerman GA, Dunn CDR, Lange RD (1980) The ultrastructural, immunocytochemical demonstration of erythropoietin receptors on developing erythrocytic cells of fetal mouse liver. Exp Hematol 8:1063–1074

Lafferty MD, Ackerman GA, Dunn CDR, Lange RD (1981) Ultrastructural, immunocytochemical localization of presumptive erythropoietin binding sites on developing erythrocytic cells of normal human bone marrow. J Histochem Cytochem 29:49–56

Lajtha LG (1979) Haemopoietic stem cells: concepts and definitions. Blood Cells 5: 447–455

Lands W, Lee R, Smith W (1971) Factors regulating the biosynthesis of various prostaglandins. Ann NY Acad Sci 180:107–122

Lange RD, Ichiki AT (1977) Immunological studies of erythropoietin. In: Fisher JW (ed) Kidney hormones, vol II, erythropoietin. Academic, London, pp 111–149

Lange RD, Gallagher NI (1962) Clinical and experimental observations on the relationship of the kidney to erythropoietin production. In: Jacobson L, Doyle M (eds) Erythropoiesis. Grune and Stratton, New York, pp 361–373
Lange RD, McDonald TP, Jordan T (1969) Antisera to erythropoietin: partial characterization of two different antibodies. J Lab Clin Med 73:78–90
Lange RD, Chen JP, Dunn CDR (1980) Erythropoietin assays: some new and different approaches. Exp Hematol 8 (Suppl 8):197–223
Lechermann B, Jelkmann W (1985) Erythropoietin production in normoxic and hypoxic rats with increased blood O_2 affinity. Respir Physiol 60:1–8
Lee-Huang S (1980) A new preparative method for isolation of human erythropoietin with hydrophobic interaction chromatography. Blood 56:620–624
Lee-Huang S (1982) Monoclonal antibodies to human erythropoietin. Fed Proc 41: 520A
Lee-Huang S (1984) Cloning and expression of human erythropoietin cDNA in *Escherichia coli.* Proc Natl Acad Sci USA 81:2708–2712
Leichtweiss H-P, Lübbers DW, Weiss C, Baumgärtl H, Reschke W (1969) The oxygen supply of the rat kidney: measurements of intrarenal pO_2. Pfluegers Arch 309: 328–349
Lertora JJL, Dargon PA, Rege AB, Fisher JW (1975) Studies on a radioimmunoassay for human erythropoietin. J Lab Clin Med 86:140–151
Leung P, Gidari AS (1982) Glucocorticoids inhibit the growth of murine fetal liver erythroid burst-forming cells. Endocrinology 111:1121:1126
Levine AS, Alter BP, Clemons G, Fried W, Weinberg R, Zanjani ED, Zuckerman KS (1983) Human erythropoietin: comparative in vivo and in vitro dose response of commercial products. Blood 62 (Suppl 1):124A
Lewis JP, Moores RR, Neal WA, Garver FA, Lutcher CL, Zucali JR, Mirand EA (1981) Inhibition of Friend virus (FVP)-induced murine erythropoiesis with prostaglandin (PGF2α): potentiation and inhibition of erythropoietin and prevention of both with PGD2. Exp Hematol 9:540–549
Lowy PH, Keighley G (1968) Inactivation of erythropoietin by Koshland's tryptophan reagent and by membrane filtration. Biochim Biophys Acta 160:413–419
Lowy PH, Keighley G, Cohen NS (1970) Stimulation by serotonin of erythropoietin-dependent erythropoiesis in mice. Br J Haematol 19:711–718
Lozzio BB, McDonald TP, Lange RD (1971) Erythropoietin formation in rats bearing congenital hydronephrosis. Isr J Med Sci 7:1001–1006
Lucarelli G, Porcellini A, Carnevali C, Carmena A, Stohlman F (1968) Fetal and neonatal erythropoiesis. Ann NY Acad Sci 149:544–559
Lukowsky W, Painter RH (1968) Molecular weight of erythropoietin from anemic sheep plasma and human urine. Can J Biochem Cell Biol 46:731–734
Lukowsky WA, Painter RH (1972) Studies on the role of sialic acid in the physical and biological properties of erythropoietin. Can J Biochem Cell Biol 50:909–917
Maeda H, Hotta T, Yamada H (1983) Murine erythroid colony-promoting factor derived from thymocyte conditioned media. Exp Hematol 11:834–840
Malgor LA, Fisher JW (1970) Effects of testosterone on erythropoietin production in isolated perfused kidneys. Am J Physiol 218:1732–1736
Mann DL, Gallagher NI, Donati RM (1967) Erythrocytosis and primary aldosteronism. Ann Intern Med 66:335–340
Markelonis G, Garbus J (1975) Alterations of intracellular oxidative metabolism as stimuli evoking prostaglandin biosynthesis. Prostaglandins 10:1087–1106
Martelo OJ, Toro EF, Hirsch J (1976) Activation of renal erythropoietic factor by phosphorylation. J Lab Clin Med 87:83–88
McCully KS, Rinehimer LA, Gillies CG, Hopfer SM, Sunderman FW (1982) Erythrocytosis, glomerulomegaly, mesangial hyperplasia, sialyl hyperplasia, and arteriosclerosis induced in rats by nickel subsulfide. Virchows Arch 394:207–220
McDonald TP (1981) Assay and site of production of thrombopoietin. Br J Haematol 49:493–499

McDonald TP, Lange RD (1967) Erythropoietin bioassays utilizing silicone rubber membrane enclosures. J Lab Clin Med 70:48–56
McDonald TP, Shadduck RK (1982) Comparative effects of thrombopoietin and colony-stimulating factors. Exp Hematol 10:544–550
McDonald TP, Martin DH, Simmons ML, Lange RD (1969) Preliminary results of erythropoietin production by bovine kidney cells in culture. Life Sci 8, Part II:949–954
McDonald TP, Zanjani ED, Lange RD, Gordon AS (1971) Immunological studies of the renal erythropoietic factor (erythrogenin). Br J Haematol 20:113–121
McGonigle RJS, Husserl F, Wallin JD, Fisher JW (1984a) Hemodialysis and continuous ambulatory peritoneal dialysis effects on erythropoiesis in renal failure. Kidney Int 25:430–436
McGonigle RJS, Wallin JD, Shadduck RK, Fisher JW (1984b) Erythropoietin deficiency and inhibition of erythropoiesis in renal insufficiency. Kidney Int 25:437–444
McGonigle RJS, Wallin JD, Husserl F, Deftos LJ, Rice JC, O'Neill WJ, Fisher JW (1984c) Potential role of parathyroid hormone as an inhibitor of erythropoiesis in the anemia of renal failure. J Lab Clin Med 104:1016–1026
McLeod DL, Shreeve MM, Axelrad AA (1974) Improved plasma culture system for production of erythrocytic colonies in vitro: quantitative assay method for CFU-E. Blood 44:517–534
Melchiorri P, Sopranzi N, Roseghini M (1976) Effects of bombesin on erythropoietin production in the anaesthetized dog. Naunyn Schmiedebergs Arch Pharmacol 294:193–197
Metcalf D (1981) Control of hemopoietic cell proliferation and differentiation. Control Cell Div Dev (A):473–486
Metcalf D, Johnson GR (1979) Interactions between purified GM-CSF, purified erythropoietin and spleen conditioned medium on hemopoietic colony formation in vitro. J Cell Physiol 99:159–174
Meytes D, Bogin E, Ma A, Dukes PP, Massry SG (1981) Effect of parathyroid hormone on erythropoiesis. J Clin Invest 67:1263–1269
Milledge JS, Cotes PM (1985) Serum erythropoietin in humans at high altitude and its relation to plasma renin. J Appl Physiol 59:360–364
Miller ME, Howard D (1979) Modulation of erythropoietin concentrations by manipulation of hypercarbia. Blood Cells 5:389–403
Miller ME, Rörth M, Parving HH, Howard D, Reddington I, Valeri CR, Stohlman F (1973) pH effect on erythropoietin response to hypoxia. N Engl J Med 288:706–710
Miller ME, Howard D, Stohlman F, Flanagan P (1974) Mechanism of erythropoietin production by cobaltous chloride. Blood 44:339–346
Miller ME, Rörth M, Stohlman F, Valeri CR, Lowrie G, Howard D, McGilvray N (1976) The effects of acute bleeding on acid-base balance, erythropoietin (Ep) production and in vivo P_{50} in the rat. Br J Haematol 33:379–385
Miller ME, Garcia JF, Cohen RA, Cronkite EP, Moccia G, Acevedo J (1981) Diurnal levels of immunoreactive erythropoietin in normal subjects and subjects with chronic lung disease. Br J Haematol 49:189–200
Miller ME, Cronkite EP, Garcia JF (1982) Plasma levels of immunoreactive erythropoietin after acute blood loss in man. Br J Haematol 52:545–549
Miller ME, Garcia JF, Shiue GG, Okula RM, Clemons GK (1983) Humoral regulation of erythropoiesis. In: Killmann SA, Cronkite EP, Muller-Berat CN (eds) Haemopoietic stem cells. Munksgaard, Copenhagen, pp 217–228
Mirand EA, Gordon AS (1966) Mechanism of estrogen action in erythropoiesis. Endocrinology 78:325–332
Mirand EA, Prentice TC (1957) Presence of plasma erythropoietin in hypoxic rats with or without kidney(s) and/or spleen. Proc Soc Exp Biol Med 96:49–51

Mirand EA, Grace JT, Johnston GS, Murphy GP (1964) Effects of hypothalamic stimulation on the erythropoietic response in the rhesus monkey. Nature 204:1163–1165
Mirand EA, Murphy GP, Bennett TB, Grace JT (1968) Erythropoietin response to repeated hemorrhage in renal allotransplanted, nephrectomized, or intact dogs. Life Sci 7, Part I:689–696
Mirand EA, Steeves RA, Groenewald JH, Van Zyl JJW, Murphy GP (1969) Extra-renal erythropoietin production in the baboon. Proc Soc Exp Biol Med 130:685–688
Mirand EA, Gordon AS, Zanjani ED, Bennett TE, Murphy GP (1972) Disappearance of exogenous erythropoietin (ESF) from the blood of germfree mice. Proc Soc Exp Biol Med 139:161–164
Misiti J, Spivak JL (1979) Erythropoiesis in vitro. Role of calcium. J Clin Invest 64: 1573–1579
Mitus WJ, Toyama K, Brauer MJ (1968) Erythrocytosis, juxtaglomerular apparatus (JGA), and erythropoietin in the course of experimental unilateral hydronephrosis in rabbits. Ann NY Acad Sci 149:107–113
Miyake T, Kung CKH, Goldwasser E (1977) Purification of human erythropoietin. J Biol Chem 252:5558–5564
Mladenovic J, Eschbach JW, Koup JR, Garcia JF, Adamson JW (1985) Erythropoietin kinetics in normal and uremic sheep. J Lab Clin Med 105:659–663
Mok M, Spivak JL (1982) Protease activity in human urine erythropoietin preparations. Exp Hematol 10:300–306
Moriyama Y, Fisher JW (1974) Effects of testosterone and erythropoietin on erythroid colony formation in rabbit bone marrow cultures. Life Sci 15:1181–1188
Morse EE (1979) Consequences of erythropoietin production by neoplasms. Ann Clin Lab Sci 9:116–120
Muirhead EE, Leach BE, Fisher JW, Kosinski M (1968) Renal transplantation and extracts and erythropoiesis. Ann NY Acad Sci 149:135–142
Mujovic VM, Fisher JW (1974) The effects of indomethacin on erythropoietin production in dogs following renal artery constriction. I. The possible role of prostaglandins in the generation of erythropoietin by the kidney. J Pharmacol Exp Ther 191: 575–580
Mujovic VM, Fisher JW (1975) The role of prostaglandins in the production of erythropoietin (ESF) by the kidney. II. Effects of indomethacin on erythropoietin production following hypoxia in dogs. Life Sci 16:463–473
Muldowney FP, Crooks J, Wayne EJ (1957) The total red cell mass in thyrotoxicosis and myxoedema. Clin Sci 16:309–314
Mulvey PF, Macaione JMR (1969) Thyroidal dysfunction during simulated altitude conditions. Fed Proc 23:1243–1246
Murphy MJ (1976) ^{125}I-Labelling of erythropoietin without loss of biological activity. Biochem J 159:287–292
Murphy GP, Brendler H, Mirand EA (1970) Erythropoietin release from renal cell carcinomas grown in tissue culture. Res Commun Chem Pathol Pharmacol 1:617–626
Murphy GP, Mirand EA, Groenewald JH, Kenny GM (1971a) Erythropoietin release from preserved kidneys. Cryobiology 8:184–189
Murphy GP, Mirand EA, Gaeta JF, Weber HW (1971b) Human renal allotransplants. Correlation of acute and chronic histologic alterations with endocrine function – a new syndrome? NY State J Med 71;2743–2746
Murphy GP, Mirand EA, Takita H, Schoonees R, Groenewald JH (1971c) The effects of hypoxia and ischemia on erythropoietin and renin release in dogs. Invest Urol 8: 521–525
Murphy MJ, MacPherson GG, Camiscoli JF, Gordon AS (1977) Erythropoietin in renal and hepatic lymph of conscious ewes. Exp Hematol 5:41–50
Myles WS, Ducker AJ (1973) The role of the sympathetic nervous system during exposure to altitude in rats. Int J Biometeor 17:51–58

Naets JP (1958) Erythropoiesis in nephrectomized dogs. Nature 181:1134–1135
Naets JP, Wittek M (1968) Erythropoiesis in anephric man. Lancet 1:941–943
Naets JP, Wittek M (1969) Erythropoietic activity of marrow and disappearance rate of erythropoietin in the rat. Am J Physiol 217:297–301
Naets JP, Wittek M (1974) A role of the kidney in the catabolism of erythropoietin in the rat. J Lab Clin Med 84:99–106
Nakao K, Takaku F, Fujioka S, Sassa S (1966) The effect of erythropoietin on hematopoietic organs of the polycythemic mouse. Blood 27:537–543
Napier JAF (1975) A comparison of methods for the initial concentration of erythropoietin from human urine. Biochem Med 13:312–318
Napier JAF (1980) Effect of age and 2, 3-DPG content of transfused blood on serum erythropoietin. Vox Sang 39:318–321
Napier JAF, Evans J (1980) Erythropoietin assay using fetal mouse liver cell cultures: a modified technique using semi-automatic harvesting of I^{125} deoxyuridine labelled erythroblasts. Clin Lab Haematol 2:13–19
Napier JAF, Dunn CDR, Ford TW, Price V (1977) Pathophysiological changes in serum erythropoiesis stimulating activity. Br J Haematol 35:403–409
Nathan DG, Chess L, Hillman DG, Clarke B, Breard J, Merler E, Housman DE (1978) Human erythroid burst-forming unit: T-cell requirement for proliferation in vitro. J Exp Med 147:324–339
Neal WA, Welch ET, Lewis JP, Lutcher CL, Wright CS, Smith LL (1978) The apparent heterogeneity of erythropoietin. Biochem Med 20:229–237
Neal WA, Lewis JP, Garver FA, Lutcher Cl (1980) The production of prostaglandins with an erythropoietin generating factor(s). Biochem Med 23:55–63
Nečas E, Neuwirt J (1970) Feedback regulation by red cell mass of the sensitivity of the erythropoietin-producing organ to hypoxia. Blood 36:754–763
Nečas E, Neuwirt J (1972) The effect of inhibitors of energy metabolism on erythropoietin production. J Lab Clin Med 79:388–396
Nečas E, Thorling EB (1972) Unresponsiveness of erythropoietin-producing cells to cyanide. Am J Physiol 222:1187–1190
Nečas E, Živný J, Neuwirt J (1972) Effect of polycythemia on erythropoietin production in the hypoxic rat. Am J Physiol 223:809–811
Needleman P, Minkes MS, Douglas JR (1974) Stimulation of prostaglandin biosynthesis by adenine nucleotides. Circ Res 34:455–460
Nelson PK, Brookins J, Fisher JW (1983) Erythropoietic effects of prostacyclin (PGI_2) and its metabolite 6-keto-prostaglandin $(PG)E_1$. J Pharmacol Exp Ther 226:493–499
Nijhof W, Wierenga PK (1983) Isolation and characterization of the erythroid progenitor cell: CFU-E. J Cell Biol 96:386–392
Ogawa M, Porter PN, Nakahata T (1983) Renewal and commitment to differentiation of hemopoietic stem cells (an interpretive review). Blood 61:823–829
Ogle JW, Lange RD, Dunn CDR (1978a) Production of erythropoietin in vitro: a review. In Vitro 14:945–950
Ogle JW, Lange RD, Dunn CDR (1978b) Erythropoiesis-stimulating factor production by rabbit kidney cultures from "programmed" rabbits. Blood 52:233–239
Ogle JW, Dunn CDR, McDonald TP, Lange RD (1978c) The in vitro production of erythropoietin and thrombopoietin. Scand J Haematol 21:188–196
Ohno Y, Rege AB, Fisher JW, Barone J (1978) Inhibitors of erythroid colony forming cells (CFU-E and BFU-E) in sera of azotemic patients with anemia of renal disease. J Lab Clin Med 92:916–923
Okabe T, Urabe A, Kato T, Chiba S, Takaku F (1985) Production of erythropoietin-like activity by human renal and hepatic carcinomas in cell culture. Cancer 55: 1918–1923
Olesen H, Fogh J (1968) Apparent molecular weight of erythropoietin determined by gel filtration. Scand J Haematol 5:211–214

Oliver WJ, Brody GL (1965) Effect of prolonged hypoxia upon granularity of renal juxtaglomerular cells. Circ Res 16:83–88

Olson RD, Skoglund ML, Nies AS, Gerber JG (1980) Prostaglandins mediate the macula densa stimulated renin release. In: Samuelsson B, Ramwell PW, Paoletti R (eds) Advances in prostaglandin and thromboxane research, vol 7. Raven, New York, pp 1135–1137

Omran N, Neumann E (1979) A haemagglutination inhibition test kit for routine analysis of erythroooietin. Blut 39:225–228

Orlic D, Gordon AS, Rhodin JAG (1968) Ultrastructural and autoradiographic studies of erythropoietin-induced red cell production. Ann NY Acad Sci 149:198–216

Ortega JA, Dukes PP, Ma A, Shore NA, Malekzadeh MH (1984) A clinical trial of prostaglandin E_2 to increase erythropoiesis in anemia of end stage renal disease. Prostaglandins Leukotrienes Med 14:411–416

Ozawa S (1967) Erythropoietin from the kidney cells cultured in vitro. Keio J Med 16:193–203

Papayannopoulou T, Finch CA (1972) On the in vivo action of erythropoietin: a quantitative analysis. J Clin Invest 51:1179–1185

Paul P, Rothmann SA, McMahon JT, Gordon AS (1984) Erythropoietin secretion by isolated rat Kupffer cells. Exp Hematol 12:825–830

Paulo LG, Roh BL, Fisher JW (1972) Antagonism of erythropoietin production in rabbits by atropine. Proc Soc Exp Biol Med 139:207–210

Paulo LG, Wilkerson RD, Roh BL, George WJ, Fisher JW (1973a) The effects of prostaglandin E_1 on erythropoietin production. Proc Soc Exp Biol Med 142:771–775

Paulo LG, Fink GD, Roh BL, Fisher JW (1973b) Influence of carotid body ablation on erythropoietin production in rabbits. Am J Physiol 224:442–444

Paulo LG, Fink GD, Roh BL, Fisher JW (1974) Effects of several androgens and steroid metabolites on erythropoietin production in the isolated perfused dog kidney. Blood 43:39–47

Pavlović-Kentera V, Sušić D, Milenković P, Biljanović-Paunović L (1980) Effects of prostaglandin synthetase inhibitors, salt overload and renomedullary dissection on the hypoxia stimulated erythropoietin production in rats. Exp Hematol 8 (Suppl 8):283–290

Penington DG (1963) Erythropoietin in tissues: features of the renal erythropoietin-producing system. In: Williams P (ed) Hormones and the kidney. Academic, New York, pp 201–219

Penington DG (1972) Haematological changes in renal disease. In: Black D (ed) Renal disease. Blackwell, Oxford, pp 739–758

Peschle C, Condorelli M (1975) Biogenesis of erythropoietin: evidence for pro-erythropoietin in a subcellular fraction of kidney. Science 190:910–912

Peschle C, Zanjani ED, Gidari AS, McLaurin WD, Gordon AS (1971a) Mechanism of thyroxine action on erythropoiesis. Endocrinology 89:609–612

Peschle C, Sasso GF, Mastroberardino G, Condorelli M (1971b) The mechanism of endocrine influences on erythropoiesis. J Lab Clin Med 78:20–29

Peschle C, Sasso GF, Rappaport IA, Condorelli M (1972a) Erythropoietin production in nephrectomized rats: possible role of the renal erythropoietic factor. J Lab Clin Med 79:950–959

Peschle C, Rappaport IA, Sasso GF, Gordon AS, Condorelli M (1972b) Mechanism of growth hormone (GH) action on erythropoiesis. Endocrinology 91:511–517

Peschle C, D'Avanzo A, Rappaport IA, Russolillo S, Marone G, Condorelli M (1973a) Role of erythrogenin from liver and spleen in erythropoietin production in the anephric rat. Nature 243:539–541

Peschle C, Rappaport IA, D'Avanzo A, Russolillo S, Marone G, Condorelli M (1973b) Renal mechanisms underlying cyclic AMP action on erythropoiesis. Br J Haematol 25:393–398

Peschle C, Rappaport IA, Sasso GF, Condorelli M, Gordon AS (1973c) The role of estrogen in the regulation of erythropoietin production. Endocrinology 92:358–362

Peschle C, Marone G, Genovese A, Rappaport IA, Condorelli M (1976) Increased erythropoietin production in anephric rats with hyperplasia of the reticuloendothelial system induced by colloidal carbon or zymosan. Blood 47:325–337

Peschle C, Rappaport IA, Magli MC, Marone G, Lettieri F, Cillo C, Gordon AS (1978) Role of the hypophysis in erythropoietin production during hypoxia. Blood 51: 1117–1124

Peschle C, Cillo C, Rappaport IA, Magli MC, Migliaccio G, Pizzella F, Mastroberardino G (1979) Early fluctuations of BFU-E pool size after transfusion or erythropoietin treatment. Exp Hematol 7:87–93

Piantadosi CA, Dickerman HW, Spivak JL (1976) Sequential activation of splenic nuclear RNA polymerases by erythropoietin. J Clin Invest 57:20–26

Popovic WJ, Brown JE, Adamson JW (1977) The influence of thyroid hormones on in vitro erythropoiesis. J Clin Invest 60:907–913

Przala F, Gross DM, Dargon PA, Fisher JW (1977) Effects of in vitro beta-adrenergic activation on rabbit bone marrow erythroid colony forming cells. Proc Soc Exp Biol Med 155:334–338

Przala F, Gross DM, Beckman B, Fisher JW (1979) Influence of albuterol on erythropoietin production and erythroid progenitor cell activation. Am J Physiol 236: H422–H426

Quesenberry P, Levitt L (1979) Hematopoietic stem cells. N Engl J Med 301:755–760, 819–823, 868–872

Radtke HW, Erbes PM, Fassbinder W, Koch KM (1978a) Serum erythropoietin measurements using the fetal mouse liver cell culture: the importance of reduction of variation in the specific activity of radioiron-transferrin. Exp Hematol 6:468–472

Radtke HW, Erbes PM, Schippers E, Koch KM (1978b) Serum erythropoietin concentration in anephric patients. Nephron 22:361–365

Radtke HW, Jubiz W, Smith JB, Fisher JW (1980) Albuterol-induced erythropoietin production and prostaglandins release in the isolated perfused dog kidney. J Pharmacol Exp Ther 214:467–471

Radtke HW, Rege AB, LaMarche MD, Bartos D, Bartos F, Campbell RA, Fisher JW (1981) Identification of spermine as an inhibitor of erythropoiesis in patients with chronic renal failure. J Clin Invest 67:1623–1629

Rambach WA, Alt HL, Cooper JAD (1957) The mode of action and nature of a heat stable plasma erythropoietic factor. Blood 12:1101–1113

Ramirez F, Gambino R, Maniatis GM, Rifkind RA, Marks PA, Bank A (1975) Changes in globin messenger RNA content during erythroid cell differentiation. J Biol Chem 250:6054–6058

Rastogi GK, Malhotra MS, Srivastava MC, Sawhney RC, Dua GL, Sridharan K, Hoon RS, Singh I (1977) Study of the pituitary-thyroid functions at high altitude in man. J Clin Endocrinol Metab 44:447–452

Recordati GM, Moss NG, Waselkov L (1978) Renal chemoreceptors in the rat. Circ Res 43:534–543

Rege AB, Brookins J, Fisher JW (1982) A radioimmunoassay for erythropoietin: serum levels in normal human subjects and patients with hemopoietic disorders. J Lab Clin Med 100:829–843

Reissman KR (1964) Protein metabolism and erythropoiesis. II. Erythropoietin formation and erythroid responsiveness in protein-deprived rats. Blood 23:146–153

Reissmann KR, Nomura T (1962) Erythropoietin formation in isolated kidneys and liver. In: Jacobson LO, Doyle M (eds) Erythropoiesis. Grune and Stratton, New York, pp 71–77

Reissmann KR, Udupa KB (1972) Effect of erythropoietin on proliferation of erythropoietin-responsive cells. Cell Tissue Kinet 5:481–489

Reissmann KR, Nomura T, Gunn RW, Brosius F (1960) Erythropoietic response to anemia or erythropoietin injection in uremic rats with or without functioning renal tissue. Blood 16:1411–1423

Reissmann KR, Diederich DA, Ito K, Schmaus JW (1965) Influence of disappearance rate and distribution space on plasma concentration of erythropoietin in normal rats. J Lab Clin Med 65:967–975
Reynafarje C, Ramos J, Faura J, Villavicencio D (1964) Humoral control of erythropoietic activity in man during and after altitude exposure. Proc Soc Exp Biol Med 116:649–650
Rich IN, Kubanek B (1979) The ontogeny of erythropoiesis in the mouse detected by the erythroid colony-forming technique. J Embryol Exp Morphol 50:57–74
Rich IN, Heit W, Kubanek B (1982) Extrarenal erythropoietin production by macrophages. Blood 60:1007–1018
Rich IN, Ihle J, Prystowsky M, Goldwasser E (1983) Interleukin-3 and its effect on hemopoiesis. Exp Hematol 11 (Suppl 14):381A
Rodgers GM, George WJ, Fisher JW (1972) Increased kidney cyclic AMP levels and erythropoietin production following cobalt administration. Proc Soc Exp Biol Med 140:977–981
Rodgers GM, Fisher JW, George WJ (1974) Lactate stimulation of renal cortical adenylate cyclase: a mechanism for erythropoietin production following cobalt treatment or hypoxia. J Pharmacol Exp Ther 190:542–550
Rodgers GM, Fisher JW, George WJ (1975a) Renal cyclic GMP and cholinergic mechanisms in erythropoietin production. Life Sci 17:1807–1814
Rodgers GM, Fisher JW, George WJ (1975b) Renal cyclic AMP accumulation and adenylate cyclase stimulation by erythropoietic agents. Am J Physiol 229:1387–1392
Rodgers GM, Fisher JW, George WJ (1975c) The role of renal adenosine 3′, 5′-monophosphate in the control of erythropoietin production. Am J Med 58:31–38
Rodgers GM, Fisher JW, George WJ (1975d) Increase in hematocrit, hemoglobin and red cell mass in normal mice after treatment with cyclic AMP. Proc Soc Exp Biol Med 148:380–382
Rodgers GM, Fisher JW, George WJ (1976) Elevated cyclic GMP concentrations in rabbit bone marrow culture and mouse spleen following erythropoietic stimulation. Biochem Biophys Res Commun 70:287–294
Roh BL, Paulo LG, Thompson J, Fisher JW (1972a) Plasma disappearance of ^{125}I-labeled erythropoietin in anesthetized rabbits. Proc Soc Exp Biol Med 141:268–270
Roh BL, Paulo LG, Fisher JW (1972b) Metabolism of erythropoietin by isolated perfused livers of dogs treated with SKF 525-A. Am J Physiol 223:1345–1348
Roodman GD, Spivak JL, Zanjani ED (1981) Stimulation of erythroid colony formation in vitro by erythropoietin immobilized on agarose-bound lectins. J Lab Clin Med 98:684–690
Rossi GB, Migliaccio AR, Migliaccio G, Lettieri F, Di Rosa M, Mastroberardino G, Peschle C (1980) In vitro interactions of PGE and cAMP with murine and human erythroid precursors. Blood 56:74–79
Rudolph W, Perretta M (1967) Effects of erythropoietin on ^{14}C-formate uptake by spleen and bone marrow nucleic acids on erythrocyte-transfused mice. Proc Soc Exp Biol Med 124:1041–1045
Russell ES, Keighley G (1972) The relation between erythropoiesis and plasma erythropoietin levels in normal and genetically anaemic mice during prolonged hypoxia or after whole-body irradiation. Br J Haematol 22:437–452
Saito T, Saito K, Trent DJ, Draganac PS, Andrews RB, Farkas WR, Dunn CDR, Etkin LD, Lange RD (1985) Translation of messenger RNA from a renal tumor into a product with the biological properties of erythropoietin. Exp Hematol 13: 23–28
Sawyer ST, Krantz SB (1984) Erythropoietin stimulates $^{45}Ca^{2+}$ uptake in Friend virus-infected erythroid cells. J Biol Chem 259:2769–2774
Scaro JL, De Levinton AM, Miranda C, De Tombolesi ARAP, Buys MC (1975a) Extraction of biologically active erythropoietin from kidneys of carbon monoxide intoxicated mice. Acta Physiol Lat Am 25:358–364

Scaro JL, Carrera MA, De Tombolesi ARAP, Miranda C (1975b) Carbon monoxide and erythropoietin production in mice. Acta Physiol Lat Am 25:204–210
Scheid CR, Fay FS (1984) β-Adrenergic effects on transmembrane ^{45}Ca fluxes in isolated smooth muscle cells. Am J Physiol 246:C431–C438
Schofield R, Lajtha LG (1976) Cellular kinetics of erythropoiesis. In: Congenital disorders of erythropoiesis. Elsevier, Amsterdam, pp 3–16
Schooley JC (1980) Erythropoietin complexed with the $F(ab')_2$ fragment of anti-erythropoietin retains its biologic activity. Exp Hematol 8:830–836
Schooley JC, Mahlmann LJ (1971a) Stimulation of erythropoiesis in plethoric mice by prostaglandins and its inhibition by antierythropoietin. Proc Soc Exp Biol Med 138:523–524
Schooley JC, Mahlmann LJ (1971b) Stimulation of erythropoiesis in the plethoric mouse by cyclic-AMP and its inhibition by antierythropoietin. Proc Soc Exp Biol Med 137:1289–1292
Schooley JC, Mahlmann LJ (1972) Evidence for the de novo synthesis of erythropoietin in hypoxic rats. Blood 40:662–670
Schooley JC, Mahlmann LJ (1975a) Hypoxia and the initiation of erythropoietin production. Blood Cells 1:429–448
Schooley JC, Mahlmann LJ (1975b) Adenosine, AMP, cyclic AMP, theophylline and the action and production of erythropoietin. Proc Soc Exp Biol Med 150:215–219
Shelton RN, Ichiki AT, Lange RD (1975) Physiochemical properties of erythropoietin: isoelectric focusing and molecular weight studies. Biochem Med 12:45–54
Sherwood JB (1984) The chemistry and physiology of erythropoietin. Vitam Horm 41:161–211
Sherwood JB, Goldwasser E (1976) Erythropoietin production by human renal carcinoma cells in culture. Endocrinology 99:504–510
Sherwood JB, Goldwasser E (1978) Extraction of erythropoietin from normal kidneys. Endocrinology 103:866–870
Sherwood JB, Goldwasser E (1979) A radioimmunoassay for erythropoietin. Blood 54:885–893
Sherwood JB, Robinson SH, Bassan LR, Rosen S, Gordon AS (1972) Production of erythrogenin by organ cultures of rat kidney. Blood 40:189–197
Sieber F (1977) Chromatography of human urinary erythropoietin and granulocyte colony-stimulating factor on insolubilized phytohaemagglutinin. Biochim Biophys Acta 496:146–154
Singer JW, Adamson JW (1976) Steroids and hematopoiesis. II. The effects of steroids on in vitro erythroid colony growth: evidence for different target cells for different classes of steroids. J Cell Physiol 88:135–144
Singer JW, Samuels AI, Adamson JW (1976) Steroids and hematopoiesis. I. The effect of steroids on in vitro erythroid colony growth: structure/activity relationships. J Cell Physiol 88:127–134
Siri WE, Van Dyke DC, Winchell HS, Pollycove M, Parker HG, Cleveland AS (1966) Early erythropoietin, blood, and physiological responses to severe hypoxia in man. J Appl Physiol 21:73–80
Skoglund RW, Finne PH (1972) Renal tubular secretion of erythropoietin. Scand J Clin Lab Invest 30:63–66
Spannhake EW, Fisher JW (1977) Partial inhibition of hypoxia-induced erythropoietin production by cholinergic blockade in the dog. Proc Soc Exp Biol Med 156: 19–23
Spear GS (1964) The glomerulus in cyanotic congenital heart disease and primary pulmonary hypertension. Nephron 1:238–248
Spear GS, Kihara I (1972) The glomerulus and serum sickness in experimental hypoxia. Br J Exp Pathol 53:265–276
Spivak JL, Graber SE (1980) Erythropoietin and the regulation of erythropoiesis. Johns Hopkins Med J 146:311–320
Spivak JL, Peck L (1979) Chemical modification of nuclear proteins by erythropoietin. Am J Hematol 7:45–51

Spivak JL, Small D, Hollenberg MD (1977) Erythropoietin: isolation by affinity chromatography with lectin-agarose derivatives. Proc Natl Acad Sci USA 74:4633–4635
Spivak JL, Small D, Shaper JH, Hollenberg MD (1978) Use of immobilized lectins and other ligands for the partial purification of erythropoietin. Blood 52:1178–1188
Stalcup SA, Lipset JS, Legant PM, Leuenberger PJ, Mellins RB (1979) Inhibition of converting enzyme activity by acute hypoxia in dogs. J Appl Physiol 46:227–234
Steinhausen M, Thederan H, Nolinski D, Dussel R, Dallenbach FD (1979) Postischämisches Akutes Nierenversagen (*Tupaia belangeri, Rattus norvegicus*). Film D 1326 des IWF, Göttingen, Publikation von Steinhausen M und Dallenbach F, Publ Wiss Film, Sekt Med, Ser 4, Nr 29/D 1326
Stephenson JR, Axelrad AA (1971) Quantitative assay method for erythropoietin in vitro. Endocrinology 88:1519–1520
Stephenson JR, Axelrad AA, McLeod DL, Shreeve MM (1971) Induction of colonies of hemoglobin-synthesizing cells by erythropoietin in vitro. Proc Natl Acad Sci USA 68:1542–1546
Stohlman F (1967) Some aspects of erythrokinetics. Semin Hematol 4:304–314
Stohlman F, Brecher G (1959) Humoral regulation of erythropoiesis. V. Relationship of plasma erythropoietine level to bone marrow activity. Proc Soc Exp Biol Med 100:40–43
Stohlman F, Howard D (1962) Humoral regulation of erythropoiesis. IX. The rate of disappearance of erythropoietine from the plasma. In: Jacobson LO, Doyle M (eds) Erythropoiesis. Grune and Stratton, New York, pp 120–124
Streeter RG, Pepelko WE, Cain SM (1975) Tissue oxygenation and splenic erythropoiesis during chronic hypoxia and hypercapnia. J Appl Physiol 38:309–314
Sue JM, Sytkowski AJ (1983) Site-specific antibodies to human erythropoietin directed toward the NH_2-terminal region. Proc Natl Acad Sci USA 80:3651–3655
Sufrin G, Mirand EA, Moore RH, Chu TM, Murphy GP (1977) Hormones in renal cancer. J Urol 117:433–438
Sunderman FW, Hopfer SM, Reid MC, Shen SK, Kevorkian CB (1982) Erythropoietin-mediated erythrocytosis in rodents after intrarenal injection of nickel subsulfide. Yale J Biol Med 55:123–136
Surks MI, Beckwitt HJ, Chidsey CA (1967) Changes in plasma thyroxine concentration and metabolism, catecholamine excretion and basal oxygen consumption in man during acute exposure to high altitude. J Clin Endocrinol Metab 27:789–799
Sušić D, Milenković P, Pavlović-Kentera V (1979) The effect of aspirin on erythropoietin formation in the rat. Proc Soc Exp Biol Med 161:476–478
Sytkowski AJ (1980) Erythropoietin forms biologically inactive complexes in solution. Biochem Biophys Res Commun 93:354–359
Sytkowski AJ, Sue JM (1984) A novel radioimmunoassay for human erythropoietin using a synthetic NH_2-terminal polypeptide and anti-peptide antibodies. J Immunol Methods 69:181–186
Sytkowski AJ, Richie JP, Bicknell KA (1983) New human renal carcinoma cell line established from a patient with erythrocytosis. Cancer Res 43:1415–1419
Sytkowski AJ, Bicknell KA, Smith GM, Garcia JF (1984) Secretion of erythropoietin-like activity by clones of human renal carcinoma cell line GKA. Cancer Res 44:51–54
Syvertsen GR, Harris JA (1973) Erythropoietin production in dogs exposed to high altitude and carbon monoxide. Am J Physiol 225:293–299
Takaku F, Hirashima K, Nakao K (1962) Studies on the mechanism of erythropoietin production. I. Effect of unilateral constriction of the renal artery. J Lab Clin Med 59:815–820
Takaku F, Nakao K, Ono T, Terayama H (1969) Changes in histone acetylation and RNA synthesis in the spleen of polycythemic mouse after erythropoietin injection. Biochim Biophys Acta 195:396–400
Takeguchi C, Kohno E, Sih CJ (1971) Mechanism of prostaglandin biosynthesis. I. Characterization and assay of bovine prostaglandin synthetase. Biochemistry 10: 2372–2376

Tambourin P, Casadevall N, Choppin J, Lacombe C, Heard JM, Fichelson S, Wendling F, Hankins WD, Varet B (1983) Production of erythropoietin-like activity by a murine erythroleukemia cell line. Proc Natl Acad Sci USA 80:6269–6273

Tarazi RC, Frohlich ED, Dustan HP, Gifford RW, Page IH (1966) Hypertension and high hematocrit. Am J Cardiol 18:855–858

Terada M, Cantor L, Metafora S, Rifkind RA, Bank A, Marks PA (1972) Globin messenger RNA activity in erythroid precursor cells and the effect of erythropoietin. Proc Natl Acad Sci USA 69:3575–3579

Thorling EB (1973) A comparison of the cobalt, methylene blue, zinc, arsenite and amino triazole effect on erythropoietin production. Br J Haematol 25:55–62

Thorn W, Bläker F, Weiland E (1981) Preparative capillary isotachophoresis: a micro method for the purification of erythropoietin. J Chromatogr 210:319–325

Thurau K (1961) Renal Na-reabsorption and O_2-uptake in dogs during hypoxia and hydrochlorothiazide infusion. Proc Soc Exp Biol Med 106:714–717

Till JE, McCulloch EA (1961) A direct measurement of the radiation sensitivity of normal mouse bone marrow cells. Radiat Res 14:213–222

Till JE, McCulloch EA, Siminovitch L (1964) A stochastic model of stem cell proliferation, based on the growth of spleen-colony forming cells. Proc Natl Acad Sci USA 51:29–36

Tong BD, Goldwasser E (1981) The formation of erythrocyte membrane proteins during erythropoietin-induced differentiation. J Biol Chem 256:12666–12672

Toyama K, Mitus WJ (1966) Experimental renal erythrocytosis. III. Relationship between the degree of hydronephrotic pressure and the production of erythrocytosis. J Lab Clin Med 68:740–752

Toyama K, Fujiyama N, Suzuki H, Chen TP, Tomaoki N, Ueyama Y (1979) Erythropoietin levels in the course of a patient with erythropoietin-producing renal cell carcinoma and transplantation of this tumor in nude mice. Blood 54:245–253

Tramezzani JH, Morita E, Chiocchio SR (1971) The carotid body as a neuroendocrine organ involved in control of erythropoiesis. Proc Natl Acad Sci 68:52–55

Trentin JJ (1970) Influence of hematopoietic organ stroma (hematopoietic inductive microenvironments) on stem cell differentiation. In: Gordon AS (ed) Regulation of hematopoiesis, vol 1. Appleton-Century Crofts, New York, pp 161–186

Udupa KB, Reissmann KR (1979) In vivo erythropoietin requirements of regenerating erythroid progenitors (BFU-e, CFU-e) in bone marrow of mice. Blood 53:1164–1171

Udupa KB, Reissmann KR (1980) Stimulation of erythroblast formation in suspension cultures of murine marrow by a factor in normal mouse serum, and its relationship to erythropoietin. Exp Hematol 8:1106–1113

Van Dyke DC, Garcia JF, Simpson ME, Huff RL, Contopoulos AN, Evans HM (1952) Maintenance of circulating red cell volume in rats after removal of the posterior and intermediate lobes of the pituitary. Blood 7:1017–1019

Van Dyke DC, Layrisse M, Larence JH, Garcia JF, Pollycove M (1961) Relation between severity of anemia and erythropoietin titer in human beings. Blood 18:187–201

Varkarakis MJ, Mirand EA, Murphy GP (1976) The response of the juxtaglomerular apparatus to stimuli effecting renin or erythropoietin release in canine renal allografts. Invest Urol 13:366–371

Von Preyss B, Goudsmit R (1976) The problem of toxicity in the bioassay for erythropoietin using mouse fetal liver cells. Exp Hematol 4:138–142

Wagemaker G (1978) Cellular and soluble factors influencing the differentiation of primitive erythroid progenitor cells (BFU-e) in vitro. In: Murphy MJ (ed) In vitro aspects of erythropoiesis. Springer, New York Heidelberg Berlin, pp 44–57

Walker BR (1982) Diuretic response to acute hypoxia in the conscious dog. Am J Physiol 243:F440–F446

Wallner SF, Vautrin RM, Kurnick JE, Ward HP (1978) The effect of serum from patients with chronic renal failure on erythroid colony growth in vitro. J Lab Clin Med 92:370–375
Wang F, Fried W (1972) Renal and extrarenal erythropoietin production in male and female rats of various ages. J Lab Clin Med 79:181–186
Wang FF, Kung CK-H, Goldwasser E (1985) Some chemical properties of human erythropoietin. Endocrinology 116:2286–2292
Ward HP (1967) An in vitro assay of erythropoietin. Proc Soc Exp Biol Med 125: 370–374
Wardle DFH, Baker I, Malpas JS, Wrigley PFM (1973) Bioassay of erythropoietin using foetal mouse liver cells. Br J Haematol 24:49–56
Weintraub AH, Gordon AS, Camiscoli JF (1963) Use of the hypoxia-induced polycythemic mouse in the assay and standardization of erythropoietin. J Lab Clin Med 62:743–752
Weiss TL, Goldwasser E (1981) The biological properties of endotoxin-free human erythropoietin. Biochem J 198:17–21
Weiss TL, Kavinsky CJ, Goldwasser E (1982) Characterization of a monoclonal antibody to human erythropoietin. Proc Natl Acad Sci USA 79:5465–5469
Wennmalm Å (1979) Prostacyclin-dependent coronary vasodilation in rabbit and guinea pig hearts. Acta Physiol Scand 106:47–52
Westbrook CA, Gasson JC, Gerber SE, Selsted ME, Golde DW (1984) Purification and characterization of human T-lymphocyte-derived erythroid-potentiating activity. J Biol Chem 259:9992–9996
White LE, George WJ (1981) Increased concentrations of cyclic GMP in fetal liver cells stimulated by erythropoietin. Proc Soc Exp Biol Med 166:186–193
Winkert J, Birchette C, Wilson M (1971) Augmentation of erythrocyte iron uptake in post-hypoxic mice by the administration of N^6, $O^{2'}$-dibutyryl cyclic adenosine 3′, 5′-monophosphate (dBcAMP). Res Commun Chem Pathol Pharmacol 2:323–329
Wolf-Priessnitz J, Schooley JC, Mahlmann LJ (1978) Inhibition of erythropoietin production in unanesthetized rabbits exposed to an acute hypoxic-hypercapnic environment. Blood 52:153–162
Wong KK, Zanjani ED, Cooper GW, Gordon AS (1968) The renal erythropoietic factor. V. Studies on its purification. Proc Soc Exp Biol Med 128:67–70
Yanagawa S, Yokoyama S, Hirade K, Sasaki R, Chiba H, Ueda M, Goto M (1984a) Hybridomas for production of monoclonal antibodies to human erythropoietin. Blood 64:357–364
Yanagawa S, Hirade K, Ohnota H, Sasaki R, Chiba H, Ueda M, Goto M (1984b) Isolation of human erythropoietin with monoclonal antibodies. J Biol Chem 259: 2707–2710
Zalman F, Maloney MA, Patt HM (1979) Differential response of early erythropoietic and granulopoietic progenitors to dexamethasone and cortisone. J Exp Med 149: 67–72
Zangheri EO, Campana H, Ponce F, Silva JC, Fernández FO, Suárez JR (1963) Production of erythropoietin by anoxic perfusion of the isolated kidney of a dog. Nature 199:572–573
Zangheri EO, Fava-De-Moraes F, Lopez OI, Marias I (1973) The role of submandibular glands in extrarenal erythropoietin production. Experientia 29:706–707
Zangheri EO, Lopez OI, Honorato LE, Puscama LO, Rodriguez ME, Reta E (1977) The role of the submandibular glands in extrarenal erythropoietin production in mice. Exp Hematol 5:237–240
Zanjani ED, Peterson EN, Gordon AS, Wasserman LR (1974) Erythropoietin production in the fetus: role of the kidney and maternal anemia. J Lab Clin Med 83:281–287

Zanjani ED, Poster J, Mann LI, Wasserman LR (1977) Regulation of erythropoiesis in the fetus. In: Fisher JW (ed) Kidney hormones, vol II, erythropoietin. Academic, London, pp 463–493
Zapf J, Waldvogel M, Froesch ER (1975) Binding of nonsuppressible insulinlike activity to human serum. Arch Biochem Biophys 168:638–645
Zaroulis CG, Hoffman BJ, Kourides IA (1981) Serum concentrations of erythropoietin measured by radioimmunoassay in hematologic disorders and chronic renal failure. Am J Hematol 11:85–92
Živný J, Travnicek T, Neuwirt J (1971) Effect of exercise on production of erythropoietin in normal and hypoxic rats. Am J Physiol 220:329–332
Živný J, Kolc J, Málek P, Neuwirt J (1972a) Renal ischaemia, hypoxic hypoxia and erythropoietin production. Scand J Haematol 9:470–476
Živný J, Neuwirt J, Borová J (1972b) The effect of aldosterone on erythropoietin production and erythropoiesis. J Lab Clin Med 80:217–223
Zucali JR, Mirand EA (1978) In vitro aspects of erythropoietin production. In: Murphy MJ (ed) In vitro aspects of erythropoiesis. Springer, New York Heidelberg Berlin, pp 218–224
Zucali JR, Stevens V, Mirand EA (1975) In vitro production of erythropoietin by mouse fetal liver. Blood 46:85–90
Zucali JR, Lee M, Mirand EA (1978) Carbon dioxide effects on erythropoietin and erythropoiesis. J Lab Clin Med 92:648–655
Zuckerman KS (1981) Human erythroid burst-forming unit. Growth in vitro is dependent on monocytes but not T lymphocytes. J Clin Invest 67:702–709
Zuckerman KS, Quesenberry PJ, Levin J, Sullivan R (1979) Contamination of erythropoietin by endotoxin: in vivo and in vitro effects on murine erythropoiesis. Blood 54:146–158

Author Index

Page numbers in *italics* refer to the bibliography

Subject Index

GPSR Compliance

The European Union's (EU) General Product Safety Regulation (GPSR) is a set of rules that requires consumer products to be safe and our obligations to ensure this.

If you have any concerns about our products, you can contact us on ProductSafety@springernature.com

In case Publisher is established outside the EU, the EU authorized representative is:

Springer Nature Customer Service Center GmbH
Europaplatz 3
69115 Heidelberg, Germany

Batch number: 10620881

Printed by Printforce, the Netherlands